P9-DXN-474

INTERFERONS: FROM MOLECULAR BIOLOGY TO CLINICAL APPLICATION

SYMPOSIA OF THE SOCIETY FOR GENERAL MICROBIOLOGY*

1 THE NATURE OF THE BACTERIAL SURFACE
2 THE NATURE OF VIRUS MULTIPLICATION
3 ADAPTATION IN MICRO-ORGANISMS
4 AUTOTROPHIC MICRO-ORGANISMS
5 MECHANISMS OF MICROBIAL PATHOGENICITY
6 BACTERIAL ANATOMY
7 MICROBIAL ECOLOGY
8 THE STRATEGY OF CHEMOTHERAPY
9 VIRUS GROWTH AND VARIATION
10 MICROBIAL GENETICS
11 MICROBIAL REACTION TO ENVIRONMENT
12 MICROBIAL CLASSIFICATION
13 SYMBIOTIC ASSOCIATIONS
14 MICROBIAL BEHAVIOUR, 'IN VIVO' AND 'IN VITRO'
15 FUNCTION AND STRUCTURE IN MICRO-ORGANISMS
16 BIOCHEMICAL STUDIES OF ANTIMICROBIAL DRUGS
17 AIRBORNE MICROBES
18 THE MOLECULAR BIOLOGY OF VIRUSES
19 MICROBIAL GROWTH
20 ORGANIZATION AND CONTROL IN PROKARYOTIC AND EUKARYOTIC CELLS
21 MICROBES AND BIOLOGICAL PRODUCTIVITY
22 MICROBIAL PATHOGENICITY IN MAN AND ANIMALS
23 MICROBIAL DIFFERENTIATION
24 EVOLUTION IN THE MICROBIAL WORLD
25 CONTROL PROCESSES IN VIRUS MULTIPLICATION
26 THE SURVIVAL OF VEGETATIVE MICROBES
27 MICROBIAL ENERGETICS
28 RELATIONS BETWEEN STRUCTURE AND FUNCTION IN THE PROKARYOTIC CELL
29 MICROBIAL TECHNOLOGY: CURRENT STATE, FUTURE PROSPECTS
30 THE EUKARYOTIC MICROBIAL CELL
31 GENETICS AS A TOOL IN MICROBIOLOGY
32 MOLECULAR AND CELLULAR ASPECTS OF MICROBIAL EVOLUTION
33 VIRUS PERSISTENCE
34 MICROBES IN THEIR NATURAL ENVIRONMENTS

*Published by the Cambridge University Press, except for the first Symposium, which was published by Blackwell's Scientific Publications Limited.

INTERFERONS: FROM MOLECULAR BIOLOGY TO CLINICAL APPLICATION

EDITED BY

D. C. BURKE AND A. G. MORRIS

THIRTY-FIFTH SYMPOSIUM OF
THE SOCIETY FOR GENERAL MICROBIOLOGY
HELD AT
THE UNIVERSITY OF LEEDS
SEPTEMBER 1983

Published for the Society for General Microbiology

CAMBRIDGE UNIVERSITY PRESS
CAMBRIDGE
LONDON NEW YORK NEW ROCHELLE
MELBOURNE SYDNEY

Published by the Press Syndicate of the University of Cambridge
The Pitt Building, Trumpington Street, Cambridge CB2 1RP
32 East 57th Street, New York, NY 10022, USA
296 Beaconsfield Parade, Middle Park, Melbourne 3206, Australia

First published 1983

Printed in Great Britain at The Pitman Press, Bath

British Library Cataloguing in Publication Data

Society for General Microbiology
Symposium (35th: 1983: University of Leeds)

Interferons.–(Society for General Microbiology Symposium; 35)
1. Interferon–Congresses
I. Burke, D. C. II. Morris, A. G. III. Series
616.07′9 QR187.5

ISBN 0 521 25069 2

CONTRIBUTORS

BILLIAU, A., Rega Institut, Katholieke Universiteit, Minderbroederstraat 10, B-3000 Leuven, Belgium

BURKE, D. C., Department of Biological Sciences, University of Warwick, Coventry CV4 7AL, UK

COLLINS, J., Gesellschaft für Biotechnologische Forschung GmbH, Mascheroder Weg 1, D 3300 Braunschweig-Stockheim, West Germany

KERR, I., Imperial Cancer Research Fund, Lincoln's Inn Fields, London WC2A 3PX, UK

KINGSMAN, A. J., Department of Biochemistry, South Parks Road, Oxford OX1 3QU, UK

KINGSMAN, S. M., Department of Biochemistry, South Parks Road, Oxford OX1 3QU, UK

MCMAHON, M., National Institute for Medical Research, Mill Hill, London, UK

MALPAS, J. S., Imperial Cancer Research Fund, Department of Medical Oncology, St Bartholomew's Hospital, West Smithfield, London EC1A 7BE, UK

MOORE, M., Christie Hospital & Holt Radium Institute, Manchester Area Health Authority (Teaching) South District, Manchester M20 9BX, UK

MORRIS, A., Department of Biological Sciences, University of Warwick, Coventry CV4 7AL, UK

SCOTT, G. M., Division of Communicable Diseases, Clinical Research Centre, Watford Road, Harrow, Middlesex HA1 3UJ, UK

STRANDER, H., Karolinska Hospital, Box 60500, S–104 01 Stockholm, Sweden

TAYLOR-PAPADIMITRIOU, M. J., Imperial Cancer Research Fund, Lincoln's Inn Fields, London WC2A 3PX, UK

TYRRELL, D. A. J., Division of Communicable Diseases, Clinical Research Centre, Watford Road, Harrow, Middlesex, HA1 3UJ, UK

WILKINSON, M. J., Department of Biological Sciences, University of Warwick, Coventry CV4 7AL, UK

CONTENTS

EDITORS' PREFACE

Twenty-five years ago, the study of interferon was an obscure part of virology. Today it has developed into a major topic for research, important not only in virology but also in interacting areas of cell biology, molecular biology, immunology and in clinical medicine. Its use in medicine, made possible on a large scale by advances in cell and molecular biology, has provoked claim and counter claim for its effectiveness against virus and malignant disease, with many unanswered questions. Partly as a result of its clinical potential, the study of the biology of interferons is very active – almost frenetically so – and will remain so for many years to come. There are many areas of profound ignorance, but we now have achieved a solid understanding of some portions of interferon biology. This Symposium spans the whole field of interferon research, with specialist reviews covering the literature up to October 1982. We feel sure it will be a useful compendium of up-to-date information for the cognoscenti of the interferon world and a stimulating overview for the tyro.

Department of Biological Sciences,
University of Warwick,
Coventry,
CV4 7AL

D. C. Burke
A. Morris

INTRODUCTION

D. A. J. TYRRELL

Division of Communicable Diseases, Clinical Research Centre, Harrow, UK

It is a great honour to be invited to contribute an introduction to this excellent volume on interferons. The phenomenon has been in my thoughts for over a quarter of a century and so I may be biased but I think I can understand why the Council of the Society thought it would be a good idea to hold a major symposium on the subject.

In the first place it is a fascinating saga of research which started in 1957 with the well-known paper of Isaacs and Lindenmann which reported on the work done in the department of a notable early member of the Society, Sir Christopher Andrewes. For a number of years until his untimely death Alick Isaacs continued to contribute experiments and ideas on what interferon was, and how it might be a response to 'foreign' nucleic acid in the cell. He showed it was produced during virus infections of animals, and showed that in some systems interferon production was associated with resistance to virus infection. He also suggested that interferon might be exploited as an antiviral agent in man, a subject on which we worked together for the rest of his life. Of course, there are many topics on which he did not have the ideas quite right, and some he did not anticipate at all, and he would have really enjoyed attending a meeting like this with its wealth of new ideas and new facts as well as some old problems with which he would have been familiar.

However, if we cannot have a scintillating Isaacs' preface, let us look at the subject as a whole and see how to pull it together, for there is no doubt that in these days reports of original work on interferon (IFN) are so numerous and diverse that no one can have read and understood them all and it can even be difficult to come to grips with expert and carefully written reviews such as we have in this volume.

When an expert historian of science comes to deal with the story of biology in the latter half of the century I think he would be well advised to give a good deal of space to the story of interferon, for it shows in so many ways how good research is actually done and how advances happen.

It all began with unexpected results to a series of experiments which had the original objective of showing that when inactivated influenza virus attached to a cell which subsequently became resistant to virus infection (in a state of interference) the nucleic acid of the virus entered the cell (Lindenmann, 1982). Instead of discarding the unexpected results Isaacs and Lindenmann followed them up and concluded that influenza virus-treated cells turned out a protein that modified other cells and made them resistant to infection, and thus was named interferon. The reactions to this report were mixed. Some doubted the phenomenon, others, particularly Isaacs himself, were enthusiastic to discover the biological significance of the phenomenon, the chemical nature of the substance and the biochemical basis of its effects; it was also foreseen that it might be applied as an antiviral agent, a successor to penicillin perhaps since it had inhibitory effects on so many diverse viruses and seemed to have no effect on the behaviour of cells in culture and therefore seemed to be non-toxic. And yet most of our naive hopes of the early days were dashed one by one. Attempts to purify the molecule by the methods then available did not succeed, partly because none were really effective in separating it from normal cell components, but even more because after enormous labour with chicken or mouse cells the amount of interferon actually present was miniscule – we marvelled more and more as we realized that material with an activity of 10^6 units per mg was grossly impure, and that therefore the active substance must be extraordinarily potent. Then studying the mode of action proved unexpectedly difficult. The first attempts ran into difficulties probably because of effects of impurities; later, when it was realized that IFN effects were mediated by the synthesis of a virus inhibitory protein (VIP), years were spent trying to detect VIP in lysates of interferon-treated cells.

Attempts at treatment with the substance were frustrated firstly by the difficulty of working and purifying material, to which I have already referred, but even more to the 'species specificity' of the effect. It is not truly species specific but at least there are enough specificities to hamper experimental and clinical work – for instance, interferon made from cells of a convenient species, such as the chicken, had no protective effect on human cells. Nevertheless it was possible to prevent infections of the skin and central nervous system in animals, and with some difficulty of the respiratory tract. Substantial clinical trials had to wait until Cantell modified and then applied methods of making human interferon from buffy coats, but then it was found, for instance, that experimental colds could be prevented by local administration and that local and systemic infections with herpes viruses could be treated with local or parenteral administration.

Perhaps the most interesting results are those which were not anticipated from the knowledge and ideas of the early days. We were clear that there must be more than one molecule, else how could one explain the 'species specificity', but the idea that there might be the great variety of interferons was beyond us. There is still a lot to learn, not merely about the chemical composition and the genetic complexity of these molecules, but also about the way they fit into the bodily economy – not merely what they can be persuaded to do in the laboratory but what their actual function is in a whole animal, whether infected or not. This area is surveyed by Dr Strander, who describes the whole wide range of effects, and who has himself pioneered the use of interferon against malignant disease.

We foresaw even less the impact of recombinant DNA technology on many aspects of the field. Firstly, the nature of the genes and their control is being rapidly unfolded as described by Dr Collins. Then sequences of cloned DNA have greatly helped our understanding of the chemical architecture of the molecules, and as described by Dr Kingsman, there is now a great range of choices open and we can decide which molecules to get made in prokaryote or eukaryote cells. This part of the book can be expected to become out of date on detail, but the principles enunciated will remain true.

It is still, of course, important for biologists to understand more about the interferons made in vertebrate cells by the genes naturally present in the nucleus and to purify them for definitive study. A particularly interesting area developed recently is the production of γ interferon, as what seems to be a natural consequence of a mitogenic stimulus to an immunocyte. Other interferons are produced and many different cells may be involved so it is worth having a chapter on the biology of the production of interferons by the immune systems (Dr Morris) as well as one on the equally complex way in which interferons can modify the many different cells and functional arms of the immune system (Dr Moore).

It was early recognized that interferon could protect a wide range of cells from virus infection, but it was not realized until much later that they modified many other functions beside that of the support of virus growth, probably because at first we were over-impressed with the fact that overall oxidative metabolism and protein synthesis were unaffected. There is now, however, an enormous literature on the effects on cell function that have been unambiguously ascribed to interferon – ranging from morphology of cells in culture to growth and differentiation of bone marrow cells. This is summarized for us by an old colleague of Dr Isaacs, namely Dr Taylor-Papadimitriou.

In modern biology we always try to push our understanding of cell behaviour to the molecular level and therefore we have two chapters in this area. One by Professor Burke deals with the control of the formation of interferons, a subject of absorbing interest since there must be very effective methods of regulating the function of the many genes which might otherwise lead to the expression of molecules that could disrupt many aspects of the body's economy. We are also fortunate in having a chapter by Dr Kerr on the biochemical changes induced by interferon – an exceedingly difficult subject on the elucidation of which he has spent more time and had more success than most.

The chapter on the pharmacokinetics and toxicology of interferon by Professor Billiau should really be regarded as an interface or transition zone between the study of the biology of interferons and their use in clinical practice. It is not enough to know how they behave in isolated systems, we need to understand how they are distributed in the body when they are applied to diseases or enter the circulation – the answers help us to understand their behaviour in life and help to select how they should be used in therapy. Furthermore, the so-called 'toxic' effects of interferons may be a nuisance to those who want to give them for treatment, but they indicate possible new actions of interferons, like producing inflammation, fever and malaise and indicate that they may well have a role in producing disease as well as curing it.

Finally, we have two reviews of the possible uses of interferon in treating infections (Dr Scott) and malignant disease (Dr Malpas). Here are areas that have been stirred into action by the large supplies of pure material provided through 'genetic engineering', but it is important for us to have a firm grasp of what had and had not been demonstrated using interferons from vertebrate cells, and to have some indications of where we can expect further advances to be made. What can be shown depends on the proper design of clinical trials, which biologists should try to understand, and what may happen depends on the fundamental properties of interferons which shows why clinicians should read the earlier chapters in the book.

It is a tribute to the quality of the books in this series that not only do they give rise to excellent expositions and discussions of contemporary science but are also taken off the shelves for reference decades later. I cannot do better than end by hoping that this is exactly what will happen in the case of this volume.

REFERENCE

Lindenmann, J. (1982). From interference to interferon: a brief historical introduction. *Philosophical Transactions of the Royal Society of London, Series B*, **299,** 3–6.

INTERFERONS AND DISEASE: A SURVEY

HANS STRANDER

Radiumhemmet, Karolinska Hospital, 10401 Stockholm, Sweden

THE ROLE OF INTERFERON IN DISEASE

Interferons (IFN) are proteins or glycoproteins capable of non-specific antiviral activity, at least in the homologous cells, through cellular metabolic processes involving RNA and protein synthesis. It has been shown that the proteins of this family share certain sequences of homology. They occur in vertebrates, and some workers assert that they should be classified as hormones, while others regard them as biological response modifiers. They are considered to be important elements of the body's defence against foreign organisms, inflammatory processes and various tumour, viral and other diseases.

Most IFNs exist in the tissue at concentrations ranging from barely detectable to extremely high levels, such as those found in the blisters developing in connection with zoster. Many of the constraints on our investigation of the role of IFN in disease have been removed, thanks to advances in the purification of IFN and the production of specific antibodies which enable precise and sensitive assays to be performed.

In 1979 human leucocyte IFN was purified to homogeneity and the cloning of Hu IFN-β in bacteria was achieved. The following year the sequence of the first 22 amino acids in human leucocyte IFN was recognized. Moreover, it was shown that one subtype produced by *Escherichia coli* consistently displays biological activity and that many monoclonal antibodies against the alpha type could be prepared. In 1982 it was reported that Hu IFN-γ can be produced from *E. coli*. In the same year crystals of a recombinant Hu IFN-α were prepared. All these achievements have furnished a basis for analysing the actions of exogenous IFN therapy and of endogenously produced IFN in greater detail, and the biological implications of the IFN system can be explored.

TYPES OF IFN

The human IFNs are assigned to three distinct types, alpha, beta and gamma, according to their biochemical properties and antigenicity. Work with bacteria, including genetic analysis and purification by use of monoclonal antibodies, has shown that there are at least 12 subtypes of IFN-α, and one or more of both beta and gamma. The biological implications of this diversity remain to be established. The IFNs detected so far are proteins secreted by cells in the extracellular compartment.

The amount of IFN available in a particular fluid is determined by an IFN assay, which is usually based on measurement of the antiviral activity in cell culture. The cells are treated for 12–24 h, after which they are exposed to a virus, the antiviral effect then being monitored. The titre is taken as the reciprocal of the highest dilution of the IFN solution that produces a 50% decrease in effect. The units so obtained are then compared to an International Reference Standard IFN Preparation, and the titre is expressed in International Units. With the recent preparation of monoclonal antibodies to various IFN species it has been possible to evolve radioimmune assay techniques, which should facilitate sensitive and precise determinations.

There is need for reference preparations that can be applied to many subtypes of Hu IFN-α. Comparisons of the various alpha preparations in different systems, which are at present difficult to perform, will be facilitated when more reference preparations for the various subtypes become available. Internal Laboratory Reference preparations could be used, but if so they should be matched to the International Reference Standards for IFN-α. We already have reference preparations of the natural human alphas and betas, and gamma preparations will shortly be available. A number of such Reference Preparations have been furnished by the United States National Institutes of Health. Specific antisera tested by strict techniques are available from the same source. The present status of the standardization of IFN was a topic of discussion by a WHO scientific group at a recent meeting in Geneva.

PRODUCTION

The extensive family of IFN inducers contains members of most virus groups, mycoplasmas, protozoa, bacteria, several substances resembling double-stranded nucleic acids, other such nucleic acids, polysaccharides, antigens and a number of prepared low-molecular-weight compounds. The type of IFN produced depends on the combination of cell and inducer. For example, lymphocyte populations exposed to tumour cells usually produce IFN-γ, whereas fibroblasts and monocytes exposed to viruses usually form the beta and gamma types, respectively. The differences and the similarities between these three types of IFN will be discussed by several contributors during this meeting.

In the course of the induction of IFN a few hours usually elapse before the release of IFN begins, after which the substance can reach remote organs in an intact state. An interesting point is that small amounts of one IFN present in various regions of the body can cause local potentiation of others, and possibly also superinduction in areas where virus or tumour cells stimulate the formation of IFN. Induction of IFN-β results in increased transcription of specific mRNA. The level of polyadenylated IFN-β RNA can be maintained longer in superinduced cells than in others.

Leucocytes are as potent producers of IFN-α in the peripheral blood of embryos as they are in that of the elderly. Monocytes seem to be the main producers of this type of IFN. Macrophage cultures, myeloblasts and B-lymphoblastoid cell lines can all produce IFN in tissue culture. Mitogens and tumour cells can induce the formation of IFN, and endotoxin elicits release of IFN. Also, it is likely that many medical agents used in the treatment of various diseases are IFN inducers. The mechanism of so-called spontaneous IFN production remains obscure.

The possible significance of glycosylation of IFN proteins is a matter of current debate. Mouse clones have been used for the production of Hu IFN-β because this type seems to be glycosylated to much the same extent in such cells as it is in human fibroblasts. Human fibroblasts are being used for mass production of IFN-β, but mouse lines can also be made to produce this type on a large scale.

Large-scale production of lymphocyte IFN-α is under way in several countries. Lymphoblastoid cells have also been used, and one myeloblast cell line is able to produce large amounts of IFN-α. Several laboratories are now engaged in attempts to make the gamma type on a large scale. As mentioned above, a variety of IFNs have recently been produced in bacteria. Monoclonal antibodies have already been employed to purify recombinant human leucocyte IFN.

GENETICS

Since 1968, when the first report on the genetics of IFN production was published, several genes coding for IFN proteins have been identified on different chromosomes. The control of IFN formation seems to occur through gene activation, gene transcription, variation of the messenger half time, and control of translation systems.

The first communication on the genetics of IFN sensitivity was issued in 1973. It was then established that the human chromosome 21 is involved in the action of IFN. The fact that trisomic 21 cells were found to be more sensitive to IFN than others has a direct bearing on the effect of IFN in disease; for the IFN system would then, at least in theory, behave in a quite different manner, in, for example, patients with Down's syndrome. The practical implications of such a difference in IFN sensitivity remain to be explored. It would seem that the action of the various types of human IFN are all dependent on the genes situated on the long arm of chromosome 21 in human cells.

The genetic control of the antiviral efficacy of IFN has been studied in the mouse. Certain genes of importance in this context may determine individual responses to various actions of the drug.

The property that the different types of IFN may possess of potentiating one another is of major importance. It is evident that the host gene type has a bearing on the action of IFN *in vivo*, and it will be interesting to see what potentiation effects can be achieved in various genetic systems. The mapping of human IFN genes is a topic of current interest. Different recombinant alphas and betas seem to have different specific molecular activities. A new area in this field of research is occupied by the search for cDNA plasmids complementary to mRNA induced by IFN therapy. These plasmids should facilitate chromosome mapping. The structure and expression of human IFN genes also call for detailed investigation.

ACTION *IN VITRO*

To judge from all the animal species so far examined the species specificity of IFN is relative. It is through their property of inhibiting the intracellular multiplication of viruses that IFNs were discovered. The process entails RNA and protein synthesis. When cells are exposed to IFN, specific proteins are induced in them. In IFN-treated cells there is a rise in the level of certain recently discovered enzymes, the best known of which are 2–5 A synthetase and a protein kinase. There is now evidence that the synthetase – RNase L system is one of the mediators of IFN action. Here, double-stranded RNA seems to be an important factor, since it can generate 2–5 A oligomers activating a cellular endononuclease.

IFN treatment can alter plasma membranes and the cytoskeleton in various cells. Also to be discussed during this Symposium are the possible effects of IFNs on virus absorption, RNA synthesis and translation, protein synthesis, and virus maturation and release. Suffice it to say at this point that IFNs exert effects at many different sites, and that their action can be rendered fairly selective.

IFNs are also capable of affecting the prostaglandin system. They can inhibit the growth of cells, especially of tumour cells. They can influence differentiation, and one of their important functions in the organism is to participate in the control of the immune response.

IFNs are produced by the immune system through mitogen, antigen and tumour-cell induction. It is of interest to note that a variety of IFNs may be produced, depending on the prevailing conditions, but at the moment we are unable to say which of them are produced most abundantly *in vivo*, and by which cell. It is clear, however, that the IFNs produced are capable of influencing both natural and adapted immunoregulation, and of playing a role in generating resistance to virus infections and perhaps also to tumour cells. IFNs can enhance the cytotoxicity of immune T cells, NK cells and macrophages, and suppress responses of lymphocytes. They can modulate and often increase antibody production and affect differentiation of B lymphocytes. The interaction between IFN and lymphocytes is a topic that will doubtless prove of the greatest importance in the context of various morbid states. The studies of the different IFN systems will be greatly facilitated when monoclonal antibodies to various sub-species of IFN are available.

IFNs influence the spontaneous cytotoxicity against cell lines *in vitro*. However, the findings on freshly separated cells from tumour biopsy specimens have been interpreted as indicating that the antitumour effect of IFN exerted through immunological mechanisms might be limited. Nonetheless IFN has been shown to induce the differentiation into NK cells of at least some populations, and also to increase the cytolytic activity of such cells for a number of targets. The stimulation of Fc receptor-mediated phagocytosis and at least a component of the inhibition of growth by IFN are mediated by cAMP. cAMP is not involved in the induction of the antiviral state. IFN can also increase phagocytosis.

In all these tested systems a number of recombinant and hybrid IFN molecules seems to share some properties and differ in respect of others.

IFNs can also exert an antiproliferative effect. Cells that are susceptible to this effect of IFN enter a G_0-like state and thus leave the active proliferative phase. Sometimes a G_1–G_2 phase delay can be detected. It is of interest to note that osteosarcoma cell lines are sensitive to the growth-inhibitory action of IFN in tissue culture. An antiproliferative activity is also exerted *in vitro* by recombinant human leucocyte IFN.

IFN-induced inhibition of the replication of murine leukaemia is due to deficiencies in the systems that result both in the formation of infectious particles and in the production of virions. Virus mRNA is impaired in IFN-treated cells. The murine leukaemia viruses produced by IFN-treated cells have been found to contain structural defects; their release in this system is inhibited by IFN. Trapping of oncorna virus particles has been found to occur at the surface of IFN-treated cells. The importance of the IFN-ganglioside interaction to the triggering of the antiviral response has been stressed. Several actions of IFN may actually involve the cell surface.

IFN has been shown to promote induced haemoglobin synthesis, but at the same time the constitutive haemoglobin production and the viral induction of leukaemia are both inhibited.

The action of IFN can be modified by butyrate. For example, a combination of IFN and butyrate may well potentiate growth inhibition in various systems, since IFN usually inhibits the passage of cells from G_0 to G_1, while butyrate inhibits events involved in the transit from G_1 to S. 2–5A seems to be the primary mediator of the inhibition of protein synthesis by dsRNA. The control mechanisms of the 2–5A system are at present being investigated in detail. From the aspect of combination therapy it is of interest that while cycloheximide and puromycin can inhibit the antiviral activity of IFNs, they can enhance their anticellular action *in vitro*.

ACTION *IN VIVO*

The genetics of the production and action of IFN have been studied in animals, and the next step is, of course, to extend this work to man. The production of IFN in response to viruses in animals is controlled by specific genes, and host genes can also influence the action of IFN. To my knowledge the relationship between production and sensitivity has not yet been extensively investigated in animal systems.

IFN is known to be capable of influencing a number of systems *in vivo*. For example, it can modify virus infections and the course of tumour diseases, and in animals it has even induced diseases. No serious toxic effects of long-term IFN therapy in man have been reported, but during the last few years the formation of homologous antibodies to IFN has been detected. The injection of IFN into newborn mice and rats can induce toxic effects in the liver and kidney.

Some IFNs are acid labile while others are stable. The practical implications of this difference *in vivo* is unknown. It has been suggested that an inhibitor of IFN action may be involved in the regulation of the whole IFN system.

IFN constitutes an autoregulatory mechanism for NK cells, but the practical significance of any such effect is not known. Perhaps the most important factor in this effect of IFN would be an influence on the growth of NK cells and their differentiation from their progenitors. The role of NK cells in the IFN patients is not yet clear. NK cell activity is raised in patients receiving either human leucocyte or recombinant IFN-α. A lowering of the NK response has been reported in some patients. No correlation has been found between the effects of IFN-α on NK cells and clinical response to the therapy. Influenza infection in man induces production of IFN and also increases the level of the NK cell activity.

It has been suggested that IFN is capable of modifying the differential gene expression involved in cell differentiation, and amniotic fluid has been shown to contain IFN.

IFN is an important factor determining the susceptibility of newborn mice to oncogenic viruses. Inhibition of the growth of human tumours in nude mice has been achieved by injecting human IFN. It has been proposed, however, that much of the antitumour action of IFN in man is host mediated. Thorough investigation of this type of action is required. In a new therapeutic method the IFN is immobilized in a perfusion chamber. However, if the effect of the substance is essentially direct, such a system should be abandoned.

PHARMACOKINETICS

Different IFNs possibly have different functions in the body; for example IFN-α might exert particular effects at a distant site while IFN-β may have short-range effects. Studies of the pharmacokinetics and of the inactivation and elimination of such different IFNs are extremely important for understanding the precise action of IFNs in the body. While the IFNs are rapidly removed from the circulation after intravenous injection, when the alpha type is injected intramuscularly a steady state can be obtained. The metabolism of IFN possibly resembles that reported for other plasma glycoproteins. The kidney seems to be involved in IFN catabolism. In this context there is need for various techniques of immune assay to be developed for estimating IFNs in tissue. Further investigation of the kinetics combined with the toxicology of IFNs is also needed, since these substances would seem to be responsible for symptoms characteristic of certain diseases.

The serum levels of IFN are difficult to determine after intramuscular or subcutaneous injections of IFN-β. Intravenous injections of this material have produced serum levels that persist for shorter times than those seen after intramuscular injections of IFN-α. The fact that IFN seems so readily to penetrate some body compartments is consistent with the glycoprotein nature of the substance. It is important to note that human IFN levels in cerebrospinal fluid are much lower (3%) than those in serum samples drawn simultaneously after parenteral injections.

In the amniotic system both alpha and beta types of IFN are present during pregnancy, and reported findings suggest that these substances are synthesized in the embryo. They do not seem to penetrate into the maternal blood.

According to clinical studies with recombinant leucocyte IFN, the pharmacokinetics and side effects would appear to be quite similar to those reported after injections of purified or semipurified Hu IFN-α obtained as a mixed population.

IFN-INDUCED SYMPTOMS

Among the main side effects of IFN therapy given by intramuscular injection are increased pulse rate and body temperature, together with thrombocytopaenia and leukopaenia. In addition, there are the various symptoms resembling those experienced during influenza, such as headache, malaise and fatigue. Local inflammatory reactions have also been reported. Injection of human lymphoblastoid IFN has been followed by pyrexia, hypertension, hypotension, myelosuppression and liver function disorders. All these important side effects, which seem to be reversible, appear to have their origin in the IFN molecule itself.

The rhesus monkey is an unreliable indicator of the toxicity of Hu IFN-α in man. When natural and recombinant IFNs are given to chimpanzees, however, the side effects are similar to those observed in man.

In comparative studies partially purified leucocyte IFN and recombinant IFN have produced various side reactions to roughly the same extent; they include fever, chills, myalgias, headache, fatigue and reversible leukopaenia.

In mice the total number of cells in the axillary lymph nodes has been found to be increased after injecting IFN. No increase in the afflux of lymphoid cells to lymphoid tissue has been reported. Multiplication of cells in the peripheral lymph nodes was not stimulated, but it was suggested that IFN inhibits release of lymphocytes from lymphoid tissue. This could, of course, have a bearing on the blood picture and account for specific symptoms in disease.

Prostaglandin A and IFN are related and it has been suggested that IFNs are involved in the action of prostaglandins. The prostaglandin system, especially the tumour prostaglandins, should now be investigated in this context. An inverse relationship between adjusted PG2 synthesis and tumour size has been noted.

ENDOGENOUS INTERFERON IN DISEASE

What kind of test can be devised to ascertain whether the level of IFN is increased in a particular disease? There is, of course, the direct measurement of IFN activity in the serum. Differences in the level of the enzyme 2–5 A synthetase can be measured, and tests and radioimmune assays for these purposes seem now to be available. The levels of IFN in the blood are known to be elevated during IFN therapy. Monitoring of the treatment can perhaps be effected to some extent by determining the synthetase activity in human peripheral leucocytes. The biological activity of IFN in the body can also be measured by examining the antiviral state in human peripheral leucocytes. It would be useful to characterize all the various IFNs being produced so as to determine their role under normal conditions and in disease.

It is important to know how IFN genes are switched on and off *in vivo*, and also which controls in the IFN system may be impaired in human disease. It would be especially valuable to know the function of the transcriptional and translational controls in various morbid conditions. Several monoclonal antibodies to human leucocyte IFN have been produced. Such antibodies are suitable for various practical purposes, and it is now time to decide what kinds of IFNs are produced, and where and under what conditions.

Among the new areas of research is the screening of some of the diseases where a virus aetiology is suspected, but where there is no evidence of such infection. Here, screening variables could be used that would indicate the activity of the IFN system. Such studies are already under way.

The IFN system is composed of a number of interacting components. IFN-γ possesses the interesting property of potentiating the effect of other IFNs. IFNs are generally potentiated by elevated temperatures. Because most of the principal actions of IFN each possesses several components the IFN system can probably be made highly selective in its action within the body. The complexity presumably also implies a great number of potential selective therapeutic applications of IFN. It has been established that IFN assumes a defensive role during viral infections. In patients treated with leucocyte IFN the serum level of beta-2-macroglobulin has been found to be increased.

We need to be able to recognize patients that have a deficient IFN system by monitoring them, especially during IFN therapy. It is possible that under certain conditions the IFN produced – which is distributed at low titres throughout the body – might potentiate other IFNs and cause superinduction of them during virus infection.

Antibodies to IFN have been detected not only in patients that have been given human IFN, but also in some that have not. The clinical significance and the frequency of such antibodies have to be determined. It is clear, however, that ones against both alpha and beta IFNs can exist; they are of the IgG type. None against gamma IFN has so far been reported. Inactivators of IFN-β have been found in human serum. Some body fluids are able to inactivate human IFN at normal temperature. Factors antagonistic to the antiviral action of IFN have been extracted from various human sarcomas.

It is of interest that prostaglandin is capable of enhancing the therapeutic efficacy of IFN inducers, and that it is also involved in the side effects of injected IFN. It is not known whether inhibitors of prostaglandin synthetase can be exploited to diminish the side effects of IFN therapy without prejudicing the efficacy of the IFN. Viruses do not promote the formation of prostaglandin. IFN induction is positively correlated to prostaglandin secretion.

It must be borne in mind that high extracellular concentrations of IFN may be required to afford protection against viruses. On the other hand, it is possible that systemic administration of IFN leads to potentiation of high-dose IFN production in patients suffering from virus infections.

Patients with trisomy 21 should be particularly sensitive to the action of IFN. Trisomic lymphocytes from such persons have proved more sensitive to the action of IFN. It is probably safe to say that trisomic 21 fibroblasts, monocytes and lymphocytes are all more sensitive to both natural and recombinant IFN-α.

IFN can exert a variety of actions on the immune system. It also modifies monocyte morphology. All tested subsets of T lymphocytes produce IFN-γ in response to phytohaemagglutinin. As the production of this type of IFN by each subset is enhanced by macrophages the lymphocyte–macrophage interaction would seem to be involved. A depressed IFN-γ response of T cells observed in patients with selective IgA deficiencies has been ascribed to lymphocyte defects. Depressed production of IFN-γ has already been observed in several diseases, and depressed production of the alpha type in various morbid states. However, what is cause and what is effect in these relationships remain to be established. The IFN system, including induction, production and so on, is extremely complicated. The type of inducer is probably an important factor. Virus-induced IFN may actually suppress immune responses.

It has been suggested that the NK-IFN system plays a critical part in defence against persistent viral infections and tumorigenesis. The role of NK cells in this context is obscure. During viral infection the immune system seems to be efficiently stimulated *in vivo* by IFN and it is possibly through this type of action that the maximum antiviral effect is to be obtained. IFN can both increase and decrease antibody production.

Two types of disorders, apart from the virus infections and tumour diseases, where IFNs have particularly been found are autoimmune connective tissue diseases and T cell chronic lymphocytic leukaemia. As has already been mentioned, IFN may normally be involved in certain of the human immunoregulatory diseases. It has been detected in a fairly large number of patients with SLE, rheumatoid arthritis, scleroderma and vasculitis. In the sera of SLE patients the predominant type of IFN has been found to be a leucocyte product, but it seems to be pH2 labile. Biologically, it resembles IFN-α, but pH2 lability is characteristic of the gamma type. The highest levels of IFNs are detected in SLE patients.

It is notable that cell-mediated immunity is decreased in SLE patients. This reaction can be inhibited by IFN. Some side effects of IFN therapy resemble symptoms associated with SLE. IFN can induce glomerulonephritis in newborn mice. It has been suggested than an increase in IFN production during certain morbid states may contribute to some of the clinical manifestations. Clearly, we need to know much more about the role of IFN in immunoregulatory disorders.

It is important to characterize the IFN-α subspecies that is inactivated at pH2 and that has been found in the sera of SLE patients. One complicating factor, however, is that the IFNs of different SLE patients seem not to be identical. The onset of autoimmune disease in NZB mice has been accelerated by the administration of IFN. No relationship between IFN and the clinical markers of SLE has been found. It has been suggested that SLE patients may have serum antibodies to IFN-α, and, moreover, that IFN–antibody complexes might be deposited in the kidneys.

Measurement of the 2–5 A synthetase levels in pregnant women has disclosed elevation of the enzyme from 12 weeks of gestation. The possibility that IFN generates resistance to infections during pregnancy has been suggested. IFN might also be involved in regulation of the maternal response to the foetal allograft. A further possibility is that IFN has an immunoregulatory role in pregnancy in general.

A technique has been worked out for estimating IFN-induced synthetase in leucocytes. The activity of the enzyme is raised in acute viral infections, immune diseases and virus-related malignancies, but not in bacterial infections or certain other non-infectious diseases. IFN-induced synthetase promises to be useful for monitoring patients. It is of interest that the level of the enzyme is increased in patients with SLE, and studies linking the IFN system with the beta-2-microglobulin system in such diseases are in progress.

While it is known that IFN can induce disease in mice and rats, the workers responsible for these findings do not consider the clinical application of IFN to be ruled out. From this research it is clear, however, that the substances being investigated are very potent, and that we need a greater fund of knowledge before IFN can be used safely in various patients.

INTERFERON AND VIRUS DISEASES

IFN has the property of inducing an antiviral state in cells which usually lasts for two to three days. The growth of most viruses tested is inhibited if enough IFN is added to the various tissue culture systems. Spontaneous leukaemia in mice caused by the murine leukaemia virus can be inhibited by injecting IFN preparations. Both therapeutic and antiviral effects have been demonstrated, and treatment of animals with virus-induced tumours has significantly reduced the tumour load and lengthened the life span.

IFN has been shown to reduce the severity of respiratory rhinovirus infections. However, the value of IFN as a therapeutic agent in such infections has yet to be substantiated. Herpes virus infections can be modified by IFN therapy. Herpes zoster patients experience fewer complications, the formation of new vesicles in primary dermatomes has been inhibited, and dissemination has been prevented. In children with varicella, visceral complications have been fewer in patients treated with IFN-α.

Forty-four child cancer patients developing varicella were treated with IFN-α. Serious dissemination developed in six out of 21 subjects composing a placebo group, compared with two out of 23 recipients of IFN. To judge from these results IFN-α therapy could benefit such patients.

Herpes labialis and herpes simplex viruses in trigeminal neuralgia have been affected by IFN treatment. The use of high concentrations of IFN applied locally in the treatment of herpetic dendritic keratitis has met with excellent results, especially when combined with tri-fluorothymidine (TFT) treatment. Both alpha and beta preparations seem to be active in herpes keratitis. Adenovirus conjunctivitis has also been treated successfully with IFN; corneal complications were less common in the IFN than in the placebo group.

Both local and systemic administration of IFN have diminished warts in many patients. There have also been reports of beneficial effects on verucca vulgaris in immunosuppressed patients, and on condyloma accuminata.

Human papilloma viruses can probably cause juvenile laryngeal papilloma in children. Prolonged IFN treatment seems to be beneficial in such patients. The results obtained in the few patients treated to date have been excellent. It is notable that though IFN-β was without effect in two patients with laryngeal papilloma, when the alpha type was subsequently given the results were clearly positive. At the moment all one can say as regards the choice of IFN preparation for use in laryngeal papilloma is that human leucocyte IFN is to be preferred.

Mouse L-cell IFN can decrease the level of transformation induced by bovine papilloma virus in mouse C127 cells. When established mouse cells transformed by papilloma virus were exposed to IFN for a long time the number of plasmid viral genomes in the transformed cells was decreased. After prolonged treatment with IFN, isolated flat revertants were found to be cured of their viral DNA sequences – as determined by hybridization experiments. This could be an elegant model for the human papilloma virus-induced diseases.

IFN therapy has been found to inhibit the activation of cytomegalovirus infection in kidney-graft recipients. In double-blind placebo-controlled trials of IFN treatment in such patients the excretion of cytomegalovirus began earlier, and viremia was more frequent in the placebo than in the IFN group. In further trials the recombinant human IFN preparations will be explored to see whether they are capable of producing similar effects. Chronic cytomegalovirus infection has also been affected by IFN treatment.

IFN therapy has been employed in chronic active hepatitis associated wih hepatitis B virus infection. While it is recognized that IFN therapy can influence this disease, it is still not known whether it is beneficial for thc paticnts. Because the results have been rather inconsistent when IFN was used alone, it is usually combined with other antiviral drugs. It is possible that this form of treatment will eventually lead to better results, and there are already indications that the combinations of IFN and ara-AMP might be helpful to these patients. IFN has also been reported to be effective in haemorrhagic dengue, which is a serious disease in some countries.

A series of analytical methods has been described in which

various factors of the antiviral IFN system have been evaluated. These methods have already been applied to some patients suffering from viral diseases. They were used to detect patients with a defective IFN system, for whom it should be considered whether or not IFN would be of benefit. Such methods could also be used for obtaining optimal pharmacokinetic information.

INTERFERON AND TUMOUR DISEASE

It has been shown in animals and in man that IFN can inhibit tumour growth. This applies to both benign and malignant tumours, including breast cancer, non-Hodgkin lymphoma and cancer of the colon. Among the tumours treated at the Karolinska Hospital in Stockholm are osteosarcoma, myelomatosis, ovarian cancer and hypernephroma. Our experience of the use of IFN in patients with these diseases will be briefly surveyed.

In an adjuvant IFN trial on osteosarcoma carried out in Sweden 51 patients have been treated with IFN since 1972; 30 have been treated without adjuvant therapy since 1972 and 21 patients have received chemotherapy from 1976 up to now. The two latter groups constitute concurrent controls, while the historical control group at the Karolinska Hospital contains 35 patients. The patients receive 3×10^6 units daily by intramuscular injection for one month, and then three times a week for a further 17 months. After three years follow-up no metastases were found in 46% of the IFN group, 44% of the chemotherapy group, 31% of the concurrent control group not receiving adjuvant therapy, and 14% of the historical control group. At five years the figures were 44, 40, 31 and 14%. The corresponding survival figures at three years were 61, 56, 34 and 17%. At five years the survival figures were 54, 40, 34 and 14%, respectively.

For seven patients developing metastases despite IFN treatment over the last three years radiotherapy (20 Gy) and daily IFN (3×10^6 units daily intramuscularly) have been given. Among the seven patients the response was complete in one and partial in one; a stationary state was recorded in five, the longest duration being more than two and a half years.

In 1979 we reported the effects of IFN therapy in a case of myelomatosis. There seemed to be an antitumour effect and this led to the first myelomatosis study in Stockholm in which human leucocyte IFN was used.

In an attempt in Stockholm to assert the value of human leucocyte IFN in the treatment of myelomatosis a pilot study on eight patients that had not previously received treatment and five patients that had was undertaken. The dose was 3×10^6 I.U. daily, injected intramuscularly. From November 1978 to January 1980 a randomized clinical trial was performed in which the results obtained with 3×10^6 I.U. of leucocyte IFN daily were compared to those yielded by high-dosage intermittent melphalan/predisolone (MP) therapy. Between May and September 1980 a study was performed in which patients received either MP or 3×10^6 I.U. of leucocyte IFN daily. From September 1980 to September 1982 patients were again assigned at random to the MP and IFN groups (dose 3×10^6 I.U. daily). From June 1981 IgG myelomatosis patients were excluded. The pooled results of these three studies are as follows.

MP was given to 53 patients and IFN to 71. There are no significant differences with respect to clinical stage or M-component subtypes. The response to initial therapy was more favourable in the MP group, with 39% responders against 14% for the IFN group ($P<0.05$). Thus, so far, MP therapy is superior to IFN for the initial treatment of myelomatosis. If, however, various immunoglobulin M-component subtypes are considered, there is no statistical difference between the IFN and the MP groups as regards the response rate of patients with IgA and Bence-Jones myelomas. Because of the very low response of IgG myelomatosis to IFN it was decided that the drug should not be prescribed for this form of tumour. In a study begun in 1982 the effects of a high-dose regimen (30×10^6 I.U.) are being compared with those obtained with a low-dose regimen (3×10^6 I.U.) and with standard MP therapy.

At the Department of Gynecological Oncology at Radiumhemmet, Karolinska Hospital in 1979 five patients with advanced seropapillary ovarian carcinoma, all of whom had previously received other types of therapy, were given human leucocyte IFN as the sole form of treatment. This consisted of daily intramuscular injections of 3×10^6 I.U. In all but one of the five patients there appeared to be some decrease in tumour volume. Ascitic fluid production was halted in two of the patients and a partial response was observed in one, while in the other two the disease was stabilized for more than a year.

The same team is at present conducting two studies on patients with advanced ovarian carcinoma. In one of the studies IFN is given to patients that have previously received other forms of treatment, but here increasing daily doses of IFN are being given (3–9–27 $\times 10^6$ I.U.) to see whether there is a dose-response correlation. In the other study five patients with advanced ovarian carcinoma (FIGO stage III or IV, seropapillary or undifferentiated) are being given 3×10^6 I.U. of IFN as the sole form of treatment.

The Urology Section of Radiumhememt and the Department of Urology at the Karolinska Hospital are carrying out a study on patients with lung metastases of hypernephroma where the primary tumour has been removed. Two alternative treatments are being given. In one group irradiation (10 Gy) is delivered to both lungs for one month, and vincristin and bleomycin are administered twice a week. In the other group IFN (3×10^6 I.U.) is given intramuscularly every day for three months, after which a fresh evaluation is made. The preliminary results suggest a beneficial effect on some of the first patients treated. Renal carcinoma has a variable natural course. Of crucial importance in this connection are staging and prognostic factors. When intramuscular injections of Hu IFN-α were given to patients with metastatic renal cell carcinoma at the M. D. Andersson Hospital partial responses accounted for 26%, while the disease progressed in 37%; in the remaining 37% the condition was stabilized. Lung or mediastinal metastases seemed to respond best.

The principal mechanism of the antitumour activity is unknown. The situation is complicated by the complexity of the actions of IFN; for example, alteration of cell metabolism, modification of the immune system, and inhibition of tumour cell growth. Priming and blocking are phenomena that remain to be thoroughly explored *in vivo* and whose practical significance must be established.

In 1981, IFN antibodies were detected in a nasopharyngeal carcinoma patient treated in Germany with Hu IFN-β. The IFN therapy led to a complete remission. Long after the IFN therapy was withdrawn antibodies were present, but there was no recurrence. In Germany 16 patients with nasopharyngeal carcinoma have received IFN treatment. Among the five for whom results are available there was complete regression in one, a stationary state in two and progression of the disease in two.

In the United States IFN-α has been given to many cancer patients in studies conducted within the Biological Response Modifiers programme in collaboration with the Burroughs Wellcome and the Roche companies. From preliminary results it appears that highly purified recombinant and non-recombinant IFN-α can exert an antitumour action *in vivo* in man.

In large trials that are being conducted in the United States, IFN is being used in the treatment of breast cancer, malignant melanoma and non-Hodgkin lymphoma. None of 23 patients with various types of tumours receiving direct intra-arterial injections of high daily doses of IFN-α exhibited a partial or complete response. Extensive studies are also being conducted on haematological disorders and melanoma in England. For melanoma the results have not yet been very promising, but for breast cancer and, especially, non-Hodgkin lymphoma they are much better. The use of combined IFN-therapy and chemotherapy has been suggested, and such combinations are at present being explored.

PROSPECTS

Rules have been drawn up for the testing and control of IFN. The need for standardizing the potency of human IFN preparations and IFN tissue content has been emphasized. IFNs have been shown to be involved in many virus infections. They are also able to affect tumours and in some cases to bring about their regression in man. During the 1980s therapeutic studies on recombinant DNA-derived IFN-β and with partially pure and recombinant IFN-γ will probably be initiated. It has been suggested that adjuvant studies with IFN may start during the present decade.

A variety of IFN preparations have so far been employed in patients suffering from a wide range of diseases. At the moment the main drawback of IFN therapy is that we do not know for certain whether the drug should be used on its own or in combination with other substances. Nor do we know the physiological role that IFN plays in the body. Increased IFN concentrations detected in a number of important diseases might be responsible for some of the observed pathological alterations. IFN might also be the cause of several symptoms in various diseases. Clearly, we are now moving ahead with refined assay techniques and purer products, and it will be exciting to follow the development within this area in the next few years. The role of IFN in disease is just beginning to be exploited.

REFERENCES

An exhaustive reference list would be outside the scope of this Survey.

There follows a list of review articles and conference reports relevant to the content of the Survey.

REFERENCES

ALLISON, A. C. (1979). Mode of action of immunological adjuvants. *Journal of the Reticuloendothelial Society*, **26,** 619–30.

ARVIN, A. M. (1980). Interferon as an antiviral and anti-tumor therapeutic agent. *Ophthalmology*, **87,** 1236–1238.

ARVIN, A. M., SCHMIDT, N., CANTELL, K. & MERIGAN, T. C. (1982). Interferon administration to infants with congenital rubella. *Antimicrobial Agents and Chemotherapy*, in press.

ATTALLAH, A. M., PETRICCIANI, J. C., GALASSO, G. J. & RABSON, A. S. (1980). Report of a workshop on standards for human interferon in clinical trials. *Journal of Infectious Diseases*, **142,** 300–1.

BAGLIONI, C. (1979). Interferon induced enzymatic activities and their role in the antiviral state. *Cell*, **17,** 255–64.

BAGLIONI, C. & NILSEN, T. W. (1981). The action of interferon at the molecular level. *American Scientist*, **69,** 392–9.

BALKWILL, F. R. (1979). Interferons as cell-regulatory molecules. *Cancer Immunology and Immunotherapy*, **7,** 7–14.

BALL, L. A. (1981). Molecular mechanisms of inferon action. *Medical and Pediatric Oncology*, **9,** 83–8.

BARON, S. (1979). The interferon system. *American Society of Microbiology News*, **45,** 358–66.

BARON, S. & DIANZANI, F. & STANTON, G. J. (1982). The interferon system. *Texas Reports on Biology and Medicine*, vol. **41,** in press.

BAUM, S. J. & LEDNEY, G. D. (1980). *Experimental Hematology Today*. New York: Springer-Verlag.

BERMAN, B. & FRANKFORT, H. M. (1982). The human interferon system. *International Journal of Dermatology*, **21,** 12–18.

BILLIAU, A. (1981). Interferon therapy: Pharmacokinetic and pharmacological aspects. *Archives of Virology*, **67,** 121–33.

BILLIAU, A. (1981). The clinical value of interferons as antitumour agents. *European Journal of Cancer and Clinical Oncology*, **17,** 949–67.

BILLIAU, A., DAMME, J. V., LEUVEN, F. V., EDY, V. G., DE LEY, M., CASSIMAN, J. J., VAN DE BERGHE, H. & DE SOMER, P. (1979). Human fibroblast interferon for clinical trials: production, partial purification, and characterization. *Antimicrobial Agents and Chemotherapy*, **16,** 49–55.

BILLIAU, A., DE SOMER, P., EDY, V. G., DECLERCQ, E. & HEREMAN, S. (1979). Human fibroblast interferon for clinical trials: pharmacokinetics and tolerability in experimental animals and humans. *Antimicrobial Agents and Chemotherapy*, **16,** 56–63.

BILLIAU, A., DE SOMER, P., SCHELLEKENS, H. & WEIMAR, W. (1981). The clinical value of interferon – a critical appraisal. *Netherlands Journal of Medicine*, **24,** 72–8.

BISHOP, D. H. L. (1980). *Rhabdoviruses*. CRC Press.

BLOOM, B. R. (1980). Interferons and the immune system. *Nature*, **284,** 593–5.

BLOOM, B. R. (1980). The unspecificity of cellular reactions. *Journal of Immunology*, **124,** 2527–9.

BOCCI, V. (1980). Possible causes of fever after interferon administration. *Biomédicine*, **32,** 159–62.

BOCCI, V. (1980). Is interferon produced in physiologic conditions? *Medical Hypotheses*, **6,** 735–45.

BOCCI, V. (1981). Which will be the most useful interferon? *Ricera in Clinica e in Laboratorio*, **11,** 27–32.

BOCCI, B. (1981). Pharmacokinetic studies of interferons. *Pharmacology and Therapeutics*, **13,** 421–40.

BOCCI, V. (1981). Production and role of interferon in physiological conditions. *Biological Reviews of the Cambridge Philosophical Society*, **56,** 49–85.

BOCCI, V. (1982). Catabolism of interferons. *Survey of Immunologic Research*, **1,** 137–43.

BORDEN, E. C. (1979). Interferons: Rationale for clinical trials in neoplastic disease. *Annals of Internal Medicine*, **91,** 472–9.

BORDEN, E. C. & BALL, L. A. (1981). Interferons: biochemical, cell growth inhibitory and immunological effects. *Progress in Hematology*, **XII,** 299–339.

BORDEN, E. C. & HAWKINS, J. J. (1980). Interferons for human neoplastic and viral diseases. *Comprehensive Therapy*, **6,** 6–15.

BORECKY, L. (1979). Twenty years after discovery of interferon: the progress and the persistant problems. *Acta Biologica et Medica Germanica*, **38,** 709–31.

BORMAN, S. A. (1981). How much interferon? *Analytical Chemistry*, **53,** 818A–20A.

BROUTY-BOYÉ, D. (1980). Inhibitory effects of interferons on cell multiplication. *Lymphokine Reports*, **1,** 99–112.

BROWN, D. & FOX, C. F. (1981). *Developmental Biology Using Purified Genes*. New York: Academic Press.

BURKE, D. C. (1981). Methods for enucleation and reconstruction of interferon-producing cells. *Methods in Enzymology*, **79,** 552–7.

BURKE, D. C. (1981). Some recent advances in the molecular biology of the interferon system. *Acta Medica Academiae Scientiarum Hungaricae*, **38,** 283–8.

CAME, P. E. & CALIGUIRI, L. A. (1982). *Chemotherapy of Viral infections*. Berlin: Springer Verlag (in press).

CAME, P. E. & CARTER, W. A. (1982). *Handbook of Experimental Pharmacology: Interferons and their Applications*. New York: Springer AG.

CANTELL, K. (1978). Towards the clinical use of interferon. *Endeavour*, **1,** 27–30.

CANTELL, K. (1979). Why is interferon not in clinical use today? *Interferon*, **1,** 1–28.

CANTELL, K. & HIRVONEN, S. (1978). Large scale production of human leukocyte interferon containing 10^8 units per ml. *Journal of General Virology*, **39,** 541–3.

CANTELL, K., HIRVONEN, S., KAUPPINEN, H. L. & MYLLYLÄ, G. (1981). Production of interferon in human leukocytes from normal donors with the use of Sendai virus. *Methods in Enzymology*, **78,** 29–38.

CANTELL, K. & STRANDER, H. (1977). Human leukocyte interferon for clinical use. In *Blood Leukocytes. Functions and Use in Therapy*, ed. C. F. Högman, K. Lindahl-Kiessling & H. Wigzell, pp. 73–5. Stockholm: Almqvist & Wicksell.

CARTER, W. A. (1979). Glycosylation, intraspecies molecular heterogeneity and trans-species activity of mammalian, interferons. *Life Sciences*, **25,** 717–28.

CARTER, W. A. & HOROSZEWICZ, J. S. (1980). Production, purification and clinical application of human fibroblast interferon. *Pharmacology and Therapeutics*, **8,** 359–77.

CHANDRA, P. (1978). *Antiviral Mechanisms in the Control of Neoplasia.* New York: Plenum Press.
CHANG, T. W. & SNYDMAN, D. R. (1979). Antiviral agents: action and clinical use. *Drugs*, **18,** 354–76.
CLEMENS, M. (1979). Interferons and cellular regulation. *Nature*, **282,** 364–5.
COHN, Z. A. (1978). The activation of mononuclear phagocytes: fact, fancy and future. *Journal of Immunology*, **121,** 813–16.
CUATRECASAS, P. & GREAVES, M. F. (1978). *Receptors & Recognition*, **A5,** 134–212.
CULLITON, B. J. & WATERFALL, W. K. (1979). War on cancer – interferon. *British Medical Journal*, **2,** 195–6.
DE CLERCQ, E. (1979). New trends in antiviral chemotherapy. *Archives Internationales de Physiologie et de Biochemie*, **87,** 353–433.
DE MAEYER, E. (1978). Interferon twenty years later. *Bulletin de l'Institut Pasteur*, **76,** 303–24.
DE MAEYER, E. (1979). L'Interféron. *Nouvelle Presse Médicale*, **8,** 287–2.
DE MAEYER, E. & DE MAEYER-GUIGNARD, J. (1979). Interferons. In *Virus-Host-Interactions: Immunity to Viruses*, ed. H. Fraenkel-Conrat & R. R. Wagner, *Comprehensive Virology*, vol. **15**. New York: Plenum Press.
DE MAEYER, E. GALASSO, G. & SCHELLEKENS, H. (1981). *The Biology of the Interferon System*. Amsterdam, New York & Oxford: Elsevier/North-Holland Biomedical Press.
DE MAEYER-GUIGNARD, J. (1979). L'interferon et ses inducteurs. *Revue du Practicien*, **29,** 2923–33.
DE VITA, V. T. & KERSHNER, L. M. (1980). Cancer, the curable disease. *American Pharmacy*, **20,** 16–22.
DE WECK, A. L., KRISTENSEN, F. & LANDY, M. (1980). *Biochemical Characterization of Lymphokines*. New York: Academic Press.
DORIA, G. & ESHKOL, A. (1980). *The Immune System: Function and Therapy of Dysfunction*. New York: Academic Press.
DRACH, J. C. (1980). Antiviral agents. *Annual Reports of Medical Chemistry*, **15,** 149–61.
DUNNICK, J. K. & GALASSO, G. J. (1979). Clinical trials with exogenous interferons: Summary of a meeting. *Journal of Infectious Diseases,* **139,** 109–23.
EDITORIAL (1979). Can interferons cure cancer? *Lancet*, **i,** 1171–2.
EINHORN, S. (1980). *Influence of Interferon on Human Lymphocytes*. Thesis, Karolinska Institute, Stockholm.
EPSTEIN, C. J. & EPSTEIN, L. B. (1982). Genetic control of the response to interferon. *Texas Reports on Biology and Medicine*, **41,** 324–31.
EPSTEIN, L. (1979). The comparative biology of immune and classical interferons. In *Biology of the Lymphokines*, ed. S. Cohen, E. Pick & J. J. Oppenheim. New York: Academic Press.
EPSTEIN, L. B. (1982). Interferon-gamma: success, structure and speculation. *Nature*, **295,** 453–4.
FINTER, N. B. (1973). *Interferons and Interferon Inducers*. Amsterdam/London: North-Holland.
FINTER, N. B. (1981). Standardization of assay of interferons. *Methods in Enzymology*, **78,** 14–22.
FINTER, N. B. & FANTES, K. H. (1980). The purity and safety of interferons prepared for clinical use: The case for lymphoblastoid interferon. *Interferon*, **2,** 65–79.
FRIEDMAN, R. M. (1977). Antiviral activity of interferon. *Bacteriological Revues*, **41,** 543–67.

Friedman, R. M. (1978). Interferons and cancer. *Journal of National Cancer Institute*, **60,** 1191–4.
Friedman, R. M. (1978). Interferon action and the cell surface. *Pharmacology and Therapeutics*, **A 2,** 425–38.
Friedman, R. M. (1979). Interferons: interaction with cell surfaces. *Interferon*, **1,** 53–74.
Friedman, R. M. (1979). Induction and production of interferon. *Methods in Enzymology*, **58,** 292–6.
Friedman, R. M. (1981). Cell surface alterations induced by interferon. *Methods in Enzymology*, **79,** 458–61.
Galasso, G. J. (1981). Antiviral agents: the road from scepticism to efficacy. *Acta Medica Academiae Scientiarum Hungaricae*, **38,** 313–24.
Galasso, G. J. (1981). An assessment of antiviral drugs for the management of infectious diseases in humans. *Antiviral Research*, **1,** 73–96.
Galasso, G. J. (1981). Antiviral agents for the control of viral diseases. *Bulletin of the World Health Organization*, **59,** 503–12.
Galasso, G. J., Merigan, T. C. & Buchanan, R. (1979). *Antiviral Agents and Viral Diseases in Man.* New York: Raven Press.
Gallmeier, W. M. (1980). Interferon gegen Krebs? *MMW. Münchener Medizinische Wochenschrift*, **122,** 1261–4.
Goffinet, D. R., Glatstein, E. J. & Merigan, T. C. (1972). Herpes Zoster – Varicella infections and lymphoma. *Annals of Internal Medicine*, **76,** 235–40.
Gordon, J. E. & Minks, M. A. (1981). The interferon renaissance: molecular aspects of induction and action. *Microbiological Reviews*, **45,** 244–66.
Gordon, M., Buchanan, R. L., Crenshaw, R. R., Kerridge, K. A., MacNintch, J. E. & Siminoff, P. (1981). A review of approaches to viral chemotherapy. *Journal of Medicine*, **12,** 289–382.
Gresser, I. (1977). On the varied biological effects of interferon. *Cellular Immunology*, **34,** 406–15.
Gresser, I. (1977). Antitumour effects of interferon. In *Cancer: a Comprehensive Treatise*, ed. F. Becker, *Chemotherapy*, **5,** 521–71.
Gresser, I. (1980). Usefulness of the results of studies on the antitumour effects of interferon in animals to interferon therapy of patients. *Recent Results of Cancer Research*, **75,** 226–8.
Gresser, I. (1981). *Interferon 1981.* New York: Academic Press.
Gresser, I. & Tovey, M. G. (1978). Antitumour effects of interferon. *Biochimica et Biophysica Acta*, **516,** 231–46.
Grieco, M. H. (1980). *Infections in the Abnormal Host.* Yorke Medical Books.
Gross, E. & Meienhofter, J. (1982). *The Peptides: Analysis, Synthesis, Biology.* New York: Academic Press.
Grossberg, S. E. (1981). On reporting interferon research. *Annals of Internal Medicine*, **95,** 115–16.
Groth, F. de S. E. & Scheidegger, S. (1980). Production of monoclonal antibodies: Strategy and tactics. *Journal of Immunology Methods*, **35,** 1–25.
Gutterman, J. U. (1981). Clinical investigation of the interferons in human cancer. *Bulletin du Cancer*, **33,** 271–8.
Gutterman, J. & Quesada, J. (1982). Clinical investigation of partially pure and recombinant DNA-derived leukocyte interferon in human cancer. *From Gene to Protein: Translation into Biotechnology*, 1982 Miami Winter Symposium (in press).
Gutterman, J. & Quesada, J. (1982). Clinical investigation of partially pure and recombinant DNA derived leukocyte interferon in human cancer. *Texas Reports on Biology and Medicine*, **41** (in press).

Harrison, E. A. (1979). *Interferon.* Springfield: National Technical Information Service.

Havanessian, A. G. (1979). Intracellular events in interferon-treated cells. *Differentiation*, **15,** 139–51.

Herberman, R. B. (1981). Natural killer (NK) cells. *Progress in Clinical and Biological Research*, **58,** 33–43.

Herberman, R. B. (1982). *Natural Cell-mediated Immunity Against Tumours* (in press).

Herberman, R. B. & Holden, H. T. (1978). Natural cell-mediated immunity. *Advances in Cancer Research*, **27,** 305–77.

Ho, M. (1979). Interferon and its therapeutic potential. *Medical Journal of Australia,* **2,** 11–12.

Ho, M. & Armstrong, J. A. (1975). Interferon. *Annual Review of Microbiology*, **29,** 131–61.

Holland, J. M. (1973). Cancer of the kidney – natural history and staging. *Cancer*, **32,** 1030–42.

Horowitz, B. (1981). Human interferon – properties, clinical application, and production. *Journal of Parenteral Science and Technology*, **35,** 223–6.

Horoszewicz, J. S., Leong, S. S., Ito, M., Di Berardino, L. & Carter, W. A. (1978). Aging in vitro and large-scale interferon production by 15 new strains of human diploid fibroblasts. *Infection and Immunity*, **14,** 720–6.

Högman, C. F. (1979). Advances in blood transfusion. *Revue Française de Transfusion et Immuno-Hematologie*, **22,** 9–15.

Ikawa, Y. (1979). *Oncogenic Viruses and Host Genes*. New York: Academic Press.

Ingimarsson, S. (1980). *Studies on Human Leukocyte Interferon in Clinical Use.* Thesis, Karolinska Institute, Stockholm.

Interferon Standards: a memorandum. *Journal of Biological Standardization*, **7,** 383–95.

International Committee: Interferon Nomenclature (1980). *Nature*, **286,** 110.

Isaacs, A. & Lindenmann, J. (1957). Virus interference. I. The interferon. *Proceedings of the Royal Society of London, Series B, Biological Sciences*, **147,** 258–67.

Johnson, H. M. (1980). Viruses, interferon and the immune response. *Illinois Medical Journal*, **157,** 294–9.

Johnson, H. M. & Baron, S. (1976). Interferon: effects on the immune response and the mechanism of activation of the cellular response. *CRC Critical Reviews in Biochemistry*, **4,** 203–27.

Johnson, H. M. & Baron, S. (1976). Regulatory role of interferons in the immune response. *International Research Communication of Systematic Medical Sciences*, **4,** 50–2.

Johnson, H. M. & Baron, S. (1977). Evaluation of effects of interferon and interferon inducers on the immune response. *Pharmacology and Therapeutics*, **A 1,** 349–67.

Joklik, W. K. (1977). The mechanism of interferon action. *Annals of the New York Academy of Sciences*, **284,** 711–17.

Karmali, R. A. (1980). Review: Prostaglandins and cancer. *Prostaglandins & Medicine*, **5,** 11–28.

Khan, A. & Hill, N. O. (1982). *Human Lymphokines* (in press).

Khan, A., Hill, N. O. & Dorn, G. L. (1980). *Interferon: Properties and Clinical Uses.* Dallas: Leland Fikes Foundation Press.

Keil, T. U. (1980). Interferon – Hysterie. *MMW Münchener Medizinische Wochenschrift*, **122,** 1088.

Kern, E. R. & Glasgow, L. A. (1981). Evaluation of interferon and interferon inducers as antiviral agents: animal studies. *Pharmacology and Therapeutics D.*, **13,** 1–38.

Kirchner, H. & Beck, J. (1980). Interferone: Proteine mit antiviralen, antiproliferativen, immunoregulatorischen und antitumoralen Eigenschaften. *Verhandlungen der Deutschen Gesellschaft für Innere Medizin*, **86,** 496–502.

Klein, E. (1981). Interpretation and in vivo relevance of lymphocytotoxicity assays. *Hämatologie und Bluttransfusion*, **26,** 345–50.

Knight, E., Jr (1978). Purification of interferons. *Pharmacology and Therapeutics*, **A 2,** 439–46.

Kock, G. & Richter, D. (1979). *Regulation of Macromolecular Weight Mediators.* New York: Academic Press.

Kohn, A. (1979). Early interactions of viruses with cellular membranes. *Advances in Virus Research*, **24,** 223–76.

Kono, R. & Vilček, J. (1982). *The Clinical Potential of Interferons.* Tokyo: University of Tokyo Press.

Krim, M. (1980). Towards tumor therapy with interferons. Part I. Interferons: production and properties. *Blood*, **55,** 711–21.

Krim, M. (1980). Towards tumor therapy with interferons. Part II. Interferons: in vivo effects. *Blood*, **55,** 875–84.

Kuznetsov, V. P., Melchedov, L. N. & Soloviev, V. D. (1979). On preparations of human leukocyte interferon for clinical use. *Acta Biologica et Medica Germanica*, **38,** 801-6.

Leanderson, T. (1982). *Antiproliferative Effect of Interferon in a Burkitt's Lymphoma Cell Line.* Thesis, Umeå University, Umeå.

Lengyel, P. (1981). Enzymology of interferon action – a short survey. *Methods in Enzymology*, **79,** 135–48.

Lengyel, P. (1982). Biochemistry of interferons and their actions. *Annual Revue of Biochemistry*, **51,** 251–82.

Leong, S. S. & Horoszewicz, J. S. (1981). Production and preparation of human fibroblast interferon for clinical trials. *Methods in Enzymology*, **78,** 87–101.

Lin, C. (1982). Antiviral drugs. *Medical Clinics of North America*, **66,** 235–44.

Luby, J. P. (1979). Antivirals with clinical potential. *Advances in Internal Medicine*, **24,** 229–54.

Marx, J. L. (1979). Interferon (I): on the threshold of clinical application. *Science*, **204,** 1183–6.

Marx, J. L. (1979). Interferon (II): learning about how it works. *Science*, **204,** 1293–5.

McCredie, K. B. & Moore, M. S. (1981). Interferon: an early evaluation. *Journal of Medical Association of Georgia*, **70,** 495–6.

McGregor, A. M. (1981). Monoclonal antibodies: production and use. *British Medical Journal (Clinical Research ed.)*, **283,** 1143–4.

Merigan, T. C. (1979). Human interferon as a therapeutic agent. *New England Journal of Medicine*, **300,** 42–3.

Merigan, T. C. (1981). Present appraisal of and future hopes for clinical utilization of human interferon. *Interferon*, **3,** 135–54.

Merigan, T. C. & Friedman, R. M. (1982). *Interferons.* New York & London: Academic Press.

Merigan, T. C., Jordan, G. W. & Fried, R. P. (1975). Clinical utilization of exogenous human interferon. *Perspectives in Virology*, **9,** 249–64.

Milstein, C. (1982). Monoclonal antibodies. *Cancer*, **49,** 1953–7.

Mogensen, K. E. & Cantell, K. (1977). Production and preparation of human leukocyte interferon. *Pharmacology and Therapeutics C*, **1,** 369–81.

Moore, G. E. (1981). Interferon. *Surgery, Gynecology and Obstetrics*, **153,** 97–102.

MUNK, K. & KIRCHNER, H. (1982). Interferon – properties, mode of action, production, clinical application. *Contributions to Oncology*, vol. 11. Basel, Munich, Paris, London, New York & Sydney: S. Karger AG.

NAGANO, Y. (1975). *Virus-Inhibiting Factor.* Tokyo: University of Tokyo Press.

NELSON, D. J. (1976). *Immunobiology of the Macrophage.* New York: Academic Press.

NICOLSON, G. L., RAFTERY, M. A. & RODBELL, M. (1976). *Cell Surface Receptors.* New York: Alan R. Liss.

NOTHKINS, A. L. (1975). *Viral Immunobiology and Immunopathology.* New York: Academic Press.

PANEM, S. (1982). Interferon emerging as hormone-like substance. *JAMA, Journal of the American Medical Association*, **247**, 418–21.

PESTKA, S. (1981). Cloning of the human interferons. *Methods in Enzymology*, **79**, 599–601.

PESTKA, S. & BARON, S. (1981). Definition and classification of the interferons. *Methods in Enzymology*, **78**, 3–14.

POHL, A., MOSER, K. & MICKSCHE, M. (1981). Human-interferone – Eigenschaften und Möglichkeiten. *Wiener Klinische Wochenschrift*, **93**, 439–57.

POLLARD, R. B. (1980). Usages of interferon and interferon inducers in man: the first half of 1980. *Medical Biology*, **58**, 293–9.

POLLARD, R. B. (1981). Usages of interferon and interferon inducers in man: the second half of 1980. *Medical Biology*, **59**, 69–76.

POLLARD, R. B. (1982). Interferons and interferon inducers: development of clinical usefulness and therapeutic promise. *Drugs*, **23**, 37–55.

PRESBER, H. W. & WASCHKE, S. R. (1981). *Zeitschrift fur Arztliche Fortbildung*, **75**, 277–85.

PRIESTMAN, T. J. (1979). Interferon: an anticancer agent? *Cancer Treatment Reviews*, **9**, 223–37.

PRIESTMAN, T. J. (1981). Interferons and cancer: the end of the beginning or the beginning of the end. *Clinical Oncology*, **7**, 271–4.

REPORT OF A WHO SCIENTIFIC GROUP: INTERFERON THERAPY (1982). *World Health Organization Technical Report Series*, 676.

REVEL, M. (1979). Molecular mechanisms involved in the antiviral effects of interferon. *Interferon*, **1**, 101–63.

REVEL, M. & GRONER, Y. (1978). Post-transcriptional and translational controls of gene expression in eukaryotes. *Annual Review of Biochemistry*, **47**, 1079–126.

RODER, J. C., KARRE, K. & KIESSLING, R. (1981). *Progress in Allergy*, **28**, 66–159.

ROSE, N. R. & FRIEDMAN, H. (1980). *Manual of Clinical Immunology*, 2nd edn. Washington D. C.: American Society of Microbiology.

RUBIN, R. H. & YOUNG, L. H. (1981). *Clinical Approach to Infection in the Immunosuppressed Host.* New York: Plenum Press.

SALMON, S. E. (1980). *Cloning of Human Tumour Stem Cells.* New York: Alam R. Liss.

SCHELLEKENS, H. & DE WILDE, G. A. (1981). Relationship between the cellular expression of the antiviral and anticellular activities of interferon. *Antiviral Research*, **1**, 135–40.

SCHESINGER, D. (1980). *Microbiology 1980.* American Society of Microbiology Publication.

SCOTT, G. M., SECHER, D. S., FLOWERS, D., BATE, J., SANTELL, K. & TYRRELL, D. A. J. (1981). Toxicity of interferon. *British Medical Journal*, **282**, 1345–8.

SCOTT, G. M. & TYRRELL, D. A. J. (1980). Interferon, therapeutic fact or fiction for the '80s? *British Medical Journal*, **280**, 1558–62.

SEHGAL, P. B. (1982). How many human interferons are there? *Interferon*, **4**, 1–22.

SEHGAL, P. B. (1982). The interferon genes. *Biochimica et Biophysica Acta*, **695,** 17–33.

SELBITZ, H. J., SCHEEL, H., VIOGT, A. & SCHÖNHER, W. (1980). The interferons and their significance for prophylaxis and treatment of humans and animals. *Zeitschrift fur die Gesamte Innere Medizin und Ihre Grenzgebiete*, **35,** 837–41.

SELIGMAN, M. & HITZIG, W. H. (1980). *Primary Immunodeficiencies*. Amsterdam: Elsevier/North-Holland.

SHANNON, W. M. & SCHABEL, F. M. (1980). Antiviral agents as adjuncts in cancer chemotherapy. *Pharmacology and Therapeutics*, **11,** 263–390.

SIKORA, K. (1980). Does interferon cure cancer? *British Medical Journal*, **281,** 855–8.

SLATE, D. L. & RUDDLE, F. H. (1979). Genetics of the interferon system. *Pharmacology and Therapeutics*, **4,** 221–30.

SLATE, D. L. & RUDDLE, F. H. (1981). Methods for mapping human interferon structural genes. *Methods in Enzymology*, **79,** 529–36.

SLOMA, A., MCCANDLISS, R. & PESTKA, S. (1982). Translation of human RNAs in *Xenopus laevis* oocytes. *Methods in Enzymology* (in press).

SOLOVIEV, V. D. & BECKTIMIROV, T. A. (1981). *Interferon in Theory and Practice of Medicine*. Moscow: Medicina.

SONNENFELD, G. & MERIGAN, T. C. (1979). A regulatory role for interferon in immunity. *Annals of the New York Academy of Sciences*, **332,** 345–55.

SONNENFELD, G. & MERIGAN, T. C. (1979). The role of interferon in viral infections. *Springer Seminars in Immunopathology*, **2,** 311–38.

STEBBING, N. (1979). Interferons and inducers *in vivo*. *Nature*, **279,** 581–2.

STEWART, II, W. E. (1977). *Interferons and Their Actions*. Cleveland: CRC Press.

STEWART, II, W. E. (1979). Varied biological effects of interferon. *Interferon*, **1,** 29–46.

STEWART, II, W. E. (1979). *The Interferon System*. Wien & New York: Springer-Verlag.

STEWART, II, W. E. (1981). Clinical status of the interferons: will their promise be kept? *Hospital Practice*, **16,** 97–101.

STINEBRING, W. R. & CHAPPLE, P. J. (1978). *Human Interferon – Production and Clinical Use*. New York & London: Plenum Press.

STIEHM, E. R., KRONENBERG, L. H., ROSENBLATT, H. M., BRYSON, Y. & MERIGAN, T. C. (1982). UCLA Conference: Interferon immunobiology and clinical significance. *Annals of Internal Medicine*, **96,** 80–93.

STRANDER, H. (1977). Interferons: anti-neoplastic drugs? *Blut*, **35,** 277–88.

STRANDER, H. (1981). Interferon therapy of tumour disease. *Microecology and Therapy*, **11,** 135–40.

STRANDER, H. & EINHORN, S. (1982). Interferon in cancer – faith, hope and reality. *American Journal of Clinical Oncology*, **5,** 297–301.

STRINGFELLOW, D. A. (1980). *Interferon and Interferon Inducers, Clinical Applications*. New York & Basel: Marcel Pekher Inc.

STRINGFELLOW, D. A. (1981). Chemotherapy of viral infections. *Archives of Dermatological Research*, **270,** 141–55.

SUNDMACHER, R. (1981). *Herpetic Eye Diseases*. München: J. F. Bergmann.

TAMM, I. & SEGHAL, P. B. (1979). Interferons. *American Journal of Medicine*, **66,** 3–5.

TAN, Y. H. (1982). Interferon: no verdict yet. *Canadian Medical Association Journal*, **126,** 9.

TAYLOR, S. (1981). Interferons as antitumour agents. *Journal of the Kansas Medical Society*, **82,** 128–30.

TAYLOR-PAPADIMITRIOU, J. (1980). Effects of interferons on cell growth and function. *Interferon*, **2,** 13–46.

TORRENCE, P. F. & DE CLERCQ, E. (1981). Interferon inducers: general survey and classification. *Methods in Enzymology*, **78,** 291–9.

TOVEY, M. G. (1980). Viral latency and its importance in human disease. *Pathologie Biologie*, **26,** 631–4.

TREUNER, J. & DANNECHER, G. (1981). Interferon – Grundlagen und bisherige Klinische Anwendung. *Fortschritte der Medizin*, **99,** 807–13.

TURKIN, D. (1980). The interferon. Newest weapon in the fight against viral diseases. *American Pharmacy*, **20,** 33–5.

TYRRELL, D. A. J. & BURKE, D. C. (1982). Interferon: twenty-five years on. *Philosophical Transactions of the Royal Society of London*, **299,** 1–144.

VAN DAMME, J. & BILLIAU, A. (1981). Large scale production of human fibroblast interferon. *Methods in Enzymology*, **78,** 101–19.

VENGRIS, V. E., FERNIE, B. F. & PILKA, P. M. (1980). The interaction between gangliosides and interferon. *Advances in Experimental Medicine and Biology*, **125,** 479–86.

VILCĒK, J. (1979). Interferon as a cell product. *Advances in Experimental Medicine and Biology*, **118,** 117–27.

VILCĒK, J., GRESSER, I. & MERIGAN, T. C. (1980). Regulatory functions of interferons. *Annals of the New York Academy of Sciences*, **350**.

WARREN, S. L. (1980). A practitioner's guide to interferon. *Annals of Allergy*, **45,** 37–42.

WELSH, R. M. (1981). Natural cell-mediated immunity during viral infections. *Current Topics in Microbiology and Immunology*, **92,** 83–106.

WILLIAMS, B. R. G. & KERR, I. M. (1980). The 2–5A system in interferon-treated and control cells. *Trends of Biochemical Sciences*, 138–40.

YABROV, A. (1979). Interferon and cell-medical immunity. *Medical Hypothesis*, **5,** 769–97.

ZSCHIESCHE, W. (1980). Actual results and problems of interferon research. *Deutsche Gesundheitswochenschrift*, **35,** 1609–18.

STRUCTURE AND EXPRESSION OF THE HUMAN INTERFERON GENES

JOHN COLLINS

G.B.F. – Gesellschaft für Biotechnologische Forschung mbH, Mascheroder Weg 1, D-3300 Braunschweig-Stöckheim, W. Germany

INTRODUCTION

The interferons are classified into three groups γ, β and α which differ in primary structure, presence or absence of glycosylation and the cell type and method of induction which yield the different species. Although accounts have appeared to the contrary, only single members of the interferon β and γ types will be referred to (for a review of the controversy, see Collins, 1983). The peptide sequence of interferon β is about 35% related with that of the consensus sequence for the IFN-α family. Interferon-γ, which will not be treated to any great extent here, is hardly related (< 15%) to interferon-α or -β at the amino acid sequence level. All the interferons share the same ability to induce an antiviral state in a target cell at femtomole concentrations, but vary in their specific activity and in the species specificity of this reaction. Human interferon-β, for example, shows almost no activity on mouse cells, whereas interferon-α1 (D) is equally active on both human and mouse cells.

To a very great extent our present knowledge of the properties of individual species comes from the cloning of individual interferon genes. It is unlikely that the picture is yet complete, but the majority of genes belonging to the three classes mentioned above are already characterized.

The isolation of individual genes has allowed the production of pure interferon peptides through genetically engineered *Escherichia coli* strains or in *Bacillus*, yeast and tissue culture cells. This represents not only the first detailed characterization of the individual species of interferon but also a realistic possibility for production of large quantities for possible clinical applications (which will be covered by other contributors to this symposium).

Another aspect of particular interest to the molecular biologist is the role played by sequences adjacent to eukaryotic genes in the regulation of the gene. Studies in this respect are well advanced and will be presented in some detail. Since this type of study has been carried out principally with the human interferon-β gene transferred into mouse cells, this has led directly to the novel possibility that correctly processed and glycosylated interferon (unobtainable from micro-organisms) could be efficiently produced from such hybrid tissue cultures. Since the significance of glycosylation is largely unknown, in particular the consequences for clinical application, this proposal could have repercussions for the production of many glycopeptides intended for medical use.

The recent reviews by Weissmann (1981) and Weissmann *et al.* (1982) have provided a rich source of information for this review, which is to some extent a shortened version of Collins (1983).

(1) INTERFERON PROTEINS

The purification of the interferons has been an arduous process which has been continuing for some 25 years. It has been complicated by the natural diversity of the gene products and subsequent post-translational modifications such as proteolytic cleavage and glycosylation. In addition, only trace amounts are normally produced, often, as for example from leucocytes, along with many other potent biologically active substances such as the lymphokines. The heterogeneity of leucocyte and lymphoblastoid interferons (mainly α-type with some β-type) has been observed on SDS gel electrophoresis (e.g. Heron *et al.*, 1981), isoelectric focusing (Törmä & Paucker, 1976; Stewart, Lin, Wiranowski-Stewart & Cantell, 1977), HPLC (Rubinstein *et al.*, 1979), affinity chromatography (Grob & Chadha, 1979) and through differential affinity to monoclonal antibodies (Staehelin *et al.*, 1981). Estimates of the molecular weight (M.W.) of IFN-α ranged between 16 and 23 kilodalton (kD). As will be discussed in detail below the M.W. of IFN-α and -β as naked peptides (without leader sequence) would be about 18.3 kD. Species lacking the 10 C-terminal amino acids, found by Levy *et al.* (1981) to be produced by chronic myelogenous leukaemia cells, would have a M.W. of 17.2 kD.

Interferons as glycoproteins

The extent of IFN-α glycosylation has been the subject of some controversy. Reports on the purification of interferons on affinity columns normally showing affinity for glycoproteins (Grob & Chadha, 1979) was an indication that IFN-α could be glycosylated (or adsorbed to glycoproteins). The reduction in overall M.W. and loss of charge heterogeneities following periodate treatment (Stewart *et al.*, 1977) was taken as a very strong indication that IFN-α was considerably glycosylated. The latter results appear not to be reproducible (M. Rubinstein, personal communication). Furthermore Allen & Fantes (1980) could find no glycopeptides in their studies on the amino sequences of the principal IFN-α species, and in fact only a very low level of amino sugars. IFN-α genes sequences indeed have only recently revealed two species out of 15 known (14 as listed in Table 1; one recently isolated by B. Lund, E. Lindgren & J. Collins, unpublished) which contain a possible site for asparagine-*N*-glycosylation, namely asparagine-X-(threonine or serine). This latter observation does not rule out the possibility that serine-*O*-glycosylation or other even rarer glycosylation sites are being used. Most authors, however, agree that less than 30% of the IFN-α species are glycosylated to any extent (see review by Berg, 1982).

(2) AMINO ACID SEQUENCES

By the summer of 1980 publications had appeared from a number of amino acid sequencing groups, which although dealing mostly with mixtures from different sources (e.g. lymphoblastoid cells or blood leucocytes), indicated that at least five different IFN-α species could be produced by human cells (Levy *et al.*, 1980; Zoon *et al.*, 1980; Allen & Fantes, 1980). As summarized by Zoon (1981) from five to ten distinct human IFN species have been reported as being induced by virus inducers in various cell types, including 'buffy coats', Namalwa cells, chronic myelogenous leukaemia cells and KG-1 cells. A summary of data on amino acid sequences of IFN-α peptides is given in Fig. 1 as compared with the IFN-α gene sequences. The comparison shows, using the consensus sequences they are 93% identical. There are some 19 polymorphisms in the peptide sequences for which no known gene exists. This may reflect artefacts in the amino acid sequencing method. Conversely a large number of gene sequence polymorphisms exist for which no corresponding amino acid has been found in the peptide sequence. This probably indicates that many of the interferon species are expressed only at a low level. IFN-α1 (D) and -α2 (A) appear to correspond to

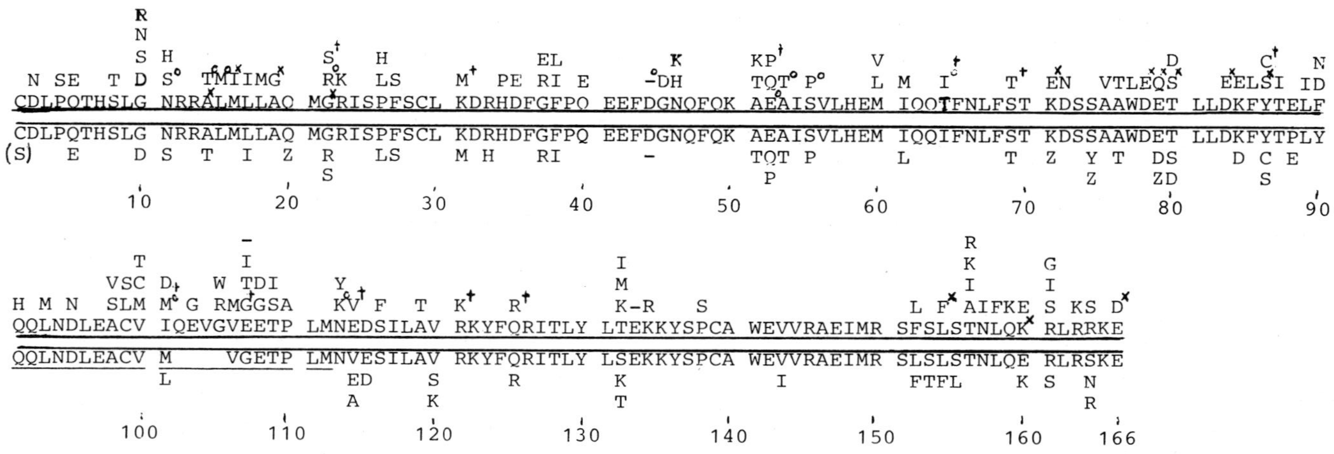

```
                R
                N
                S H              S        H                   EL              K          KP          V            +            D            C   N
  N SE    T D S°  TMIIMG    RK    LS      M  PE  RI E       -DH        TQT° P°     L M   I         T   EN    VTLEQS      EELSI ID
CDLPQTHSLG NRRALMLLAQ MGRISPFSCL KDRHDFGFPQ EEFDGNQFQK AEAISVLHEM IQQTFNLFST KDSSAAWDET LLDKFYTELF
====================================================================================================
CDLPQTHSLG NRRALMLLAQ MGRISPFSCL KDRHDFGFPQ EEFDGNQFQK AEAISVLHEM IQQIFNLFST KDSSAAWDET LLDKFYTPLY
(S)   E     D S  T  I  Z   R    LS       M H     RI         -         TQT  P        L          T   Z   Y T   DS      D C E
                           S                                        P                              Z     ZD        S
          10        20        30        40        50        60        70        80        90

                      -                                                       R
                T     I                     I                                 K      G
              VSC D   W TDI       Y                         M                 I      I
H M N    SLM M  G RMGGSA      KV F   T  K    R         K-R    S               L F AIFKE S KS D
QQLNDLEACV IQEVGVEETP LMNEDSILAV RKYFQRITLY LTEKKYSPCA WEVVRAEIMR SFSLSTNLQK RLRRKE
====================================================================================================
QQLNDLEACV M     VGETP LMNVESILAV RKYFQRITLY LSEKKYSPCA WEVVRAEIMR SLSLSTNLQE RLRSKE
           L               ED     S      R       K                I       FTFL      K S  N
                           A      K              T                                     R
         100       110       120       130       140       150       160     166
```

Fig. 1. A comparison of sequence heterogeneities found in the IFN-α genes and in leucocytes and lymphoblastoid IFN peptides. The consensus sequence for the IFN-α gene family, plus the observed divergences are shown above the double line (compiled from data referred to in Weissmann, 1981; see also Table 1) for IFN-α1 through 8, $\psi\alpha$10, IFN-A, -B, -C, -D, -F, -G, -H, λ2h and λ2cl, and for three additional IFN-H and -C-like genes from Björn Lund (University of Umeå; personal communication). The consensus peptide sequence, derived from the sequences of three lymphoblastoid IFN fractions (Allen & Fantes, 1980; Zoon, 1981) and for leucocyte IFN (γl, γ2 and β1 according to Levy *et al.* (1981) but which should not be confused with the standard nomenclature) is shown, again with sequence divergences below the double line. Parentheses indicate uncertainty in amino acid designation. Z designates an unspecified amino acid heterogeneity. The region underlined is derived from the sequence of only one of the three fractions. Amino acid abbreviations: A, alanine; C, cysteine; D, aspartic acid; E, glutamic acid; F, phenylalanine; G, glycine; H, histidine; I, isoleucine; K, lysine; L, leucine; M, methionine; N, asparagine; P, proline; Q, glutamine; R, arginine; S, serine; T, threonine; V, valine; W, tryptophan; Y, tyrosine. Heterogeneities characteristic for IFN-α1 are designated with^{+}, for IFN-α2 with °, and those specific for the group IFN-α4, -7, -10 and -C with an x.

the predominant interferon species expressed in virus-induced leucocytes and lymphoblastoid cells. As will be discussed in more detail below, interferon gene clones have been obtained either by making copy DNA (cDNA) using mRNA from interferon-producing cells as template, or by directly isolating genomic DNA fragments. The isolation of a genomic clone without identification of a corresponding clone amongst cDNA clones is insufficient evidence that the genomic gene can be functionally expressed. Thus although 15 interferon-α genes have been characterized only 10 have been found amongst cDNA sequences.

Knight *et al.* (1980) found a unique amino acid sequence for IFN-β which is identical to cloned cDNA and genomic sequences (see below). M. Revel (personal communication) has evidence that a second IFN-β species is also induced by the action of poly rI : rC on human fibroblasts. Van Damme *et al.* (1981) have reported the presence of an IFN-β-type (immunologically cross-reacting) interferon species in the supernatant of Con A-stimulated human leucocytes.

Although a number of reports exist as to the heterogeneity of interferon-γ species (e.g. Goldstein *et al.* (1981); Yip, Barrowclough, Urban & Vilcek (1982) only a single mRNA species (Ullrich, Gray, Goeddel & Dull, 1982) and a unique gene have been found (Gray *et al.*, 1982). The predicted amino acid sequence (146 a.a.) contains two possible glycosylation sites, but the naked peptide would have a molecular weight of 17.1 kD. Since Yip *et al.* (1982) report two species with molecular weights of 20 and 24 kD it is likely that differential post-translational glycosylation and/or proteolytic cleavage has taken place. Earlier reports of higher molecular weight proteins could imply the formation of multimers or possibly association with carrier proteins in the serum.

Comparison of IFN protein sequences

A comparison is made in Fig. 2 between individual interferon sequences (omitting leader peptides). The amino acid sequence of IFN-α1 is taken as representative for the α-family. Homology between IFN-α and -β (with judicious gaps to optimize the frame of comparison) is about 35%. The region 120–150 shows the highest homology (122–152 = 56% homology). The comparison with IFN-γ shows only seven and nine per cent homology to IFN-α and -β respectively.

(*a*)

```
           1       10         20        30          40          50        60         70          80
            L       N        LL Q    R      CL DR   F   P E      QFQK  A      E  Q IF  F     SS  W E
IFN-α 1    CDLPETHSLDNRRTLMLLAQM-SRISPSSCLMDRHDFGFPQEEFDGNQFQKAPAISVLHELIQQIFNLFTTKDSSAAWDEDLLD-
IFN-β    MSYNLLGFLQRSSNFQCQKLLWQLNGR--LEYCLKDRMNFDIPEEIKQLQQFQKEDAALTIYEMLQNIFAIFRQDSSSTGWNETIVE-
IFN-γ      CYCQDPYVKEAENLKKYFNAGHSD-VADNGTLFLGILKNWKEESDRKIMQSQIVSFYFKLFKNFKDDQSINKSVETIKEDMNVK-
           C                      S                  EE D    Q                         K
                     K                               E       Q                          S  T     V

               90         100         110         120         130         140         150         160
               Y Q N L                      S L    Y  RI   YL   K YS CAW  VR EI R          L   LR
          -KFCTELYQQLNDLEACVMQEERVGETPLMNADSILAVKKYFRRITLYLTEKKYSPCAWEVVRAEIMRSLSLSTNLQERLRRKE
          -WLLANVYHQINHLKTVLEEKLEKEDFTRGKLMSSLHLKRYYGRILHYLKAKEYSHCAWTIVRVEILRNFYFINRLTGYLRN
          -FF--NSNKKKRDDFEKLTNYSVTDLNVQRKAIKELIQVMAELSPAAKTGKRKRSQMLFQGRRASQ
            F         D                 A    L                K S        RA
               N          L            K    L                  S        R
```

(*b*)

```
         1      10        20        30         40             50        60         70        80
IFN-β    MSYNLLGFLQRSSNFQCQKLLWQLNGRLEYCLKDRMNFDIPEE----IKQLQQFQKEDAALTIYEMLQNIFAIFRQDSSSTGWNETIVE-
IFN-γ     CYCQDPYVKEAENLKKYFNAGHSDVADNGTLFLGILKNWKEESDRKIMQSQ----------IVSFYFKLFKNF-KDQSINKSVETI---
          Y           N                  L         EE     I Q Q              I         F  F  D S    ETI

              90         100        110        120        130        140        150        160
         -WLLANV-YHQINHLKTVLEEKLEKEDFTRGKLMSSLHLKRYYGRILHYLKAKEYSHCAWTIVRVEILRNFYFINRLTGYLRN
         -KEDMNVKFFNSNKKKRDDFEKLTNYSVTDLNVQRKAIKELIQVMAELSPAAKTGKRKRSQMLFQGRRASQ
              NV      N  K     EKL       T                   AK
```

Fig. 2. Comparison of the amino acid sequences of IFN-α1, -β and -γ. (*a*) The homologies between IFN-α1 and -β (uppermost), IFN-α and -γ (just below the sequences) and IFN-β and -γ (lowest row). A maximum of three gaps per sequence have been introduced to optimize alignment. (*b*) The IFN-β and -γ sequences are compared with up to 13 gaps per sequence having been introduced to optimize homology. Between IFN-β amino acids 81 and 107 there is some 37% homology between IFN-β and IFN-γ (with three inserts). Sequences are from Ohno & Taniguchi (1981), Mantei *et al.* (1980) and Gray *et al.* (1982). Amino acid abbreviations are as used in Fig. 1.

(3) CLONING OF IFN cDNA

IFN-β

The pioneering work of Taniguchi *et al.* (1979) showed the basic principle by which interferon genes can be cloned in the absence of knowledge of the amino acid sequence of the product. mRNA was isolated from poly rI : rC-induced and non-induced human fibroblast cell cultures (DIP-2). This RNA was fractionated according to size in a sucrose gradient. Individual fractions were tested for their IFN mRNA content by injecting them into *Xenopus laevis* oocytes, which after two days incubation were tested for the production of human interferon in a bioassay using the virus cytotoxic inhibition test. This arduous indirect method showed that IFN-β mRNA was contained in the 12S mRNA fraction. Using the now classic reactions the mRNA was converted to double stranded copy DNA (cDNA), tailed with homopolymers and cloned into the plasmid pBR322. The *Escherichia coli* clones were then screened, firstly by a hybridization method which demonstrated whether individual clones contained sequences complementary to induced mRNA. It should be noted that at least 15 other proteins are induced coordinately with IFN-β (Raj & Pitha, 1980a,b) and that the genes for some of these have also been cloned (Gross, Mayr & Collins, 1981b; Weissenbach *et al.*, 1980; Content *et al.*, 1982). A hybridization-translation assay was used to identify the IFN-β clones amongst those colonies selected by the colony hybridization method. In this method DNA from individual or groups of clones is immobilized on filter paper and hybridized with IFN-β mRNA-containing samples. mRNA which had specifically hybridized is eluted and tested for its capacity to code for IFN-β in the oocyte injection method discussed above. Positive clones were then sequenced. One clone contained a 770 bp insert (not counting A : T tails) which could code for peptides of 187 and 166 amino acids. When the amino-terminal amino acid sequence of huIFN-β became known a few months later (Knight *et al.*, 1980), Taniguchi was able to conclude that his cDNA clone was indeed for IFN-β and that either the 166 amino acid long peptide was produced directly or a leader peptide of 21 amino acids was cleaved off the 187 a.a. long peptide to produce the mature, active interferon (Taniguchi *et al.*, 1980a,b). By comparison with other interferons it seems likely that all have a leader peptide of similar length which is probably cleaved off during transport out of the cell. Other groups (Derynck *et al.*, 1980a; Goeddel *et al.*, 1980b) subsequently obtained IFN-*β* cDNA clones which had identical coding regions.

Using IFN-*β* cDNA as a hybridization probe only a single mRNA species can be identified in both poly rI:rC induced human fibroblasts and Sendai virus-induced lymphoblastoid cells (Gross, Bruns, Mayr & Collins, 1982) indicating that no closely related gene is expressed under these conditions. As discussed below only a single IFN-*β* allele has been found amongst human genomic clones.

The final demonstration that the cDNA cloned was indeed an IFN-*β* gene was the expression of the gene, devoid of its leader peptide in the bacterium *E.coli* under the control of various prokaryotic promoters (Goeddel *et al.*, 1980b; Derynck *et al.*, 1980b; Taniguchi *et al.*, 1980c).

IFN-α cDNA

Nagata *et al.* (1980a) constructed a cDNA colony bank using mRNA from induced blood leucocytes as template. IFN-*α* clones were enriched in a manner similar to that described above for the IFN-*β* clone isolation. The search for positive clones was accelerated by the fortuitous discovery that some of the *E. coli* clones were producing low levels of active human interferon, as assayed by the virus cytopathic effect reduction test on human cells. A detailed and entertaining account of how this work developed is presented by Weissmann (1981).

The sequence of the first IFN-*α* cDNA from this group (Mantei *et al.*, 1980) showed a peptide coding region which differed in only nine out of 35 amino acids from the consensus amino-terminal sequence of human lymphoblastoid IFN-*α* (Zoon *et al.*, 1980), and had a good overall correlation with the total amino acid composition reported for leucocyte interferon by Rubinstein *et al.* (1979). The first AUG, or start codon for translation was found 22 amino acids before the first amino acid (cysteine) of the mature interferon sequence. The presence of such a leader peptide sequence, which contains some 50% hydrophobic amino acids, is a common feature of all the interferons.

Combining the data on cDNA clones from the groups of Weissmann (Nagata *et al.*, 1980a,b) and Goeddel (Goeddel *et al.*, 1981) yielded a total of 10 different sequences for normal IFN-*α* peptides and a pseudo-IFN-*α* gene (IFN-E) transcript which contains a single base insert early in the coding sequence which destroys the trans-

Table 1. *Interferon-α gene family*

Genes: coding for a full length peptide (166 amino acids) $\alpha 1$(D); $\alpha 2$(A, $\lambda\alpha 2$); $\alpha 5$(G); $\alpha 6$(K, λ5K); $\alpha 8$; B; $\alpha 7$(J, λ1J); α4A(α4B); G-chr 27(H, λ2H); F; I; C; λ2C1; G-chr 19L+
Total of 14 genes.

Genes designated in parentheses are either identical or allelic with three or less amino acid differences.

Pseudogenes: not coding for a full length peptide $\alpha 10$(L)*; $\alpha 9$; $\alpha 11$; $\alpha 12$; E; M(G-chr-10LL); G-chr-26R+; G-chr-19R+; G-chr-16L+
Total of 6 to 9 pseudogenes.

* Point mutation from an otherwise normal gene.
+ Not sequenced.

lational reading frame. The genes were designated $\alpha 1$, $\alpha 2$, $\alpha 4 \rightarrow 8$ by Weissmann's group and IFNA → H by Goeddel's group. The identical or nearly identical alleles are listed in Table 1, along with genomic alleles which were discovered subsequently.

IFN-γ cDNA

The isolation of cDNA clones using gel-fractionated 18S mRNA from enterotoxin-stimulated peripheral blood lymphocytes as template did not lead immediately to the certain identification of interferon-γ cDNA clones. The 'plus-minus' hybridization method, again similar to that described by Taniguchi *et al.* (1979), showed up positive clones (Gray *et al.*, 1982). Sequence analysis, however, revealed an open translational reading frame for a peptide of 166 amino acids, and containing the sequence Ser-Leu-Gly-Cys (amino acids 18 to 21) which is identical to the leader peptide cleavage site for most of the IFN-α genes which had been sequenced. No other significant homology with the other interferon genes was present. In particular, at that time, the molecular weight of IFN-γ was estimated to be over 40 000 daltons. Undaunted by this, Gray *et al.* (1982) fused a synthetic DNA fragment to the putative mature IFN-γ gene fragment such that translation would be initiated at a new methionine codon immediately preceeding the cysteine (formerly amino acid 21). When this construct was inserted downstream from a trytophan operon promoter, interferon activity was synthesized, albeit at a low level, in *E.coli* strains carrying the recombinant plasmid. Furthermore the insertion of the total cDNA downstream from the late SV40 late promoter led to IFN synthesis in monkey cells (cos-7 cells) transformed with the virus vector hybrid DNA. The interferon synthesized by the hybrid *E.coli* and monkey cells was not neutralized by anti-IFN-α or -β antisera, but by an anti-serum against partially purified interferon from peripheral blood lymphocytes induced by bacterial toxin. The interferon was also, like normal IFN-γ, pH 2 and SDS-labile.

According to these criteria the cloned DNA appears to be the structural gene for huIFN-γ.

The identification and cloning of a single IFN-γ chromosomal gene (Gray *et al.*, 1982; Gray & Goeddel, 1982) and the demonstration by Derynck *et al.* (1982) that only a single homologous mRNA species was to be found in lymphocyte and spleen cells making IFN-γ, make it extremely likely that only a single IFN-γ gene exists.

(4) GENOMIC CLONES

The isolation of human genomic DNA fragments homologous to radioactively labelled cDNA probes followed classic methodologies such as those described in volume 68 of *Methods in Enzymology*. Collections of hybrid molecules either in λ-bacteriophage or cosmid vectors have been described (e.g. Lawn *et al.*, 1978; Gross *et al.*, 1981a). Using these systems hybrid molecules are obtained containing, for Lambda vectors, 10 to 20 kilobase pairs, and, for cosmids, 30 to 45 kilobase pairs of contiguous human DNA per hybrid. Thus by the screening of some 500 000 Lambda clones or 200 000 cosmid clones a high probability of finding 'single copy' gene sequences can be achieved. As will be seen below since very large neighbouring regions are isolated at the same time, overlapping clones have been found which indicate that many of the IFN-α genes are located together in the human genome.

IFN-α genomic clones

Nagata *et al.* (1981) reported the isolation of ten Lambda clones carrying regions which hybridized to IFN-α cDNA probes. Subsequent analysis and the isolation of further clones defined 11 distinct regions homologous to IFN-α-mRNA. A linkage between six members of this group within a region of 36 kb (IFN α7, α4, α10, α9, G-chr-10LL and G-chr-26R; Fig. 3) could be defined by overlapping restriction maps and heteroduplex studies. The hetero-

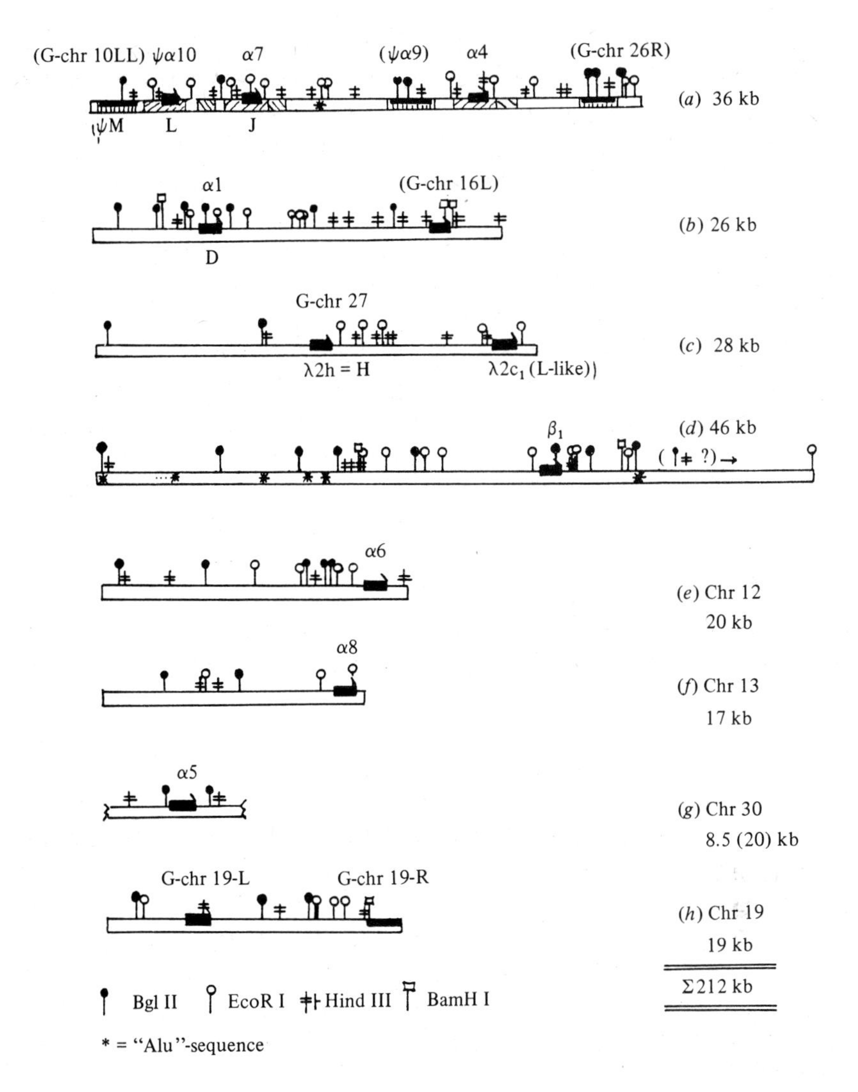

Fig. 3. Structure of genomic clones containing IFN-β and IFN-α genes. References are as follows; IFN-β, Gross *et al.* (1981b), Mory *et al.* (1981) and personal communication Y. Mory and M. Revel; IFN-α1 and -α4 through – 10, Nagata *et al.* (1981), Brack *et al.* (1981), Weissmann (1981); G-Chr 10LL, -chr 26R, -chr 16L, -chr 19L, -19R, Brack *et al.* (1981), Weissmann (1981); λ2h and $\lambda 2c_1$, Lawn *et al.* (1981); IFN-D, -H, Goeddel *et al.* (1981); IFN-J, ψ-L, M, Ullrich *et al.* (1982); IFN-A (α2), Lawn *et al.* (1981b). Genes designated with 'eared' black boxes are IFN-α1-like genes (see Table 1). In (*a*) the shaded boxes show areas having sequence homology as defined either by direct sequencing (Ullrich *et al.*, 1982) or by heteroduplex studies (Brack *et al.*, 1981). The greek letter ψ is used to designate 'pseudo'-genes in which coding regions more or less homologous to other IFN-genes are interrupted by stop codons. Low flat black boxes designate IFN-α-like genes showing less than 80% homology to IFN-α1. All maps are drawn with the direction of IFN gene transcription from left to right. At least 10 IFN-α genes, two pseudo IFN-α and four pseudo IFN-α-like genes are distinctly defined through this mapping as non-allelic loci.

duplexing studies showed, furthermore, that some 4500 bp regions including these genes were present as homologous tandemly repeated blocks. Sequencing of the chromosomal regions enable a correlation between some of the genomic clones and cDNA clones (Mantei *et al.*, 1980; K. Henco, A. Schamböck and J. Fiyisowa, unpublished results; Weissmann, 1981; Weissmann *et al.*, 1982).

Independently Lawn *et al.* (1981b,c) reported a further clone showing another linkage group (λ 2h and λ 2c1; Fig. 3*c*) and the sequence around the IFN-α2 gene. Ullrich *et al.* (1982) sequenced 9937 bp of the region containing $\psi\alpha$10, α7 and G-chr-10LL (L, J and M in their nomenclature).

The genomic regions so far cloned and mapped in unlinked groups covering at least 230 kb are presented in Fig. 3. All the IFN-α genomic clones described above were isolated from the λ bank of Lawn *et al.* (1978). Recently a cosmid clone has been isolated containing three linked IFN-α genes which appears to be an extension of the map in Fig. 3*c* but contains an additional novel IFN-α gene (B. Lund, E. Lindgrun & J. Collins, unpublished results).

IFN-β genomic clones

Using the same human genomic lambda library (foetal liver DNA) as was used for the IFN-α gene cloning, Houghton *et al.* (1981), Lawn *et al.* (1981a), Ohno & Taniguchi (1981) and Tavernier, Derynck & Fiers (1981) isolated IFN-β gene-containing clones. Independently Mory *et al.* (1981) constructed a bank from peripheral blood cells from a thallassaemia patient and isolated the IFN-β gene plus some 15 kb downstream of the gene.

Gross *et al.* (1981a,c) isolated a 36 kb region including the IFN-β gene from a cosmid bank made with human placenta DNA.

Taking the data of Gross *et al.* (1981c) and Mory *et al.* (1981), a complete restriction map of 46 kb surrounding the IFN-β gene can be constructed (Fig. 3*d*).

The sequences obtained by all these groups for the IFN-β gene and closely neighbouring regions agree to the base pair. This lack of polymorphism stands in stark contrast to the overwhelming variety found for the IFN-α genes. The significance of pseudogenes for the IFN-α family is at present unknown but it can be speculated that a non-functional gene located between two genes for which there is a strong selective advantage cannot easily be lost from the population.

(5) CHROMOSOME LOCALIZATION OF THE IFN GENES

The most commonly applied method for designation of a gene to a particular chromosome is the study of expression of that gene in somatic cell hybrids between human and animal cells, usually hamster or rat, in which a number of the human chromosomes have been lost. The human chromosome content of the hybrid cells is defined by the presence of diagnostic iso-enzymes coded for by genes on particular chromosome arms. Using this method Tan, Creagan & Ruddle (1974) and Slate & Ruddle (1979) correlated the presence of human chromosomes 2 and 5 with the ability of such cell hybrids to produce interferon (probably IFN-β). Meager *et al.* (1979b) however found that the production of human IFN-β correlated only with the presence of human chromosome 9, and that when this chromosome was specifically deleted from the hybrid (by fusion, with chromosome X and selection in HAT medium) then all interferon producing capacity was lost (Meager *et al.*, 1979a). Recently Sehgal, Sagar, Braude & Smith (1981) proposed that mRNAs for IFN-β and IFN-β-related peptides were inducible in cells containing human chromosomes 2, 5 and 9. Using labelled cDNA fragments as hybridization probes Owerbach *et al.* (1981) correlated only the presence of chromosome 9 with the α and β genes. D. Goeddel (personal communication) reported that the human IFN-γ gene is located on chromosome 12. The significance of the earlier designations of IFN genes to chromosomes 2 and 5 is at present unclear, but may imply the presence of regulatory elements on these chromosomes. A further possibility is that the designation of IFN genes to chromosomes 2 and 5 is due to an artefact in which fragments of chromosome 9 may have translocated to other chromosomes in the cell hybrids.

The clustering of genes of related function has been found for a number of human gene families, e.g. the transplantation antigen genes family, the globin genes family, and the growth hormone and related genes family.

(6) COMPARISON OF IFN GENE SEQUENCES

The lack of introns in IFN-α and -β genes is at present unique amongst vertebrate nuclear genes. Yeast, sea urchin and *Drosophila* histone genes, some virus and mitochondrial genes also lack introns (see references in Lawn *et al.*, 1981a). The significance of 'splicing' the process in which the introns are removed from the primary transcript, in terms of the evolution of genes, the regulation of the gene expression or the kinetics of induction and/or mRNA stability is unknown,

Comparing the sequences obtained for IFN-α genes by Weissmann's group and Goeddel's group it is surprising to note that only two alleles show absolute identity. Considering, $\alpha 8$ and B to be allelic variants, then at least 15 non-allelic genes can be defined, for seven of which closely related alleles exist. This high incidence of allelic variants would lead to the prediction that a high polymorphism in the population as a whole is to be expected. Whether or not this implies a redundancy in certain regions of the IFN peptide or if there is some deeper significance (hybrid virulence?) or selective advantage is still unclear.

The amino acid sequences for IFN-α peptides are shown in Fig. 4. The number of amino acid differences between different IFN alleles ranges between 12 (for $\alpha 4$B and $\alpha 7$) and 43 ($\alpha 1$ and $\alpha 8$). Weissmann *et al.* (1982) have proposed that the IFN-αs fall into two major subclasses, one comprising $\alpha 1$ (D), $\alpha 2$ (A), $\alpha 6$ and $\alpha 5$ (G), the other, $\alpha 4$ (A and B), C ($\psi\alpha$ 10), λ C1 and $\alpha 7$ (also called subgroup C by Ullrich *et al.*, 1982), with an intermediate group comprising H (λ2H), $\alpha 8$, B and F. The closely related members of subgroup C are also found as a gene cluster.

(7) STUDIES ON EUKARYOTIC GENE REGULATION SEQUENCES

The definition of a eukaryotic promoter appears to vary according to whether studies were carried out *in vitro* or *in vivo* (i.e. reintroducing the genes into a living cell). In general *in vitro* experiments (Corden *et al.*, 1980; Hu & Manley, 1981; Wasylyk *et al.*, 1980; Grosveld, Shewmaker, Tat & Flavell, 1981; Mathis & Chambon, 1981) have shown a dependence on the TATA box (Gannon *et al.*, 1979) whereas *in vivo* experiments have shown that absence of this region causes only a slight decrease in the levels of transcription along with a concommitant loss of specificity in the transcription start (e.g. Grosveld, De Boer, Shewmaker & Flavell, 1982; Benoist & Chambon, 1981; McKnight, Gavis & Kingsbury, 1981).

The extent of the region largely responsible for efficient initiation of transcription *in vivo* (i.e. the equivalent to the prokaryotic promoter) is for rabbit β-globin -100 to -58 (Grosveld *et al.*, 1981); for HSV thymidine kinase -100 to -40 (McKnight *et al.*, 1981) and for mouse metallothionein-I gene -90 to -1 (Brinster *et al.*, 1982) with respect to the initiation start. This latter example is also particularly interesting in that the responsiveness of the metallothionein gene to induction by cadmium is also located within this region but indistinguishable from the 'promoter'.

Regions involved in *inducible* transcription have been investigated for the following genes, by studying the expression in foreign cells: mouse mammary tumour virus (MMTV), inducible with glucocorticoid hormones (Hynes *et al.*, 1981; Lee, Mulligan, Berg & Ringold, 1981); rat $\alpha_{2\mu}$ globulin (Kurtz, 1981) and the heat-shock protein genes of *Drosophila* (Corces, Pellicer, Axel & Meselson, 1981). These were all studied in mouse fibroblast cell cultures. The studies of Brinster *et al.* (1982) on the metallothionein gene 'promoter' were carried out by injecting DNA into mouse eggs. Apart from the metallothionein gene, described above, 5′-distal regions required for inducibility have been only roughly located to within a few hundred to one thousand base pairs before the transcription start.

Interferon gene expression in transformed cells

Experimental design

This section is confined to experiments which shed a light on normal interferon gene regulation. Experiments where the IFN gene has been expressed under control of a foreign promoter are not considered in this review. In all the studies discussed here, human chromosomal genes are transferred into mouse, hamster or rabbit cells. The inducibility of the gene is then investigated either within a few days ('transient expression') in a population of transformed cells, or in established cell-clones which had previously been selected, subcloned and kept in culture for several weeks or months. It has been observed in several systems that transferred genes, on entering the cells may be expressed transiently, but are not further inducible, e.g. chicken ovalbumin genes (Breathneck, Mantei & Chambon, 1980) and rabbit β-globin genes (Wold *et al.*, 1979). Established lines then usually show no further synthesis using the foreign genes. Recent exceptions to this rule are the MMTV genes, rat α_2-microglobulin and metallothionein genes.

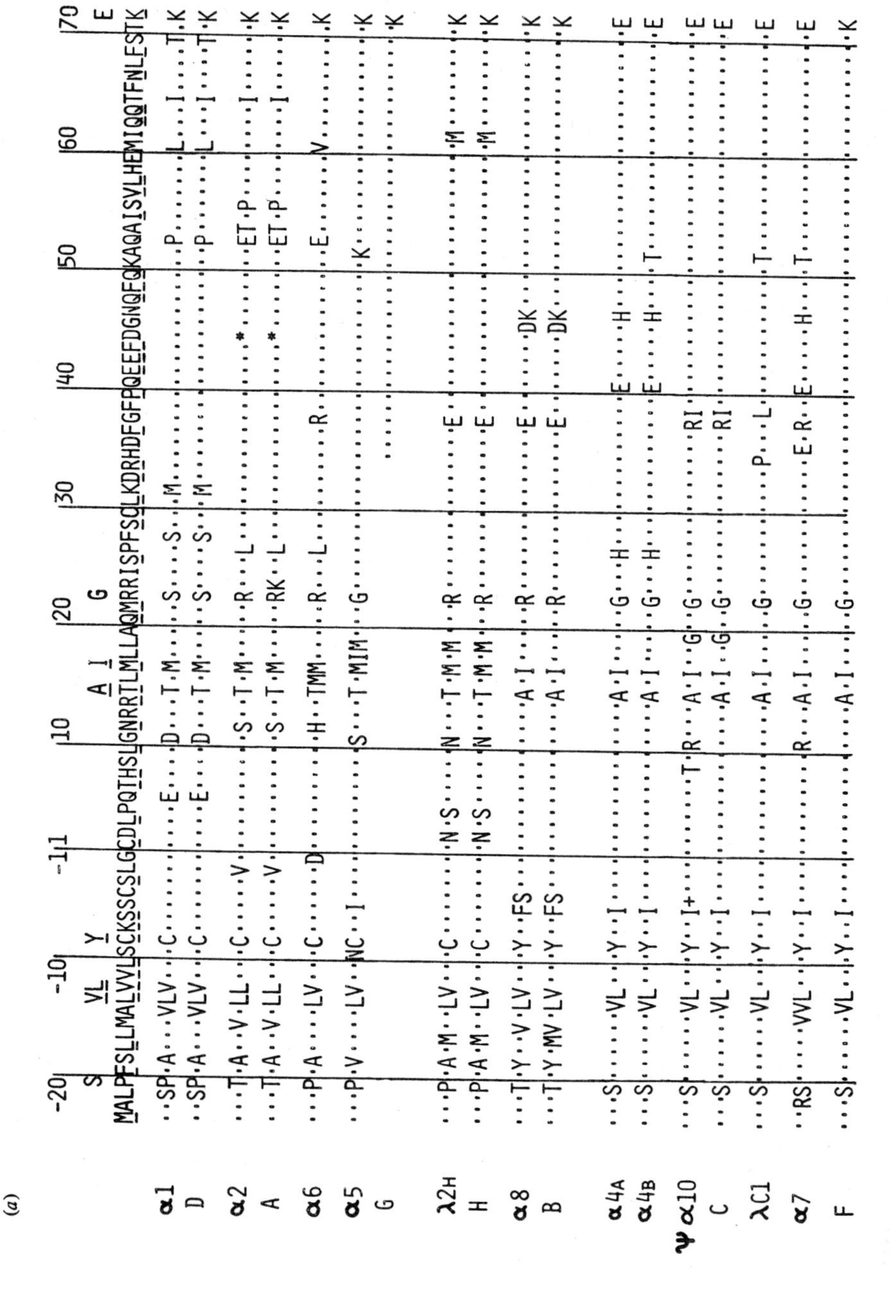

Fig. 4. Amino acid sequences of α-interferons. All sequences have been derived from the nucleotide sequences of cloned cDNAs or chromosomal DNAs. Sequences referred to in Weissmann *et al.* (1982) are from A. Schamböcke (IFN-α4a). K. Henco (IFN-α4b), J. Fujisawa (IFN-α5), J. Schmid (IFN-α6), T. Koracic (IFN-α7), M. Pasek (IFN-α8) and J. Hochstedt and J. Brosius ($\psi\alpha$10). Other sequence references are given in the main text. The

(*b*)

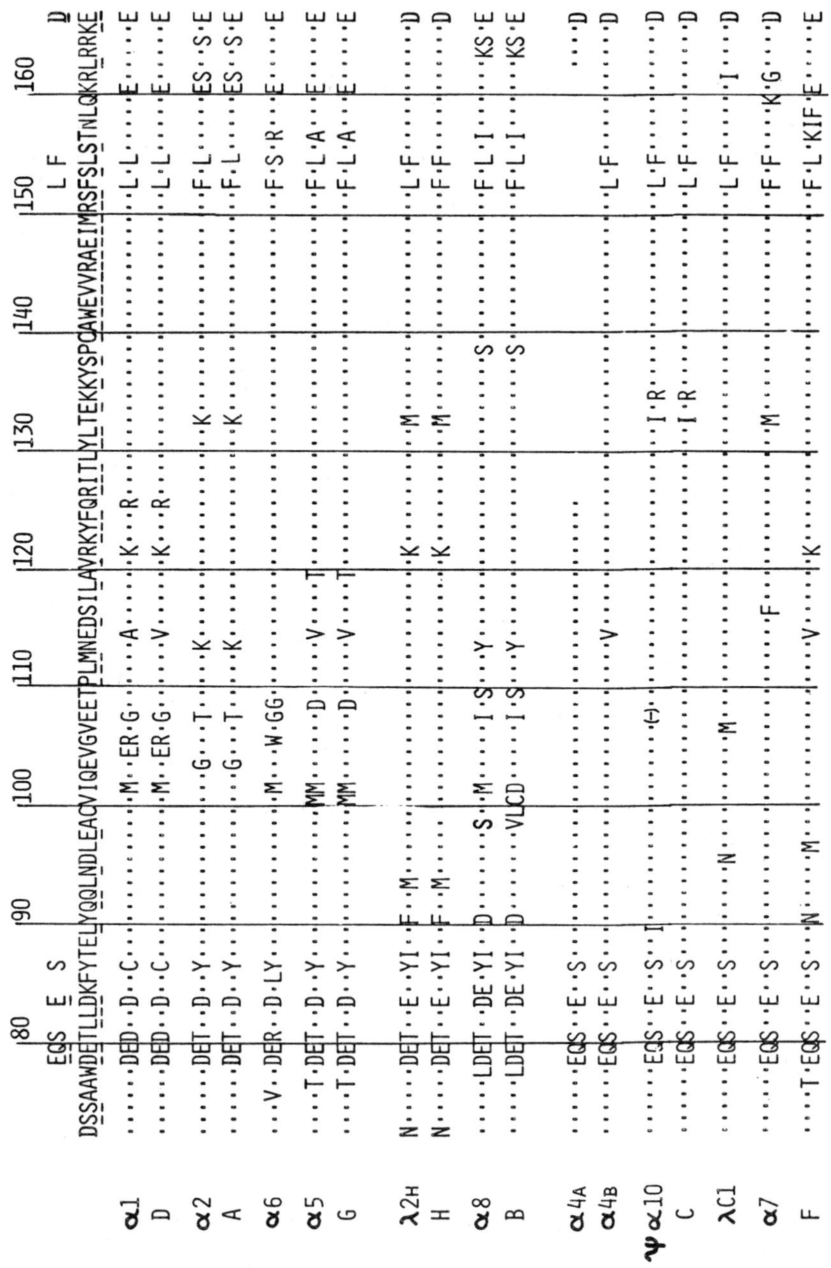

```
              |80        |90        |100       |110       |120       |130       |140       |150       |160
              EQS  E  S                                                                   L F        D
       DSSAAWDETLLDKFYTELYQQLNDLEACVIQEVGVEETPLMNEDSILAVRKYFQRITLYLTEKKYSPCAWEVVRAEIMRSFSLSTNLQKRLRRKE
α1     ······DED··D··C············M··ER·G·····A······K···R··························L·L·····E·····E
D      ······DED··D··C············M··ER·G·····V······K···R··························L·L·····E·····E
α2     ······DET··D··Y··············G···T····K·················K·····················F·L·····ES··S·E
A      ······DET··D··Y··············G···T····K·················K·····················F·L·····ES··S·E
α6     ···V··DER··D·LY············M···W·GG···························································F·S·R···E·····E
α5     ····T·DET··D··Y···········MM······D····V····T······························F·L·A···E·····E
G      ····T·DET··D··Y···········MM······D····V····T······························F·L·A···E·····E
λ2H    N·····DET··E··YI··F··M·······················K·········M·····················L·F···········D
H      N·····DET··E··YI··F··M·······················K·········M·····················F·F···········D
α8     ·····LDET··DE·YI··D······S··M·····I·S···Y·················S·················F·L·I······KS·E
B      ·····LDET··DE·YI··D······VLCD·····I·S···Y·················S·················F·L·I······KS·E
α4A    ······EQS··E··S·········································                                ···D
α4B    ······EQS··E··S···················V··························L·F···········D
ψα10   ······EQS··E··S··I··············(-)·················I·R···················L·F···········D
C      ······EQS··E··S·························I·R·········································L·F···········D
λC1    ······EQS··E··S·······N········M······································L·F······I···D
α7     ······EQS··E··S··························F············M·····················F·F····K·G···D
F      ····T·EQS··E··S··N·····M··········V·····K·············································F·L·KIF·E·····E
```

top line shows the consensus sequence with common amino acids underlined. * indicates, no amino acid. In $\psi\alpha10$ an insertion (+) and a deletion (−) have been introduced to restore the reading frame. This figure is reproduced by kind permission of Academic Press from Weissmann *et al.* (1982).

DNA on entering the cell is usually found to recombine to form large (concatameric?) polymers referred to as the transgenomic form (Scangos & Ruddle, 1981). The establishment of stable lines occurs at less than one thousandth of this efficiency through the integration of these long concatamers. Where a virus replication origin has been coupled to the transferred gene, it is not at all clear that its presence makes any difference to the subsequent fate of the DNA. No authors have convincingly demonstrated that SV40 or polyoma origins in these recombinant viruses are actually functional in mouse cell lines (e.g. Mulligan & Berg, 1981) or that they play any role in the efficiency with which stable clones can be established. An exception to this is the use of bovine papilloma virus BPV which is maintained as an extra-chromosomal plasmid-like molecule in the cell nucleus (e.g. Zinn, Mellon, Ptashne & Maniatis, 1982). It has been shown by a number of groups that transferred DNA may integrate at varied positions on many chromosomes and that in the absence of selection may be unstable (see Robins, Axel & Anderson, 1981).

(8) HUMAN IFN-β GENE IN HETEROLOGOUS CELLS

From studies on the control of IFN-β mRNA synthesis and specific degradation in normal human fibroblasts the main conclusions are that regulation is exerted (i) at the level of transcription and (ii) at later times by a process of specific mRNA degradation which is dependent on *de novo* RNA/protein synthesis (Cavalieri, Havell, Vilcek & Pestka, 1977; Sehgal, Dobberstein & Tamm, 1977; Raj & Pitha, 1980a,b; Raj & Pitha, 1981, in contrast to Raj & Pitha, 1977). In connection with this last conclusion, it should be pointed out that the half-life of IFN-β mRNA was not directly measured and that the possibility that the superinduction effect is in fact due to lack of shut-off of transcription has not been ruled out. The concommitant induction of other mRNA species from at least 15 other genes has also been reported (Raj & Pitha, 1980a,b; Content *et al.*, 1982).

The discovery that some of these co-inducible genes are located near the IFN-β gene on a genomic clone (Gross, Mayr & Collins, 1981b; Fig. 5) gave rise to the speculation that these neighbouring genes might be involved in IFN-β gene regulation at one of these two levels. To test this, the whole pCos IFN-β (36 kb human genome; Fig. 5) was co-transformed with a tk^+ gene plasmid into

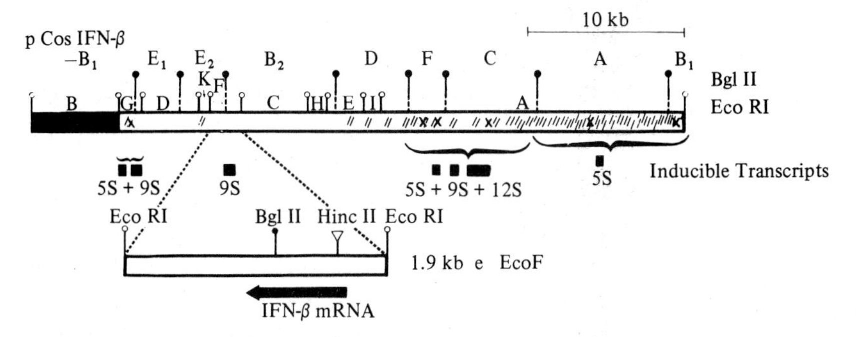

Fig. 5. Physical map of the cosmid hybrid pCosIFN-β showing the location of regions inducibly transcribed in human fibroblasts. The black box designates the location of the vector DNA pJB8. The crosses indicate the location of 'Alu-I' family sequences. Shading indicates the distribution of DNA sequences which occur in multicopies in the human genome.

mouse Ltk$^-$ cells (Hauser *et al.*, 1982). Tk$^+$ clones which had a low constitutive HuIFN-β production were found with a co-transfer frequency of over 90%; 90% of these were further inducible to very high levels of HuIFN synthesis (in a few clones, using either NDV or poly(rI):(rC) DEAE-Dextran as inducer, higher than in super-induced human fibroblasts). It was found that the inducible transcripts were identical with normal HuIFN-β gene transcripts. In the non-induced cells the constitutive HuIFN-β production was correlated with the presence of transcripts starting several hundred or a thousand bases 5′-distal to the normal start point but terminating at the normal termination site some 200 bp 3′-distal to the coding region. Transcripts specifically induced by poly (rI):(rC) or NDV showed normal 5′ and 3′ ends, i.e. transcription was initiated at the normal transcription start site and termination (also polyadenylation) took place as in normal induced human cells. The HuIFN-β activity was characterized according to the spectrum of antiviral activity on human and mouse cells, and according to molecular weight through monoclonal antibody – 'Western blotting'. It was found to be similar to normal glycosylated HuIFN-β. Since the human and mouse interferons did not cross-react their relative induction could be studied. The induction of both followed the pattern expected for a mouse IFN-β gene. An analysis of the integrated DNA allowed the conclusion that in the two clones producing high levels of HuIFN-β, 20 and 50 copies of the entire pCosIFN-β plasmid had integrated, mostly in tandem, in the mouse genome. The poly (rI):(rC) or NDV induction thus appeared to induce the transcription of the human IFN-β gene in the same way as in normal human fibroblasts.

Semi-continuous IFN-β production from the hybrid mouse cell lines could be obtained over several cycles of induction, interspersed by two days in fresh recovery medium. It was shown that HuIFN-β mRNA was accumulated stably for at least 12 h after induction. These findings, taken together with the absence of any super-induction effect (characteristic of mouse L cells but not for mouse cells in general), indicated that specific breakdown of the HuIFN-β mRNA was not taking place. It was found that genes neighbouring IFN-β on the human DNA were also specifically induced in both normal human fibroblasts and in the hybrid mouse clones. The HuIFN-β gene was not found to be inducible during the 'transient' expression, i.e. during the first few days after DNA transfer. The inducibility appeared only after the establishment of stable clones, i.e. presumably after stable integration into the genome. The magnitude of induction over the background constitutive level was in the order of 10- to 100-fold.

Hauser *et al.* (1982) estimated the yield of IFN per gene (IFN units/day/gene) on induction with NDV was about 5% of that of super-induced human fibroblasts but higher than that in human fibroblasts simply induced by poly (rI) : (rC). Ohno & Taniguchi (1981) using mouse mammary tumour cells transformed with SV40tk-IFN-β (EcoF fragment, Fig. 5) were able to produce about 80 units/10^6 cells/24 h in the highest producer (2 or 3 genes per cell?), compared to 25 000 units under similar conditions obtained by Hauser *et al.* (1982) for a clone with some 50 gene copies per cell. The extent of induction in this latter case was about a 200-fold increase over the constitutive level. A further difference observed was that the mouse fibroblasts but not the mouse mammary tumour hybrids produced a constitutive level of HuIFN-β. This may not be a significant difference since it is possible that with the general low level of expression, the constitutive level was just below the level of detection. Both groups found that HuIFN-β mRNA accumulated for at least 12 h implying a long half life for the IFN-β mRNA in the mouse hybrids (although it was not directly measured). Canaani & Berg (1982) who used an SV40 *neo*^{R} vector were able to obtain 2- to 5-fold stimulations of interferon formation in mouse fibroblasts but only a little stimulation in rabbit RK13 kidney cells. Zinn *et al.* (1982) using a fragment of the bovine papilloma virus as vector DNA, which replicates in the mouse fibroblast nucleus as an extrachromosomal element, showed a 2- to 20-fold increase in the amount of IFN mRNA, although the efficiency in terms of amounts of mRNA/gene was only 0.5% of that in human osteosarcoma cells. Both of these latter groups found that the transcripts were normal.

The most recent results of experiments carried out in our group to narrow down the control region indicate that the 'promoter' region, −20 to −90, can be distinguished from the 'inducible response' region which is −67 to −107 (H. Hauser, H. Dinter, personal communication; Fig. 6) using experiments carried out as described by Hauser *et al.* (1982) but using successive deletions from both the 3′ and 5′ ends of the gene. In Fig. 6 the 5′-distal sequence before the IFN-β transcription initiation point is shown. Deletions extending from the 5′ side (left side) to −107 are all still inducible. Deletions extending further, up to −67 and beyond, are no longer inducible. This defines a region which we would term an 'inducibility response' region, for which it is tempting (by analogy with prokaryotic systems) to postulate the interaction with effector proteins. The experiments cited in this section represent the first example of a human gene being under normal control in a foreign cellular environment. The varying levels of gene activity reported by different groups may be largely dependent on the particular recipient cell lines.

In summary, it can be concluded that should the coinduced (IA) genes be involved in IFN-β gene regulation then (i) they may play a role in specific mRNA breakdown but be expressed too weakly in mouse cells to be effective and (ii) if they are involved in stimulating transcription (and translation?) then the mouse cell can replace all the required functions in this respect. Finally a specific DNA region can be defined, separate from the 'promoter', which determines inducibility of the IFN-β gene in cis.

The laboratory of P. Pitha provides a set of data which appear largely incompatible with the experiments described above, for Pitha *et al.* (1982a,b) and Reyes *et al.* (1982) described experiments in which IFN cDNA (840/560 bp) constructs completely lacking the 5′-distal regions and even lacking 5′-non-translated or leader peptide coding sequences are transferred into mouse fibroblasts to yield (albeit unstable) hybrid clones in which the IFN-β gene is inducible.

In these experiments the IFN cDNA is in some hybrids fused to the tk gene promoter. Inducible expression is obtained with or without the tk promoter being present. When the tk promoter was

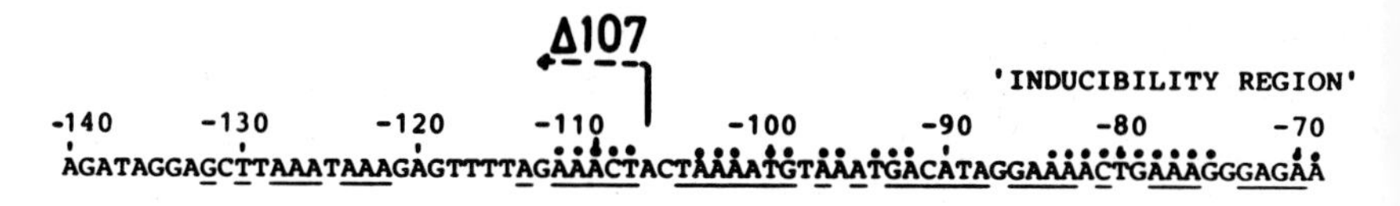

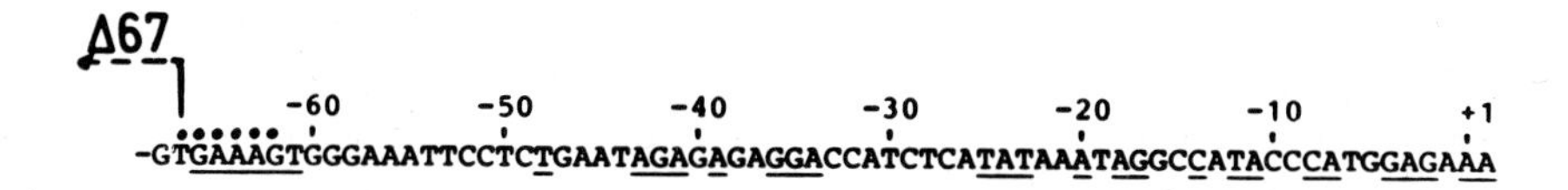

Fig. 6. DNA sequence 5′ distal to HuIFN-β gene. The sequence immediately distal to the initiation of transcription of the HuIFN-β gene is shown numbered from the first base transcribed. The TATAAAT sequence around −20 is a feature common to the consensus sequence for eukaryotic polymerase II-promoters. Deletions extending in from the left (5′ -end) up to −107 do not affect the inducibility of the IFN-β gene when introduced into mouse L cells. The additional deletion of the adjacent region to −67 (labelled 'inducibility') leads to a loss of inducibility in this system (H. Hauser, H. Dinter *et al.*, unpublished results). Direct sequence homology based on the consensus sequence AAACTGAAAG is shown by dots above the bases. The thin underlining indicates regions that are homologous with the IFN-α1 'promoter' sequence (see Gross *et al.*, 1981b).

present, Herpes simplex virus could be used to induce IFN-β synthesis, and this virus has a known direct effect on the tk promoter. One possibility proposed to explain these phenomena is that a sequence *within* the IFN-β structural gene is responsible for promoter and regulation activity. Such a situation occurs in mammals for transcripts made by RNA polymerase III (e.g. for the AluI-family gene; see Shen & Maniatis, 1982). These transcripts differ from normal polII transcripts in that they are not 'capped' (i.e. no inverted 7-methyl G at the 5Ä end) and are not polyadenylated. In view of the sequences 5′-distal to the IFN-β gene a polII promoter is to be expected, and although 'capping' has not been looked for, the normal transcript is certainly polyadenylated. This makes it unlikely that polIII and/or internal sites within the gene are being used. The final test of these models, in which the IFN-β 5′-distal regions are coupled to another gene have not yet been carried out. If specific mRNA degradation is taking place during the turn-off of normal expression, then a sequence within the transcript must be recognized by the breakdown machinery. Such a sequence has not so far been defined. A final suggestion by Pitha *et al.* (1982b) is that the DNA transferred into the mouse cells may have integrated into regions which themselves are inducibly transcribed by poly (rI) : (rC). This would require that such regions occur with high frequency throughout the genome. This has also been suggested by Gross *et al.* (1981b) who found highly repetitive genes inducible by poly (rI) : (rC), but not by NDV. It is interesting that in contrast to the results obtained with intact HuIFN-β gene the 'cDNA' clones (Pitha *et al.*, 1982a) were not in general inducible to any great extent with NDV although Hauser *et al.* (1982) have shown that this is by far the best inducer for mouse Ltk$^-$ cells. Reports in Pitha *et al.* (1982b) that mouse IFN-β did not seem to be induced to any appreciable extent in mouse Ltk cells and the hybrids and that hybrid clones producing HuIFN-β were obtained at a rate of 10% by co-transferring total human genomic DNA with the tk$^+$ gene (expected co-transfer is about 0.01%) are again in direct contradiction to reports from other laboratories. In view of the low levels of IFN-β synthesis, and the instability of the cell lines obtained by Pitha *et al.* (1982a,b) in comparison with the studies reported above one is at present inclined to accept a simple model of IFN-β gene regulation in which the immediate 5'-distal 150 bp before the transcription start contain the essential cis-acting information for the control of transcription.

(9) HUMAN IFN-α GENE IN HETEROLOGOUS CELLS

Mantei & Weissmann (1982) have reported the transfer of a genomic fragment containing the IFN-α 1 gene and 5.4 kb of 5'-flanking and 1.2 kb of 3'-flanking sequences together with a tk$^+$ plasmid into house Ltk$^-$ fibroblasts to obtain hybrids containing up to 20 copies per genome. The HuIFN-α gene was transcribed from start points 5'-distal to the normal start point. Induction by NDV caused initiation of transcription at the correct start point with a frequency of about 0.5 to 10 transcripts per cell in 11 h. This level of specific transcription is about a thousand-fold lower than reported for IFN-β (Hauser *et al.*, 1982), and did not lead to detectable levels of HuIFN-α. However, the principle that specific induction could be detected was established. Most of the transcripts, as for IFN-β, were correctly terminated and polyadenylated. The kinetics of induction of the HuIFN-α 1 transcripts paralleled the synthesis of mouse IFN ($\alpha + \beta$) transcripts.

Weissmann *et al.* (1982) have carried out experiments in which the IFN-α 5 gene has been transferred to hamster DHFR$^-$ cells (lacking dihydrofolate reductase gene) linked to an SV40 vector containing the mouse DHFR gene. The normal IFN promoter had been replaced by a viral promoter in the recombinant DNA used to transform these cells. Using selection over several months in methotrexate, which selects for DHFR gene amplification, it was found that the IFN-α 5 gene was also amplified in some cell lines, leading to constitutive production of some 3×10^4 U/ml/day (cell density not given), comparable to induced lymphoblastoid cells. This experiment, although it does not shed light on the normal regulation mechanisms of the IFN-α gene, demonstrates a methodology for gene amplification which will lead to practical application of tissue cultures for studies on gene regulation and the glycosylation of the IFN peptides, and finally for large-scale production of correctly processed interferons or glycoproteins in general.

(10) TISSUE CULTURE AS AN ALTERNATIVE SOURCE OF GLYCOPROTEINS SUCH AS INTERFERON

It has been pointed out in the previous sections that tissue cultures can be manipulated with defined DNA fragments, in much the same way as micro-organisms, in order to create cell lines which can produce large amounts of human interferons (Hauser *et al.*, 1982; Weissmann *et al.*, 1982). Initial analysis of the molecular weight of HuIFN-β by SDS-PAGE 'Western' blotting indicated that the product (identified through a monoclonal antibody) had the correct size for a normally glycosylated peptide. The process of glycosylation is a complicated multistep process which is carried out in eukaryotes during translation on the rough endoplasmic reticulum. The process involves the formation of long branched modified sugar chains containing, for example, sialic acid, *N*-acetyl glucosamine, mannose, galactosamine and fucose (Gibson, Kornfeld & Schlesinger, 1980). It is unlikely in the near future that a process can be developed for the post-translational modification of the naked peptides produced cheaply in recombinant DNA micro-organisms. This may well prevent the widespread use of such non-glycosylated products in the clinic, since the altered structure may, after fairly short periods of treatment, lead to immunological or allergic reactions. In some cases the glycosylation of particular proteins (and enzymes) has been shown to be an integral part of the stability of the peptide and may even influence the active centre (Kuo & Lampen,

1974). In some cases the glycosylation appears to be required for peptide transport out of the cell (Housley *et al.*, 1980). The full significance of glycosylation of blood proteins is not yet established. For many enzymes, including ribonuclease B, glycosylation appears to be used as a method of establishing the half life of the peptide in blood. Through the action of neuraminidase the rate of cleavage of the terminal sugar residues can be controlled. This terminal cleavage in turn exposes a sugar which causes the whole to be bound by lectins on the surface of for example macrophage or liver cells, where it is normally degraded (Stahl & Schlesinger, 1980). One possibility is that glycosylation also determines tissue target specificity, e.g. phosphoryl mannose in the targeting of lysosomal enzymes. In some cases both glycosylated and non-glycosylated forms of the same peptide may be produced from different precursors (e.g. ACTH, Gasson, 1980).

In the light of the speculations on the role of glycosylation it may well be that many glycoproteins when produced as naked peptides in recombinant micro-organisms will not be suitable for clinical use. This is currently a critical point with respect to IFN-β which is being used at this moment in the first clinical trials. To make interferon-β economically from large-scale tissue culture a number of improvements are required. Firstly tissue culture medium must be much cheaper than that normally used. Less fastidious cell lines can be used which can grow well with a cheap bovine serum fraction instead of foetal calf serum, and a semi-continuous high-level production has already been described (Hauser *et al.*, 1982). Using further gene amplification (see for example Brown, Kaufman, Haber & Schimke, 1982; Weissmann *et al.*, 1982) perhaps in conjunction with promoter deletions or other eukaryotic promoters it is possible that continuous high-level production in the absence of inducers can be obtained. These considerations along with recent dramatic improvements in large-scale purification methods and techniques of tissue culture indicate that tissue culture will find a new role in the biotechnological production of pharmaceuticals.

My gratitude is due to those who provided preprints of work in press and for other forms of personal communication particularly Walter Fiers, Björn Lund, Erik Lundgren, Paula Pitha, Michael Revel and Charles Weissmann.

Many thanks also to members of our interferon group, namely Wolfgang Bruns, Gerhard Gross, Hansjörg Hauser, Werner Lindenmaier and Ulrich Mayr for their support. Particularly thanks to Bärbel Seeger-Kunth for typing the manuscript.

REFERENCES

ALLEN, G. & FANTES, K. H. (1980). A family of structural genes for human lymphoblastoid (leukocyte-type) interferon. *Nature*, **287,** 408–11.

BENOIST, C. & CHAMBON, P. (1981). In vivo sequence requirements of the SV40 early promoter region. *Nature*, **290,** 304–10.

BENOIST, C., O'HARE, K., BREATHNECK, R. & CHAMBON, P. (1980). The ovalbumen gene-sequence of putative control elements. *Nucleic Acids Research*, **8,** 127–42.

BERG, K. (1982). Purification and characterisation of murine and human interferons. A review of the literature of the 1970's. *Acta Pathologica Microbiologica et Immunologica Scandinavica*, Section C, Supplement 279. Copenhagen: Munksgaard.

BRACK, C., NAGATA, S., MANTEI, N. & WEISSMANN, C. (1981). Molecular analysis of the human interferon-α gene family. *Gene*, **15,** 379–94.

BREATHNECK, R., MANTEI, N. & CHAMBON, P. (1980). Correct splicing of a chicken ovalbumin gene transcript in mouse L-cells. *Proceedings of the National Academy of Sciences of the USA*, **77,** 740–4.

BRINSTER, R. L, CHEN, H. Y., WARREN, R., SARTHY, A. & PALMITER, R. D. (1982). Regulation of metallothionein-thymidine kinase fusion plasmids injected into mouse eggs. *Nature*, **296,** 39–42.

BROWN, P. C., KAUFMAN, R. J., HABER, D. & SCHIMKE, R. T. (1982). Characteristic of dihydrofolate reductase gene amplification in murine and chinese hamster ovary cell lines. In *Gene Amplification*, ed. R. T. Schimke, pp. 9–13. New York: Cold Spring Harbor.

CANAANI, D. & BERG, P. (1982). Regulated expression of human interferon β1 gene after transduction into cultured mouse and rabbit cells. *Proceedings of the National Academy of Sciences of the USA*, **79,** 5166–70.

CAVALIERI, R. L., HAVELL, E. A., VILCEK, J. & PESTKA, S. (1977). Induction and decay of human fibroblast interferon mRNA. *Proceedings of the National Academy of Sciences of the USA*, **74,** 4415–19.

COLLINS, J. (1983). Interferon genes: gene structure and elements involved in gene regulation. In *Interferons and interferon inducers*, vol. 3, ed. N. Finter (in press).

CONTENT, J., DE WIT, L., PIERARD, D., DERYNCK R., DE CLERCQ, E. & FIERS, W. (1982). Secretory proteins induced in human fibroblasts under conditions used for the production of interferon β. *Proceedings of the National Academy of Sciences of the USA*, **79,** 2768–72.

CORCES, V., PELLICER, A., AXEL, R. & MESELSON, M. (1981). Integration and control of a drosophila heat shock gene in mouse cells. *Proceedings of the National Academy of Sciences of the USA*, **78,** 7038–42.

CORDEN, J., WASYLYK, B., BUCHWALDER, A., CORSI, P. S., KEDINGER, C. & CHAMBON, P. (1980). Promoter sequences of eukaryotic protein coding genes. *Science*, **209,** 1406–14.

DERYNCK, R., CONTENT, J., DE CLERCQ, E., VOLCKAERT, G., TAVERNIER, J., DEVOS, R. & FIERS, W. (1980a). Isolation and structure of a human fibroblast interferon gene. *Nature*, **285,** 542–7.

DERYNCK, R., REMAUT, E., SAMAN, E., STANSSENS, P., DE CLERCQ, E., CONTENT, J. & FIERS, W. (1980b). Expression of human fibroblast interferon gene in *E. coli. Nature*, **287,** 193–7.

GANNON, F., O'HARE, K., PERRIN, F., LE PENNEC, J. P., BENOIST, C., COCHET, M., BREATHNECK, R., ROYAL, A., GARAPIN, A., CAMI, B. & CHAMBON, P. (1979). Organisation and sequences at the 5′-end of a cloned ovalbumin gene. *Nature*, **278,** 428–34.

GASSON, J. (1980). Steroidogenic activity of high molecular weight forms of corticotropin. *Biochemistry*, **18,** 4215–24.

GIBSON, R., KORNFELD, S. & SCHLESINGER, S. (1980). A role for oligosaccharides in glycoprotein biosynthesis. *Trends in Biochemical Sciences*, **5,** 290–7.

GOEDDEL, D. V., LEUNG, D. W., DULL, T. J., GROSS, M., LAWN, R. M., MCCANDLISS, R., SEEBURG, P. H., ULLRICH, A., YELVERTON, E. & GRAY, P. W. (1981). The structure of eight distinct cloned human leukocyte interferon cDNAs. *Nature*, **290,** 20–6.

GOEDDEL, D. V., SHEPARD, H. M., YELVERTON, E., LEUNG, D., CREA, R., SLOMA, A. & PESTKA, S. (1980b). Synthesis of human fibroblast interferon by *E. coli. Nucleic Acids Research*, **8,** 4057–73.

GOEDDEL, D. V., YELVERTON, E., ULLRICH, A., HEYNECKER, H. L., MIOZZARI, G., HOLMES, W., SEEBURG, P. H., DULL, T., MAY, L., STEBBING, N., CREA, R., MOADA, S., CANDLISS, R. M., SLOMA, A., TABOR, J. M., GROSS, M., FAMILLETTI, P. C. & PESTKA, S. (1980a). Human leukocyte interferon produced by *E. coli* is biologically active. *Nature*, **287,** 411–15.

GOLDSTEIN, L. D., LANGFORD, M. P., STANTON, G. J., DE LEY, M. & GEORGIADES, J. A. (1981). Human gamma interferon: different molecular species. In *The biology of the interferon system*, ed. E. De Maeyer, G. Galasso & H. Schellekens, p. 313. London, Amsterdam: Elsevier/North-Holland Biomedical Press.

GRAY, P. W. & GOEDDEL, D. V. (1982). Structure of the human interferon gene. *Nature*, **298,** 859–63.

GRAY, P. W., LEUNG, D. W., PENNICA, D., YELVERTON, E., NAJARIAN, R., SIMONSEN, C. C., DERYNCK, R., SHERWOOD, P. J., WALLACE, D. M., BERGER, S. L., LEVINSON, A. D. & GOEDDEL, D. V. (1982). Expression of human immune interferon cDNA in *E. coli* and monkey cells. *Nature*, **295,** 503–7.

GROB, P. M. & CHADHA, K. C. (1979). Separation of human leukocyte interferon components by concanavalin A-agarose affinity chromatography and their characterization. *Biochemistry*, **18,** 5782–6.

GROSS, G., BRUNS, W., MAYR, U. & COLLINS, (1982). The identity of interferon-β1 mRNA transcripts in human fibroblasts and Namalva cells. *FEBS Letters*, **139,** 201–4.

GROSS, G., MAYR, U., BRUNS, W., GROSVELD, F., DAHL, H. H. M. & COLLINS, J. (1981c). The structure of a thirty-six kilobase region of the human chromosome including the fibroblast interferon gene IFN-β. *Nucleic Acids Research*, **9,** 2495–507.

GROSS, G., MAYR, U. & COLLINS, J. (1981b). New poly I-C inducible transcribed regions are linked to the human IFNβ gene in a genomic clone. In *The biology of the interferon system*, ed. E. De Maeyer, G. Galasso & H. Schellekens, pp. 85–90. Amsterdam: Elsevier/North-Holland Biomedical Press.

GROSS, G., MAYR, E., GROSVELD, F., DAHL, H. M., FLAVELL, R. A. & COLLINS, J. (1981a). Isolation and analysis of a cosmid hybrid containing the human genomic interferon gene, HuIFNβ1. In *Molecular Biology, Pathogenicity, and Ecology of Bacterial Plasmids*, ed. S. B. Levy, R. C. Clowes & E. L. König, pp. 429–38. New York: Plenum.

GROSVELD, G., SHEWMAKER, C. K., TAT, P. & FLAVELL, R. A. (1981). Localization of the DNA sequences necessary for transcription of the rabbit β-globin gene in vitro. *Cell*, **25,** 215–26.

GROSVELD, G. C., DE BOER, E., SHEWMAKER, C. K. & FLAVELL, R. A. (1982). DNA sequences necessary for the transcription of the rabbit β-globin gene in vivo. *Nature*, **295,** 120–6.

HAUSER, H., GROSS, G., BRUNS, W., HOCHKEPPEL, H. K., MAYR, U. & COLLINS, J.

(1982). Inducibility of human β-Interferon gene in mouse L-cell clones. *Nature*, **297,** 650–4.

Heron, I., Spargsbjerg, N., Hokland, M. & Berg, K. (1981). Effects of electrophoretically pure interferon on human lymphocytes functions. In *The biology of the interferon system*, ed. E. De Maeyer, G. Galasso & H. Schellekens, pp. 235–40. Amsterdam: Elsevier/North-Holland Biomedical Press.

Houghton, H., Jackson, I. J., Porter, A. G., Doel, S. M., Catlin, G. H., Barber, C. & Carey, N. H. (1981). The absence of introns within a human fibroblast interferon gene. *Nucleic Acids Research*, **9,** 247–66.

Housley, T. J., Rowland, F. N., Ledger, P. W., Kaplan, J. & Tanzer, M. L. (1980). Effects of tunicamycin on the biosynthesis of procollagen by human fibroblasts. *Journal of Biological Chemistry*, **255,** 121–8.

Hu, S. L. & Manley, J. (1981). DNA sequence required for the initiation of transcription in vitro from the major late promoter of adenovirus – 2. *Proceedings of the National Academy of Sciences of the USA*, **78,** 820–4.

Hynes, N. E., Kennedy, N., Rahmsdorf, U. & Groner, B. (1981). Hormone responsive expression of an endogenous proviral gene of mouse mammary-tumour virus after molecular cloning and gene transfer into cultured cells. *Proceedings of the National Academy of Sciences of the USA*, **78,** 2038–42.

Knight, E., Jr., Hunkapiller, M. W., Korant, B. D., Hardy, R. W. F. & Hood, L. E. (1980). Human fibroblast interferon: amino acid and amino terminal amino acid sequence. *Science*, **207,** 525–6.

Kuo, S. C. & Lampen, J. O. (1974). Tunicamycin – an inhibitor of yeast glycoprotein synthesis. *Biochemical and Biophysical Research Communications*, **58,** 287–95.

Kurtz, D. T. (1981). Hormonal inducibility of rat $_{2U}$globulin genes in transfected mouse cells. *Nature*, **291,** 629–31.

Lawn, R. M., Adelman, J., Dull, T. J., Gross, M., Goeddel, D. & Ullrich, A. (1981c). DNA sequence of two closely linked human leukocyte interferon genes. *Science*, **212,** 1159–62.

Lawn, R. M., Adelman, J., Franke, A. E., Houck, C. M., Gross, M., Najarian, R. & Goeddel, D. V. (1981a). Human fibroblast gene lacks introns. *Nucleic Acids Research*, **9,** 1045–52.

Lawn, R. M., Fritsch, E. F., Parker, R. C., Blake, G. & Maniatis, T. (1978). The isolation and characterization of linked α- and β-globin genes from a cloned library of human DNA. *Cell*, **15,** 1157–74.

Lawn, R. M., Gross, M., Houck, C. M., Franke, A. E., Gray, P. V. & Goeddel, D. V. (1981b). DNA sequence of a major human leukocyte interferon gene. *Proceedings of the National Academy of Sciences of the USA*, **78,** 5435–9.

Lee, F., Mulligan, R., Berg, P. & Ringold, G. (1981). Glucocorticoids regulate expression of dihydrofolate reductase cDNA in mouse mammary tumour virus chimaeric plasmids. *Nature*, **294,** 228–32.

Levy, W. P., Shively, J., Rubinstein, M., Del Valle, U. & Pestka, S. (1980). Amino-terminal amino acid sequence of human leukocyte interferon. *Proceedings of the National Academy of Sciences of the USA*, **77,** 5102–4.

Levy, W. P., Rubinstein, M., Shively, J., Del Valle, U., Lai, C.-Y., Moschera, J., Brink, L., Gerber, L., Stein, S. & Pestka, S. (1981). Amino acid sequence of a human leukocyte interferon. *Proceedings of the National Academy of Sciences of the USA*, **78,** 6186–90.

Mantei, M., Schwarzstein, M., Streuli, M., Panem, S., Nagata, S. & Weissmann, C. (1980). The nucleotide sequence of a cloned human leukocyte interferon cDNA. *Gene*, **10,** 1–10.

MANTEI, N. & WEISSMANN, C. (1982). Controlled transcription of a human α-interferon gene introduced into mouse L-cells. *Nature*, **297,** 128–32.

MATHIS, D. J. & CHAMBON, P. (1981). The SV40 early region TATA box is required for accurate in vitro initiation of transcription. *Nature*, **290,** 310–15.

MCKNIGHT, S. L., GAVIS, E. R. & KINGSBURY, R. (1981). Analysis of transcriptional regulatory signals of the HSV thymidine kinase gene: identification of an upstream control region. *Cell*, **25,** 385–98.

MEAGER, A., GRAVES, H., BURKE, D. C. & SWALLOW, D. M. (1979a). Involvement of a gene on chromosome 9 in human fibroblast interferon formation. *Nature*, **280,** 493–5.

MEAGER, A., GRAVES, H. E., WALKER, J. R., BURKE, D. C., SWALLOW, D. M. & WESTERVELD, A. (1979b). Somatic cell genetics of human interferon production in human-rodent cell hybrids. *Journal of General Virology*, **45,** 309–21.

MORY, Y., CHERNAJOVSKY, Y., FEINSTEIN, S. I., CHEN, L., NIR, U., WEISSENBACH, J., MALPIECE, Y., TIOLLAIS, P., MARKS, D., LADNER, M., COLBY, C. & REVEL, M. (1981). Synthesis of human interferon $\beta1$ in *Escherichia coli* infected by a lambda phage recombinant containing a human genomic fragment. *European Journal of Biochemistry*, **120,** 197–202.

MULLIGAN, R. C. & BERG, P. (1981). Selection for animal cells that express the *E. coli* gene coding for xanthine-guanine phosphoribosyl transferase. *Proceedings of the National Academy of Sciences of the USA*, **78,** 2072–6.

NAGATA, S., BRACK, C., HENCO, K., SCHAMBÖCK, A. & WEISSMANN, C. (1981). Partial mapping of ten genes of the human interferon-α family. *Interferon Research*, **1,** 333–6.

NAGATA, S., MANTEI, N. & WEISSMANN, C. (1980b). The structure of one of the eight or more distinct chromosomal genes for human interferon-α. *Nature*, **287,** 401–8.

NAGATA, S., TAIRA, H., HALL, A., JOHNSRUD, I., STREULI, M., ECDOSI, J., BOLL, W., CANTELL, K. & WEISSMANN, C. (1980a). Synthesis in *E. coli* of a polypeptide with human leukocyte activity. *Nature*, **284,** 316–20.

OHNO, S. & TANIGUCHI, T. (1981). Structure of a chromosomal gene for human interferon β. *Proceedings of the National Academy of Sciences of the USA*, **78,** 5305–9.

OWERBACH, D., RUTTER, W. J., SHOWS, T. B., GRAY, P., GOEDDEL, D. V. & LAWN, R. M. (1981). Leukocyte and fibroblast interferon genes are located on human chromosome 9. *Proceedings of the National Academy of Sciences of the USA*, **78,** 3123–7.

PITHA, P. M., CIUFO, D. M., KELLUM, M., RAJ, N. B. K., REYES, G. R. & HAYWARD, G. S. (1982b). Induction of human β-interferon synthesis with poly rI : rC in mouse cells transfected with cloned cDNA plasmids. *Proceedings of the National Academy of Sciences of the USA*, **79,** 4337–40.

PITHA, P. M., REYES, G. R., RAJ, N. B. K. & HAYWARD, G. S. (1982a). Expression of human β-interferon gene in heterologous cells. In *UCLA Symposia on Molecular and Cellular Biology*, vol. 25, ed. T. C. Merigan & R. M. Friedman. New York: Academic Press (in press).

RAJ, N. B. K. & PITHA, P. M. (1981). Analysis of interferon mRNA in human fibroblast cells induced to produce interferon. *Proceedings of the National Academy of Sciences of the USA*, **78,** 7426–30.

RAJ, N. B. K. & PITHA, P. M. (1980a). The messenger RNA sequences in human fibroblast cells induced with poly rI : rC to produce interferon. *Nucleic Acids Research*, **8,** 3427–37.

RAJ, N. B. K. & PITHA, P. M. (1980b). Synthesis of new proteins associated with the induction of interferon in human fibroblast cells. *Proceedings of the National Academy of Sciences of the USA*, **77,** 4918–22.

RAJ, N. B. K. & PITHA, P. M. (1977). Relationship between interferon production and interferon messenger RNA synthesis in human fibroblasts. *Proceedings of the National Academy of Sciences of the USA*, **74,** 1483–7.

REYES, G. R., GOVIS, E. R., BUCHAN, A., RAJ, N. B. K., HAYWARD, G. S. & PITHA, P. M. (1982). Expression of human β-interferon cDNA under the control of the thymidine kinase promoter from Herpes simplex virus. *Nature*, **297,** 598–601.

ROBINS, D. M., AXEL, R. & HENDERSON, A. S. (1981). Chromosome structure and DNA sequence alterations associated with mutation of transformed genes. *Journal of Molecular and Applied Genetics*, **1,** 191–203.

RUBINSTEIN, M., RUBINSTEIN, S., FOMILLETTE, P. L., MILLER, R. S., WALDMAN, A. A. & PESTKA, S. (1979). Human leukocyte interferon: production, purification to homogeneity and initial characterisation. *Proceedings of the National Academy of Sciences of the USA*, **76,** 640–4.

SCANGOS, G. & RUDDLE, F. H. (1981). Mechanisms and applications of DNA mediated gene transfer in mammalian cells – a review. *Gene*, **14,** 1–10.

SEHGAL, P. B., DOBBERSTEIN, B. & TAMM, I. (1977). Interferon messenger RNA content of human fibroblast during induction, shut-off, and super induction of interferon production. *Proceedings of the National Academy of Sciences of the USA*, **74,** 3409–13.

SEHGAL, P. B., SAGAR, A. D., BRAUDE, I. A. & SMITH, D. (1981). Heterogeneity of human α and β interferon mRNA species. In *The Biology of the Interferon System*, ed. E. De Maeyer, G. Galasso & H. Schellekens, pp. 43–6. Amsterdam: Elsevier/North-Holland Biomedical Press.

SHEN, C.-K. J. & MANIATIS, T. (1982). The organization, structure and *in vitro* transcription of Alu family RNA polymerase III transcription units in the human α-like globin gene cluster: precipitation of in vitro transcripts by Lupus anti-La antibodies. *Journal of Molecular and Applied Genetics*, **1,** 343–60.

SLATE, D. L. & RUDDLE, F. H. (1979). Fibroblast interferon in man is coded by two loci on separate chromosomes. *Cell*, **16,** 171–80.

STAEHELIN, T., DURRER, B., SCHMIDT, J., TAKACS, B., STOCKER, J., MIGGIANO, V., STÄHLI, C., RUBINSTEIN, M., LEVY, W. P., HERSHBERG, R. & PESTKA, S. (1981). Production of hybridomas secreting monoclonal antibodies to the human leukocyte interferons. *Proceedings of the National Academy of Sciences of the USA*, **78,** 1848–52.

STAHL, P. D. & SCHLESINGER, P. H. (1980). Receptor-mediated pinocytosis of mannose/N-acetyl glucosamine – terminated glycoproteins and lysosomal enzymes by macrophages. *Trends in Biochemical Sciences*, **5,** 194–6.

STEWART, W. E. II, LIN, L. S., WIRANOWSKI-STEWART, M. & CANTELL, K. (1977). Elimination of size and charge heterogeneities of human leukocyte interferons by chemical cleavage. *Proceedings of the National Academy of Sciences of the USA*, **74,** 4200–4.

TAN, Y. H., CREAGAN, R. P. & RUDDLE, F. H. (1974). The somatic cell genetics of human interferon: assignment of human interferon loci to chromosomes 2 and 5. *Proceedings of the National Academy of Sciences of the USA*, **71,** 2251–5.

TANIGUCHI, T., GUARENTE, L., ROBERTS, T. M., KIMELMAN, D., DOUHAN III, J. & PTASHNE, M. (1980c). Expression of the human fibroblast interferon gene in E.coli. *Proceedings of the National Academy of Sciences of the USA*, **77,** 5230–3.

TANIGUCHI, T., MANTEI, N., SCHWARZSTEIN, M., NAGATA, S., MURAMATSU, M. & WEISSMANN, C. (1980b). Human leukocyte and fibroblast interferons are structurally related. *Nature*, **285,** 547–9.

TANIGUCHI, T., OHNO, S., FUJII-KURIYAMA, Y. & MURAMATSU, M. (1980a). The nucleotide sequence of human fibroblast interferon cDNA. *Gene*, **10,** 11–15.

TANIGUCHI, T., SAKAI, M., FUJII-KURIYAMA, Y., MURAMATSU, M., KOBAYASHI, S. & SUDO, T. (1979). Construction and identification of a bacterial plasmid containing the human fibroblast interferon gene sequences. *Proceedings of the Japan Academy*, **55,** 464–9.

TAVERNIER, J., DERYNCK, R. & FIERS, W. (1981). Evidence for a unique human fibroblast interferon (IFN-β1) chromosomal gene, devoid of intervening sequences. *Nucleic Acids Research*, **9,** 461–71.

TÖRMÄ, E. T. & PAUCKER, K. J. (1976). Purification and characterisation of human leukocyte interferon components. *Journal of Biological Chemistry*, **251,** 4810.

ULLRICH, A., GRAY, A., GOEDDEL, D. V. & DULL, T. J. (1982). Nucleotide sequence of a portion of human chromosome 9 containing a leukocyte interferon gene cluster. *Journal of Molecular Biology*, **156,** 467–86.

VAN DAMME, J., DE LEY, M., CLAEYS, H., BILLIAU, A., VERMYLEN, C. & DE SOMER, P. (1980). Isolation of a β-type interferon from concanavalin A-induced human leukocytes: non-identity with fibroblast interferon. In *The biology of the interferon system*, ed. E. De Maeyer, G. Galasso & H. Schellekens, pp. 331–4. Amsterdam: Elsevier/North-Holland Biomedical Press.

WASYLYK, B., KEDINGER, C., CORDEN, J., BRISON, O. & CHAMBON, P. (1980). In vitro initiation of transcription on conalbumin and ovalbumin genes and comparison with adenovirus – 2 early and late genes. *Nature*, **285,** 388–90.

WEISSENBACH, J., CHERNAJOVSKY, Y., ZEEVI, M., SHULMAN, L., SOREQ, H., NIR, U., WALLACH, D., PERRICAUDET, M., TIOLLAIS, P. & REVEL, M. (1980). Two interferon mRNAs in human fibroblasts: In vitro translation and *Escherichia coli* cloning studies. *Proceedings of the National Academy of Sciences of the USA*, **77,** 7152–6.

WEISSMANN, C. (1981). The cloning of interferon and other mistakes. In *Interferon*, vol. 3, ed. I. Gresser, pp. 101–34. New York & London: Academic Press.

WEISSMANN, C., NAGATA, S., BOLL, W., FOUNTOULAKIS, M., FUJISAWA, A., FUJISAWA, J.-I., HAYNES, J., HENCO, K., MANTEI, N., RAGG, H., SCHEIN, C., SCHMID, J., SHAW, G., STREULI, M., TAIRA, H., TODOKORO, K. & WEIDLE, U. (1982). Structure and expression of human alpha-interferon genes. *ICN-UCLA Symposium (March 1982), Chemistry and Biology of Interferons: Relationship to Therapeutics*, ed. T. C. Merigan & R. M. Friedman, **35,** 295–326.

WOLD, B., WIGLER, M., LACY, E., MANIATIS, T., SILVERSTEIN, S. & AXEL, R. (1979). Introduction and expression of a rabbit β-globin gene in mouse fibroblasts. *Proceedings of the National Academy of Sciences of the USA*, **76,** 5684–8.

YIP, Y. K., BARROWCLOUGH, B. S., URBAN, C. & VILCEK, J. (1982). Molecular weight of gamma interferon is similar to that of other human interferons. *Science*, **215,** 411–13.

ZINN, K., MELLON, P., PTASHNE, M. & MANIATIS, T. (1982). Regulated expression of an extra chromosomal human β-interferon gene in mouse cells. *Proceedings of the National Academy of Sciences of the USA*, **79,** 4897–901.

ZOON, K. (1981). Purification, sequencing and properties of human lymphoblastoid and leukocyte interferon. In *The Biology of the Interferon System*, ed. E. De Maeyer, G. Galasso & H. Schellekens, pp. 47–56. London, Amsterdam: Elsevier/North-Holland Biomedical Press.

ZOON, K., SMITH, M. E., BRIDGEN, P. J., ANFINSEN, C. B., HUNKAPILLER, W. E. & HOOD, L. E. (1980). Amino terminal sequence of the major component of human lymphoblastoid interferon. *Science*, **207,** 527–8.

THE CONTROL OF INTERFERON FORMATION

D. C. BURKE

Department of Biological Sciences, University of Warwick, Coventry, CV4 7AL, England

INTRODUCTION

The cells of probably all vertebrate species contain the genes coding for IFN-α,β and presumably, also for IFN-γ. These genes are transiently activated after exposure of the cells to viruses, double-stranded nucleic acids, mitogens or antigens, activation leading to the production of one or more of the three classes of IFN. The types of IFN produced depend on the nature of the cell and the inducing agent and the IFN genes in some cell systems cannot be activated at all. What controls whether the IFN genes can be activated or not: and what controls whether IFN-α,β or γ or a mixture is produced? Gene activation leads to the formation of IFN mRNA by transcription with RNA polymerase II, rather than by activation of pre-existing IFN mRNA, and the molecular basis of this process is discussed by Dr Collins earlier in this volume (p. 35). However, the amount of IFN mRNA found in induced cells, and the rate of mRNA synthesis can be affected by pretreatment of the cells with various reagents. How is this effect brought about? The mRNA moves from the nucleus to the cytoplasm, but there is little processing of the IFN-α and -β mRNAs since neither the genes for interferon-α or -β contain introns, although the IFN-γ gene does. Translation on polysomes leads to synthesis of IFN-α, -β or -γ but the lifetime of the IFN mRNA varies widely – varying from cell to cell but also varying in the same cell depending on the inducer or the way in which the cells have been treated. How is the life-time of the IFN mRNA varied? The IFN mRNA is competing with other cellular mRNAs for initiation factors and ribosomes, and this competition is probably controlled by nucleotide sequences in the mRNA. There is some evidence that the efficiency of translation of IFN mRNA can be varied by suitable treatment of the cells. How is this brought about? Transcription of the IFN-α and -β gene(s) stops after a few hours, and a second treatment of the cells with the inducer does not lead to renewed IFN formation, although the cells gradually become inducible again. Why is the gene no longer inducible after IFN mRNA transcription has ceased, and is this the same mechanism as that which brings about cessation of transcription? Finally the IFN-β and -γ polypeptides are glycosylated as they leave the cell.

There are thus many questions still unanswered about the control of IFN production – questions whose answers may be relevant not only to this genetic system but to inducible eukaryotic genome systems in general. The purpose of this review is to describe critically what is known about these control processes, but because of our lack of knowledge about the mechanism of IFN-γ production, the review will be restricted almost completely to IFN-α and -β.

THE COMPONENTS OF THE INTERFERON SYSTEM CAN BE ACCURATELY MEASURED

Interferon genes and their organization

The number of IFN genes and their distribution about the human chromosomes can now be determined by use of the cloned IFN genes and somatic cell hybrids. When a human genomic library is hybridized with a radioactive IFN cDNA clone, the genomic IFN genes can be readily detected. Restriction enzyme analysis of such genomic clones can show whether they are identical to each other or not. Since it is sometimes possible to isolate a genomic clone containing more than one IFN gene, their distance apart can be directly determined. Further, by analysis of mouse : human somatic cell hybrids which contain all the mouse chromosomes but only a limited number of identified human chromosomes, it is possible to determine which chromosomes carry the IFN genes. This information is discussed elsewhere in this volume (pp. 46, 47), but suffice it to say that while there are single IFN-β and -γ genes the IFN-α family is complex: 13 non-allelic authentic α genes and 6 pseudogenes have been distinguished, and in addition nine genes believed to be allelic to the 13 authentic genes have been sequenced. All these α genes lie on human chromosome 9, and they are currently being mapped in detail. Analysis of IFN-α production must therefore take into account this diverse family: an increase in IFN-α production could, for example, be due either to an increase in the rate of transcription of a few genes or to new transcription from genes which were previously inactive. There is only a single genomic copy of the IFN-β gene, which is also on chromosome 9, and also a single copy of the IFN-γ gene on chromosome 12 (see Weissmann *et al.*, 1982, and Fiers *et al.*, 1982, for recent reviews).

Interferon mRNA

The amounts of IFN mRNA can be determined either functionally by translation to form biologically active IFN or by hybridization using a cDNA interferon clone. These two methods do not necessarily give the same information. There have been reports of IFN-α and -β mRNA derived from genes on chromosomes 2 and 5 (Seghal & Sagar, 1980). Such mRNAs may or may not hybridize with the IFN-α and -β genes from chromosome 9 that have recently been cloned. Conversely, an IFN mRNA derived from a pseudogene would not be translatable to give biologically active IFN but might hybridize with a cDNA clone. Mouse IFN mRNA may be measured by translation in a cell-free system (Le Bleu *et al.*, 1978) although there have been no consistent reports of such translation of human IFN mRNA preparations (Sehgal, Dobberstein & Tamm, 1977). The amount of translatable IFN-mRNA may be measured more readily by injection into *Xenopus* oocytes, which translate IFN mRNA efficiently and stoichiometrically and export the product from the oocyte into the surrounding fluid (Reynolds, Premkumar & Pitha, 1975; Sehgal *et al.*, 1977; Cavalieri, Havell, Vilcek & Pestka, 1977a; Colman & Morser, 1979; Shuttleworth, Morser & Burke, 1982a). The amount of hybridizable IFN mRNA may be measured by Northern blot transfer and quantitated by the dot-blot procedure. All these methods measure the *amount* of IFN mRNA present in the cells: but they do not measure its *rate of synthesis*. This can be done in principle by hybridizing very briefly labelled nuclear RNA to the DNA sequences corresponding to the mRNA. As pointed out by Darnell (1982), this is best done by exposure of the cells to radioactive precursors for up to five minutes. This is often not practicable because so little of the total RNA output (usually $5\times10^{-6}-5\times10^{-5}$ of the total) from the nucleus is due to transcription of a single gene, that the amount of labelling would be negligible. An alternative method is to use isolated nuclei containing previously initiated RNA polymerase II nucleus, and adding labelled nucleoside triphosphates when about 500 nucleotides are incorporated in 10–20 min of incubation. This labels the RNA much more efficiently than *in vivo*, and such *in vitro* synthesis appears to be a true reflection of the processes occurring *in vivo*. However no such studies have been carried out for interferon.

Interferon

Interferon itself may be measured either by determining its antiviral activity or by determining the amount of IFN protein. The antiviral activity may be determined on a variety of cells using a number of different challenge viruses, but none of the assays are more accurate than $\pm 0.3 \log_{10}$ and some no more accurate than $\pm \log_{10}$. Antiviral units may then be transformed into weight units by using a value for the specific activity, but it has not been easy to reach agreement on such a figure, and the possibility of inactivation during purification always makes such figures minimum ones. Further, it is now known that the different interferon-α species have different specific activities (Franke *et al.*, 1982). The alternative method of determining interferon is by determining the amount of IFN protein – either directly by a physical procedure using highly purified IFN (Fantes & Allen, 1981; see Rubinstein, 1982, for a review), or by an immunoradiometric assay using monoclonal antibody (Secher, 1981), although the immunoradiometric methods have to be calibrated with a pure sample of IFN. As yet only monoclonal antibodies to IFN-α are freely available, although there are recent reports of preparation of such antibodies to IFN-β and IFN-γ.

Thus all the components of the IFN system, the gene, the mRNA and the protein, can be quantitated. However the agents which switch transcription of the genes on and off have not been identified, let alone quantitated.

CONTROL AT THE LEVEL OF AVAILABILITY OF THE IFN GENES FOR TRANSCRIPTION

Interferon production in mouse embryonal carcinoma cells

All vertebrate cells appear to contain the genes for IFN-α and -β (Wilson *et al.*, 1983), but these genes may not necessarily be in a state where they can be expressed. The best example of this is mouse embryonal carcinoma (EC) cells. Burke, Graham & Lehman

(1978) found that undifferentiated mouse EC cells could neither make nor respond to IFN but that upon differentiation the cells become both inducible and responsive to IFN. EC cells are derived from teratocarcinomas, which are tumours composed of both pluripotent embryonal carcinoma cells and a variety of differentiated cell types (see Illmensee & Stevens, 1979, for a review). EC cells differentiate, either spontaneously or after treatment with retinoic acid, to form endoderm-like cells, and this change in morphology is accompanied by changes in a number of cell markers, and the cells also become permissive to infection with either RNA or DNA tumour viruses. It is not known why EC cells fail to make or respond to IFN: it might be either because the putative inducer of IFN formation cannot be formed in these cells, or because of differences in the control of gene expression between EC cells and their differentiated progeny. Two such differences have been identified: it has been suggested that the reason that the SV40 T antigen is not detected in infected EC cells is that the cells are unable to carry out the splicing reactions that are essential for transport of the virus RNA transcripts from nucleus to cytoplasm (Sehgal, Levine & Khoury, 1979; Sehgal & Khoury, 1979), but that differentiated cells can perform this reaction. This suggests that changes in splicing may be involved in changes in gene expression during differentiation. However this is unlikely to be the explanation of the failure to produce IFN since human IFN-α and -β, and probably mouse IFN-α and -β, lack introns and splicing may therefore not be necessary for gene expression. The second difference is in the methylation of the cytosine residues. Stewart, Stuhlman, Jahner & Jaenish (1982) showed that after infection of EC cells with Moloney murine leukaemia virus, up to 100 integrated proviral genomes were present, but that these integrated genomes were highly methylated. In contrast, the integrated genomes in differentiated cells, where the virus is expressed, were not methylated. The proviral genomes from EC cells were only infectious when the cells were treated with 5-aza-cytidine, a drug known to interfere with DNA methylation. They concluded that there was a correlation between the suppression of proviral genome expression and DNA methylation, and suggested that *de novo* methylation may be a characteristic of early embryonic cells. Such a mechanism could explain the suppression of IFN expression. Either mechanism could explain why the characteristic is recessive in intraspecific cell hybrids (Veomett, Hekman & Veomett, 1980).

Interferon production in the developing mouse embryo

Pluripotent EC cells are considered to be analogous to pluripotent cells of the early mouse embryo (Martin, 1980). Thus the finding that the IFN system is not expressed in the pluripotent EC cells implies that there might be a similar absence in the pluripotent cells of the mouse embryo: IFN production and sensitivity only being acquired after differentiation has occurred. In order to test this, Barlow, Randle, Burke & Graham (1982) developed an IFN assay suitable for very small pieces of tissue (~50 cells), and found that no mouse embryo tissue could be induced to produce IFN before day seven, but by day thirteen all the major embryonic and extra-embryonic tissues had become competent. Interferon inducibility arose throughout the embryo during day seven to eight, and although initially confined to the trophoblast, developed progressively throughout the embryo independent of cell lineage. Rather similar kinetics of appearance have been reported for alphafetoprotein (Dziadek & Adamson, 1978) and plasminogen activator (Bode & Dziadek, 1979) and this suggests that all three markers are controlled by a similar process, such as nucleic acid methylation (Jahner *et al.*, 1982). It is too early yet to assess the significance of the findings for the pathology of virus infection of the embryo, but it is clear that the whole IFN system, which in mouse as in man, is located on at least two chromosomes, can be activated during a developmental process.

Interferon production in tissue culture cells

Some tissue culture cells also fail to produce IFN after induction, and in one case at least the gene is known to be present but is not expressible. Meager *et al.* (1979) found that although a Chinese hamster cell line, W^{g3h}, could not produce any hamster IFN, human–Chinese hamster hybrids derived from this cell could do so. This was true whether the hybrid could produce human IFN or not (because the human structural gene on chromosome 9 was present or not), showing that the Chinese hamster cells contain the IFN gene but failed to express it, probably because of some repressor substance present in the hamster cytoplasm. Similar situations have been described for other cell hybrids. Several other tissue culture cell lines fail to produce IFN–BHK/21, VERO and mouse myeloma cells – and it has been reported that the IFN gene is absent or defective in VERO cells (Emeny & Morgan, 1979).

It is known that the production of either IFN-α or -γ by peripheral blood lymphocytes is limited to distinct sub-populations of lymphocytes (see the review by Wilkinson & Morris, p. 149). This implies that other cells in this population are unable to produce interferon in response to the inducer, but almost nothing is known as yet about the control of interferon formation in such somatic cells.

CONTROL AT THE TRANSCRIPTIONAL AND POST-TRANSCRIPTIONAL LEVELS

The induction of IFN mRNA synthesis

The IFN-α and -β genes are activated by treatment of the cells with viruses or double-stranded RNA, and it is likely that the formation of double-stranded RNA is the key common process with these two inducers (see review by Burke, 1982). The IFN-γ gene is activated by mitogens or the appropriate antigen, and unlike IFN-α and -β, IFN-γ is produced only by lymphocytes. However, in none of these cases do we know the nature of the proximal inducer or the mechanism of induction. Since the IFN-α and -β genes are often coordinately expressed (see later), it is assumed that a common mechanism is involved and that sequences adjacent to the IFN-α and -β genes are activated by some substance whose nature is unknown. It is unlikely that double-stranded RNA would have the specificity to interact with such sequences, so it is postulated that the proximal inducer is not double-stranded RNA, but some product of its interaction with the cell. Double-stranded RNA is known to elicit the formation of several new proteins in the interferon-treated cell. Might it produce a protein in the untreated cell that is the proximal inducer?

Several other mechanisms have been proposed (Stewart, 1979, pp. 75–6). The 'repressor-depletion' hypothesis was suggested as an explanation of why cycloheximide and other reversible inhibitors of macromolecular synthesis could induce IFN synthesis (Tan & Berthold, 1977). The ability of the inducer to initiate IFN synthesis was suggested to be due to inhibition of synthesis of a short-lived repressor molecule, which normally repressed IFN gene activity. Indeed many virus and double-stranded RNA do depress macromolecular synthesis, but conversely several effective inducers do not. The 'basal-level IFN' hypothesis is related to the repressor-duplication hypothesis and suggests that the repressor is in fact IFN which is synthesized constitutively at a low level. This suggestion is supported by the surprising observation that IFN is bound selectively to both poly rI . poly rC and poly rI . poly rU immobilized on columns (De Maeyer-Guignard, Thang & De Maeyer, 1977). The selective binding of basal-levels of IFN by infectious virus RNA, replicative intermediates or synthetic polynucleotides would then remove the repression of the IFN genes. There is no recent evidence to decide between these various hypotheses, and we know even less about the mechanism of IFN-γ production (see review by Wilkinson & Morris, p. 149).

Whatever the mechanism, interferon production requires the *de novo* synthesis of new mRNA rather than the activation of an inactive precursor. This was first shown by use of the inhibitor actinomycin which blocks interferon production in a large number of systems by virtue of its inhibitory effect on DNA-directed RNA synthesis. Interferon production is also inhibited by other inhibitors of RNA synthesis, inducing α-amanitin which inhibits DNA polymerase II, the enzyme involved in mRNA synthesis (Atherton & Burke, 1978).

IFN mRNA is transported, after its synthesis, into the cytoplasm as shown by the fact that induced mouse L^{929} cells, enucleated 10 h after induction, continue to synthesize IFN for a further 8–10 h despite the lack of a nucleus (Burke & Veomett, 1977). It is interesting that the half-life of the IFN mRNA was unaffected by enucleation.

The amount of IFN mRNA has been measured in a number of systems using injection into *Xenopus* oocytes. In either human or mouse fibroblasts induced with poly rI . poly rC the amount of interferon mRNA correlated well with the rate of IFN synthesis (Sehgal *et al.*, 1977; Cavalieri *et al.*, 1977a; Raj, Fernie & Pitha, 1979; Pang, Hayes & Vilcek, 1980). A similar result was obtained with induced human lymphoblastoid (Namalwa) cells; the amount of IFN mRNA measured either as total intracellular RNA or as polysomal RNA, rose in parallel with the rate of IFN production, and then fell as IFN production ceased (Morser *et al.*, 1979). This suggested that the rate of IFN production was directly determined by the amount of IFN mRNA present in the cells.

The amount of IFN mRNA has recently been measured in human

fibroblasts treated with poly rI . poly rC using the cloned IFN-β gene as a hybridization probe (Raj & Pitha, 1981), and also in induced lymphoblastoid cells using both the cloned IFN-α and -β genes as hybridization probes (Shuttleworth, Morser & Burke, 1983). Both groups found that the time course of either IFN-β mRNA or of both IFN-α and IFN-β mRNA production was similar to that of IFN production, and that the time courses of IFN-α and -β mRNA production were similar to each other. By calibration of the dot-blot hybridization assay, Shuttleworth *et al.* (1983) found that IFN-α mRNA accounted for 0.015% and IFN-β mRNA 0.02% of the total poly(A)$^+$ RNA.

Control of transcription in cells treated with 5-bromodeoxyuridine or butyrate

The yield of IFN from a given virus–cell system can often be increased by suitable treatment of the cells, and it is pertinent to ask whether this effect is mediated at the transcriptional or translational level, or both. Significant enhancement of IFN production can, for example, be achieved by treatment of Namalwa cells with butyrate or 5-bromodeoxyuridine before induction (Adolf & Swetley, 1979; Johnston, 1980; Baker, Morser & Burke, 1980; Tovey, Begon-Lours, Gresser & Morris, 1977; Baker, Bradshaw, Morser & Burke, 1979). This increase is not due to the production of an IFN-α of higher specific activity since there was a corresponding increase in the amount of interferon protein as measured by the immuno-radiometric assay. Nor was there any change in the activity of this interferon on heterologous cells, suggesting that the increase was not due to production of interferons with differing activities on different cells. Measurement of the amount of IFN mRNA in treated and untreated cells by injection into *Xenopus* oocytes showed that both treatments caused an increase in the amount of IFN mRNA, without any effect on the time course of IFN production. However the increase in mRNA levels (26-fold for butyrate treatment and 3-fold for 5-bromodeoxyuridine) did not completely account for the increase in IFN yield (77-fold for butyrate treatment and 11-fold for 5-bromodeoxyuridine), and it was concluded that although the main effect of these treatments was operating at the level of transcription or processing of IFN mRNA, there was some increase in translational efficiency (Morser, Meager & Colman, 1980; Shuttleworth *et al.*, 1982a). The effect of treatment with butyrate and 5-bromodeoxyuridine on IFN mRNA levels has also been measured by dot-blot hybridization. Shuttleworth *et al.* (1983) found that the levels of IFN-α and -β mRNA were increased by the treatments, but that they were still present in the same ratio, and that the size of the mRNAs, and the heterogeneity of the IFN-α mRNA, were unaltered.

There is thus very strong evidence that such treatments increase the amount of IFN mRNA in the cells, and although there have been no measurements of the effect of these treatments on the *rate* of IFN mRNA synthesis, for the reasons discussed earlier, it is very likely that because of the unchanged kinetics of IFN mRNA formation and decay, the effect is due to an increase in the rate of transcription. However the mechanism of this effect is not known, although both reagents cause similar effects in other systems (see references quoted in Shuttleworth *et al.*, 1982a).

Recently Shuttleworth, Morser & Burke (1982b) have compared the pattern of protein synthesis in Namalwa after treatment with either butyrate or 5-bromodeoxyuridine. Athough these treatments caused a substantial increase in interferon yield, there was little effect on synthesis of other cellular proteins, a 3–4-fold enhancement of a protein of 35 000 molecular weight being the only significant effect. However those proteins that made up less than 0.002% of the radioactivity of labelled cellular proteins were undetectable in the system used and since interferon represents no more than 0.001% of such labelled protein, changes in the levels of IFN or of other low-abundance proteins would not have been detected. It was concluded that these treatments have a relatively specific effect on protein synthesis in Namalwa cells.

Control of transcription in primed cells

Both the rate of IFN production and the final yield of IFN are increased when the cells are induced after pretreatment with interferon – the priming phenomenon (see Stewart, 1979, pp. 233–6). Several groups have compared the IFN mRNA levels in primed with those in unprimed cells, using the *Xenopus* oocyte innoculation technique to measure the IFN mRNA levels. Abreu, Bancroft & Stewart (1979) found that although no more IFN mRNA was produced in primed cells, it was produced earlier after induction, and also disappeared more rapidly. However Saito *et al.* (1976) and Fujita, Saito & Kohno (1979) found more IFN mRNA in primed cells than in unprimed cells although they did not look at the decay phase. Sehgal & Gupta (1980) found that more IFN mRNA was produced in human fibroblasts treated with poly rI . poly rC, and that the rate of decay was similar to that in untreated cells. Raj & Pitha (1981), using a nucleic acid hybridization assay, found 10-fold more IFN-β mRNA in primed cells than in unprimed cells, while the IFN mRNA appeared and decayed with rather similar kinetics. Thus there is good evidence that IFN mRNA is made earlier or in larger amounts, and this has usually been interpreted as being due to a stimulation in the rate of transcription, although there has been no direct measurement of such rates.

Control of transcription in superinduced cells

A number of tissue culture cell lines produce more IFN if the cells are treated with cycloheximide immediately after induction with poly rI . poly rC, then with actinomycin before incubation in inhibitor-free medium. This same effect can be obtained by ultraviolet irradiation of the cells before induction. This is the so-called 'superinduction' phenomenon (for a review see Stewart, 1979, pp. 97–103). The effect has been widely interpreted by the repressor model, described by Tomkins *et al.* (1966) to explain a similar effect of antimetabolites on the production of tyrosine amino-transferase. They proposed that control was exerted through the action of a second, repressor, gene whose product prevented continued translation of the mRNA for tyrosine amino-transferase. Anti-metabolites blocked the formation of this repressor molecule, with the result that synthesis of the enzyme continued. A series of experiments with different metabolic inhibitors supports this interpretation of the superinduction of interferon formation. Recently more direct evidence has been sought by measurement of the levels of IFN mRNA in superinduced cells, since the Tomkins model predicts that there will be more mRNA present in the cells and for longer after induction. A number of groups have measured the levels of IFN mRNA in superinduced cells and all agree that levels are raised (Raj & Pitha, 1977; Sehgal *et al.*, 1977; Cavalieri, Havell, Vilcek & Pestka, 1977b). Normally induced cells showed a rapid production of IFN mRNA which peaked by three hours after induction followed by its decay, while in superinduced cells the IFN mRNA was produced with rather similar kinetics but decayed much more slowly. Further analysis has shown that the increased levels are due to two distinct mechanisms, a small (3–4-fold) increase in the initial rate of transcription plus a much larger (14-fold) effect on the stability of the mRNA (Sehgal, Lyles & Tamm, 1978; Sehgal & Tamm, 1979). Superinduction increased the half-life of the IFN mRNA during the decay phase, which increased from 0.5 h to 6–8 h. Thus the rapid decrease in interferon production in induced cells is not due to the cessation of transcription, but rather to the existence of a post-transcriptional mechanism that degrades or inactivates IFN mRNA. This post-transcriptional mechanism requires protein synthesis, is accompanied by decreased size of the mRNA (Raj & Pitha, 1981), is unaffected by whether the cells are grown in suspension or attached to a solid support (Sehgal, 1981) and does not affect the bulk of the cellular RNA. Nor is it mediated by any 2,′5′,-oligo A which might be formed, and which then could activate an endonuclease to inactivate IFN mRNA (Sehgal & Gupta, 1980). A similar effect of IFN mRNA present in cells is seen when virus-induced human lymphoblastoid (Namalwa) cells are incubated at 28 °C, 3.5 h after induction (Morser & Shuttleworth, 1981). Again increased yields of IFN were due to prolonged translation of IFN mRNA. Similar situations have been found in the regulation of other eukaryotic mRNAs: for example Guyette, Matusik & Rosen (1979) found that treatment of rat mammary gland organ cultures with prolactin increased the rate of casein mRNA transcription 2- to 4-fold but increased the half-life of casein mRNA by 17- to 25-fold. The mechanisms of these interesting effects on the stability of the mRNA are thus unknown, but their selectivity, and the possibility that such mechanisms could be involved in the control of expression of other eukaryotic genes, make it an interesting problem to study.

Control of transcription in hyporesponsive cells

After cells have produced interferon, they are generally unresponsive to a second treatment with the same or another inducer: a situation usually called hyporesponsiveness (see review by Stewart, 1979, pp. 103–6). Little or nothing is known about the molecular basis of this hyporesponsiveness: in particular it is not known whether the mechanism that leads to the decay of the IFN mRNA is the same as that which prevents interferon being formed when the cells are reinduced.

Thus in summary, the rate of IFN-α and -β production depends on the amount of intracellular IFN mRNA, but this depends on both the rate of transcription and the life-time of the IFN mRNA, and both these can be changed by suitable treatment of the cells.

The differential control of transcription of interferon genes

It is clear that the induction of the IFN-α and -β genes, which are both on chromosome 9, is linked in some way, for both human and mouse tissue culture cells produce mixtures of IFN-α and IFN-β on induction (Hayes, Yip & Vilcek, 1979; Yamamoto, 1981; Vonk, Hekman & Trapman, 1981). Virus-induced lymphoblastoid cells contain both IFN-α and IFN-β mRNA (Cavalieri *et al.*, 1977b; Sagar *et al.*, 1981; Gross *et al.*, 1981), and produce both IFN-α and -β. However polynucleotide-induced FS4 cells contain only IFN-β mRNA (Cavalieri *et al.*, 1977b; Raj & Pitha, 1981; Sehgal & Sagar, 1980) and produce only IFN-β.

Yamamoto (1981) reported that mouse spleen cells from a mouse which had been sensitized with mouse L cell produced both IFN-β and IFN-γ after coculture with L cells, although of course it is not known whether both interferons are produced by the same cell. The relative amounts of the different interferons produced can vary widely. Matsuyama, Hinuma, Watanabe & Kawade (1982) found that at least some non-T–non-B lymphoblastoid cells produced IFN-β as the major or sole component in response to virus infection, whereas cloned T lymphocytes growing in TCGF produced IFN-α when induced by Sendai virus, but IFN-γ when treated with mitogens (Marcucci, Waller, Kirchner & Krammer, 1981) or the appropriate antigen (Morris, Lin & Askonas, 1982). Using a variety of cell strains, Hayes *et al.* (1979) showed that the relative proportions of IFN-α and IFN-β produced by the cells depended on both the cell type and the inducer. For example, GM 258 human cells, which are trisomic for human chromosome 21, produced a substantial proportion of IFN-α when induced with Newcastle disease virus (NDV) or vesicular stomatitis virus (VSV), but none when induced with poly rI . poly rC. Using GM-258 cells induced with NDV, it was possible to change the relative proportions of IFN-α and IFN-β and the time at which they were made by altering the conditions of induction. The proportion of IFN-α was increased by decreasing the multiplicity of infection or by using virus irradiated with ultraviolet light. Lebon, Commoy-Chevalier, Robert-Galliot & Chany (1982) have shown a difference in the mechanism of production of IFN-α and IFN-β in lymphocytes. When treated with Sendai virus or glutaraldehyde-fixed cells, infected by herpes virus, they released IFN-α. Its synthesis was inhibited by oubain (5×10^{-7} M) without any effect on RNA and protein synthesis. In contrast, IFN-β induction in human fibroblasts is not affected by 100 times higher concentrations of oubain. The differential effect of oubain suggests a different mechanism for the production of IFN-α in lymphocytes, where a functional membrane Na^+ K^+ pump is essential, and where the inducer does not need to enter the cell (Ito *et al.*, 1978), from the production of IFN-β in fibroblasts where the Na^+ K^+ pump is not necessary and the inducer is probably double-stranded RNA in the cytoplasm. The functional membrane Na^+ K^+ pump is also needed for some cell-mediated human immune functions such as T-cell- or antibody-dependent cellular cytotoxicity.

The human interferon produced by a number of rodent : human cell hybrids has in every case been found to be IFN-β, whether the rodent parent was mouse or Chinese hamster (Meager *et al.*, 1979; Slate & Ruddle, 1979; Chany *et al.*, 1981). This was true whether virus or double-stranded RNA was used to induce the cells, and despite the fact that the parental mouse cell produced both IFN-α and -β (Yamamoto, 1981). Working on the hypothesis that IFN-β production was a characteristic of fibroblastic cells, Meager *et al.* (1982a) characterized the interferon made by a number of non-adherent mouse–human hybrids derived from human lymphoblastoid cells. The interferon produced was only IFN-β, but traces of IFN-α were detectable in the interferon produced by some adherent cell hybrids by using a sensitive assay for IFM-α. Thus the product is largely but not completely IFN-β.

How specific is the induction of the IFN-genes?

Are just the IFN genes induced? Or are there a number of other genes induced, but because of its high biological activity, is IFN the only product that is detected? It is certain that when lymphocytes are treated with mitogens to produce IFN-γ that many other substances are also produced, but what about the more specific inducers like viruses and double-stranded RNA? There have been no analyses at the level of mRNA, but several reports of production of proteins additional to interferon upon induction. Raj & Pitha (1980) translated the mRNA extracted from human fibroblasts treated with poly rI . poly rC in cell-free systems before analysis of the labelled product by two-dimensional gel electrophoresis. They found that 23 new proteins were translatable from the mRNAs of the induced cells but not from the mRNAs of the controls. These polypeptides had molecular weights from 15 000 to 70 000, and 13 of them were detected in induced cells labelled with ^{35}S-methionine. Content *et al.* (1982) characterized one of such proteins in detail. The initial product of cell-free protein synthesis had a molecular weight of 26 000. It could be processed in a reticulocyte system in the presence of dog pancreas microsomes to give a protein of molecular weight of 19 000 that had presumably lost its signal sequence, and a protein of 24 000 molecular weight which was probably a glycosylated product, while after secretion the proteins had molecular weights of 27 000 and 22 000. None of these proteins had any of the biological activities of interferon, and their function is unknown. Content *et al.* (1982) have also suggested that the human IFN-β_2 gene described by Weissenbach *et al.* (1980) was not the gene for an IFN but for the 26 000 molecular weight protein described above. Recently Gross *et al.* (1981) have cloned a 36-kilobase region of the human genome, including and surrounding an intact IFN-β gene. Transfer of this human DNA into mouse Ltk$^-$ cells led to stable transformants which could be induced to make human IFN-β by treatment with either double-stranded RNA or NDV (Hauser *et al.*, 1982). Neighbouring genes on the transferred DNA were co-induced showing that IFN induction does lead to the expression of other human genes, whose role is at present unknown.

Whatever is the mechanism that controls the transcription of the interferon and related genes, it is clear that IFN-α, -β or -γ can be produced as a single product or as a mixture of varying proportions, depending upon the cell and the inducer. We do not know how this loosely coordinated system is controlled. Nor do we know whether the different IFN-α preparations contain the same proportions of the different members of the family, or whether the relative proportions of the components can be varied.

CONTROL AT THE TRANSLATIONAL AND POST-TRANSLATIONAL LEVEL

Human IFN-α, -β and -γ mRNAs can be translated in *Xenopus* oocytes but despite early reports to the contrary, neither human IFN-α nor -β mRNAs yield biologically active interferon in cell-free systems (Sehgal *et al.*, 1977; J. Morser, personal communication).

However mouse IFN mRNA can be translated in both the oocyte and cell-free systems (Le Bleu *et al.*, 1978). The reason for this difference is not obvious. It does not appear to be a difference in post-translational processing, for both types of mRNA contain signal sequences which would not be removed in the cell-free systems, and there is no reason why the mouse mRNA product should be processed and the human mRNA product not. In any case, human interferons with signal or other sequences at the N-terminal end are biologically active (Slocombe, Easton, Boseley & Burke, 1982; Tuite *et al.*, 1982). There is evidence of an intracellular processing step which has not been identified (Morser & Colman, 1980). Human IFN-β and -γ are also glycosylated during the process of secretion. Active secretion of IFN also occurs in *Xenopus* oocytes (Colman & Morser, 1979).

The effect of removal of the poly A tail and also of the 3′ non-coding region has been studied by Soreq, Sagar & Sehgal (1981) by digestion of the IFN mRNA with polynucleotide phosphorylase. Removal of the poly A tail (~100 residues) and ~100 more residues from human IFN-β mRNA, which has a 3′ non-coding region of 203 residues, had no affect on its translational activity or its functional stability in *Xenopus* oocytes, whereas removal of another 100 residues decreased its translational efficiency.

There is some evidence for control at the translational level. When Namalwa cells are treated with butyrate or 5-bromodeoxyuridine, the IFN mRNA levels rise less than the IFN yields, and it appears that this is due to increased efficiency of translation of IFN mRNA in Namalwa cells (Shuttleworth *et al.*, 1982a). It was suggested that the effect was due to an increased rate of elongation and/or termination or to a larger number of ribosomes on each IFN mRNA molecule. This is some evidence for a similar pleiotrophic effect of added reagents on the production of tyrosine aminotransferase in hepatoma cells (Snoek, Voorma & Van Vijk, 1981). It has also been suggested that IFN mRNA behaves like some virus RNAs in that it can be translated more efficiently than other cellular mRNAs under suboptimal conditions such as hypertonic salt (Garry & Waite, 1979). However, Meager *et al.* (1982b) found this not to be so for production of IFN-β in human MG63 cells.

SUMMARY

The formation of interferon is controlled at at least three levels:

(1) At the level of accessibility of the interferon genes for transcription.
(2) At the transcriptional and post-transcriptional level.
(3) At the translational level.

The system shows much of the subtlety of other eukaryotic systems. It is thus a worthwhile system with which to study the general problem control of gene expression as well as being a system whose product we hope will be of some practical use.

The author is much indebted to his colleagues John Morser and Alan Morris and to his students Denise Barlow and John Shuttleworth for many hours of stimulating discussion, and to the Medical Research Council for financial support. This review was written while enjoying the kind hospitality of Professor M. Azuma, in the Department of Microbiology, Asahikawa Medical College, Asahikawa, Hokkaido, Japan.

REFERENCES

Abreu, S. L., Bancroft, F. C. & Stewart, W. E. II. (1979). Interferon priming: effects on interferon messenger RNA. *Journal of Biological Chemistry*, **254,** 4114–18.

Adolf, G. R. & Swetly, P. (1979). Interferon production by human lymphoblastoid cells is stimulated by inducers of Friend cell differentiation. *Virology*, **99,** 158–66.

Atherton, K. T. & Burke, D. C. (1978). The effects of some different metabolic inhibitors on interferon superinduction. *Journal of General Virology*, **41,** 229–37.

Baker, P. N., Bradshaw, T. K., Morser, J. & Burke, D. C. (1979). The effect of 5-bromodeoxyuridine on interferon production in human cells. *Journal of General Virology*, **45,** 177–84.

Baker, P. N., Morser, J. & Burke, D. C. (1980). Effects of sodium butyrate on a human lymphoblastoid cell line (Namalwa) and its interferon production. *Journal of Interferon Research*, **1,** 71–7.

Barlow, D. P., Randle, B., Burke, D. C. & Graham, C. F. (1982). The appearance of virus-induced interferon production in the tissue of the early post-implantation mouse embryo. (Submitted for publication.)

Bode, V. C. & Dziadek, M. (1979). Plasminogen activator secretion during mouse embryogenesis. *Developmental Biology*, **73,** 272–89.

Burke D. C. (1982). The mechanism of interferon production. *Philosophical Transactions of the Royal Society of London Series B*, **299,** 51–7.

Burke, D. C., Graham, C. F. & Lehman, J. M. (1978). Appearance of interferon inducibility and sensitivity during differentiation of murine teratorcarcinoma cells *in vitro*. *Cell*, **13,** 243–8.

BURKE, D. C. & VEOMETT, G. (1977). Enucleation and reconstruction of interferon-producing cells. *Proceedings of the National Academy of Sciences of the USA*, **74,** 3391–5.

CAVALIERI, R. L., HAVELL, E. A., VILCEK, J. & PESTKA, S. (1977a). Induction and decay of human fibroblast interferon messenger RNA. *Proceedings of the National Academy of Sciences of the USA*, **74,** 4415–19.

CAVALIERI, R. L., HAVELL, E. A., VILCEK, J. & PESTKA, S. (1977b). Induction and decay of human fibroblast interferon messenger RNA. *Proceedings of the National Academy of Sciences of the USA*, **74,** 4415–19.

CHANY, C., FINAZ, C., WEIL, D., VIGNAL, M., VAN CONG, N. & GROUCHY, J. (1981). Investigations on the chromosomal localizations of human and chimpanzee interferon genes: possible role of chromosome 9. *Annales de Genétique*, **23,** 201–7.

COLMAN, A. & MORSER, J. (1979). Export of proteins from oocytes of *Xenopus laevis*. *Cell*, **17,** 517–26.

CONTENT, J. (1982). The human fibroblast and human immune interferon genes and their expression in homologous and heterologous cells. *Philosophical Transactions of the Royal Society of London, Series B*, **299,** 29–38.

CONTENT, J., DE WIT, L., PIERARD, D., DERYNCK, R., DE CLERQ, E. & FIERS, W. (1982). Secretory proteins induced in human fibroblasts under conditions used for the production of interferon β. *Proceedings of the National Academy of Sciences of the USA*, **79,** 2768–72.

DARNELL, J. E. JR. (1982). Variety in the level of gene control in eukaryotic cells. *Nature*, **297,** 365–71.

DE MAEYER-GUIGNARD, J., THANG, M. N. & DE MAEYER, E. (1977). Binding of mouse interferon to polynucleotides. *Proceedings of the National Academy of Sciences of the USA*, **74,** 3787–90.

DZIADEK, M. & ADAMSON, E. D. (1978). Localization and synthesis of alphafetoprotein in postimplantation mouse embryos. *Journal of Experimental Embryology and Morphology*, **43,** 289–313.

EMENY, J. M. & MORGAN, M. J. (1979). Regulation of the interferon system: evidence that VERO cells have a genetic defect in interferon production. *Journal of General Virology*, **43,** 247–52.

FANTES, K. H. & ALLEN, G. (1981). Specific activity of pure human interferons and a non-biological method for estimating the purity of highly purified interferon preparations. *Journal of Interferon Research*, **1,** 465–72.

FIERS, W., REMANT, E., DEVOS, R., CHEROUTRE, H., CONTRERAS, R., DEGRAVE, W., GHEYSEN, D., STANSSENS, P., TOVEMIER, J., TOYA, Y. & CONTENT, J. (1982). The human fibroblast and human immune interferon genes and their expression in homologous and heterologous cells. *Philosophical Transactions of the Royal Society of London, Series B*, **299,** 29–38.

FRANKE, A. E., SHEPARD, H. M., HOUCK, C. M., LEUNG, D. W., GOEDDEL, D. V. & LAWN, R. M. (1982). Carboxy terminal region of hybrid leukocyte interferons affects antiviral specificity. (In press.)

FUJITA, T., SAITO, S. & KOHNO, S. (1979). Priming increases the amount of interferon mRNA in poly (rI) . poly (rC)-treated L cells. *Journal of General Virology*, **45,** 301–8.

GARRY, R. F. & WAITE, M. R. F. (1979). Na^+ and K^+ concentrations and the regulation of interferon system in chick cells. *Virology*, **96,** 121–8.

GROSS, G., MAYR, U., BRUNS, W., GROSVELD, F., DAHL, H. M. & COLLINS, J. (1981). The structure of a thirty-six kilobase region of the human chromosome including the fibroblast interferon gene IFN-β. *Nucleic Acids Research*, **9,** 2495–507.

GUYETTE, W. A., MATUSIK, R. J. & ROSEN, J. M. (1979). Prolactin-mediated transcriptional and post-transcriptional control of casein gene expression. *Cell*, **17,** 1013–23.
HAUSER, H., GROSS, G., BRUNS, W., HOCHKEPPEL, H.-K., MAYR, U. & COLLINS, J. (1982). Inducibility of human β-interferon gene in mouse L-cell clones. *Nature*, **297,** 650–4.
HAYES, T. G., YIP, Y. K. & VILCEK, J. (1979). Interferon production by human fibroblasts. *Virology*, **98,** 351–63.
ILLMENSEE, K. & STEVENS, L. C. (1979). Teratomas and chimeras. *Scientific American*, **240,** April, 86–98.
ITO, Y., NISHIYAMA, Y., SHIMAKATA, K., NOGATA, I., TAKEYAMA, H. & KUNII, A. (1978). The mechanism of interferon induction in mouse spleen cells stimulated with HVJ. *Virology*, **88,** 128–32.
JAHNER, D., STUHLMAN, H., STEWART, C. L., HORBERS, K., LOHLER, J., SIMON, I. & JAENISH, R. (1982). *De novo* methylation and expression of retroviral genomes during mouse embryogenesis. *Nature*, **298,** 623–8.
JOHNSTON, M. D. (1980). Enhanced production of interferon from human lymphoblastoid (Namalwa) cells pre-treated with sodium butyrate. *Journal of General Virology*, **50,** 191–4.
LE BLEU, B., HUBERT, E., CONTENT, J., DE WIT, L., BRAIDE, I. A. & DE CLERCQ, E. (1978). Translation of mouse interferon in *Xenopus laevis* oocytes and in rabbit reticulocyte lysates. *Biochemical and Biophysical Research Communications*, **82,** 665–73.
LEBON, P., COMMOY-CHEVALIER, M. J., ROBERT-GALLIOT, B. & CHANY, C. (1982). Different mechanisms for α and β interferon induction. *Virology*, **119,** 504–7.
MARCUCCI, F., WALLER, M., KIRCHNER, H. & KRAMMER, P. (1981). Production of immune interferon by murine-T-cell clones from long-term cultures. *Nature*, **291,** 79–81.
MARTIN, G. R. (1980). Teratocarcinomas and mammalian embryogenesis. *Science*, **209,** 768–76.
MATSUYAMA, M., HINUMA, Y., WATANABE, Y. & KAWADE, Y. (1982). Production of interferon-β by human lymphoblastoid cells of T and non-T-non-B lineages. *Journal of General Virology*, **60,** 191–4.
MEAGER, A., BUCHANAN, P., SIMMONS, J. G., HAYES, T. G. & VILCEK, J. (1982a). Production of human alpha- and beta-interferons by human–rodent hybrids. *Journal of Interferon Research*, **2,** 167–76.
MEAGER, A., GRAVES, H. E., WALKER, J. R., BURKE, D. C., SWALLOW, D. M. & WESTERVELD, A. (1979). Somatic cell genetics of human interferon production in human-rodent cell hybrids. *Journal of General Virology*, **45,** 309–21.
MEAGER, A., SHUTTLEWORTH, J., JUST, M. D., BOSELEY, P & MORSER, J. (1982b). The effect of hypertonic salt on interferon and interferon mRNA synthesis in human MG63 cells. *Journal of General Virology,* **59,** 177–81.
MORRIS, A. G., LIN, Y.-L. & ASKONAS, B. A. (1982). Immune interferon release when a cloned cytotoxic T-cell line meets its correct influenza-infected target cell. *Nature*, **295,** 150–2.
MORSER, J. & COLMAN, A. (1980). Post-translational events in the production of human lymphoblastoid interferon. *Journal of General Virology*, **51,** 117–24.
MORSER, J., FLINT, J., MEAGER, A., GRAVES, H., BAKER, P. N., COLMAN, A. & BURKE, D. C. (1979). Characterization of interferon messenger RNA from human lymphoblastoid cells. *Journal of General Virology*, **44,** 231–4.
MORSER, J., MEAGER, A. & COLMAN, A. (1980). Enhancement of interferon mRNA levels in butyric acid treated Namalwa cells. *FEBS Letters*, **112,** 203–6.
MORSER, J. & SHUTTLEWORTH, J. (1981). Low temperature treatment of Namalwa

cells causes superproduction of interferon. *Journal of General Virology*, **56,** 163–74.

PANG, R. H. L., HAYES, T. G. & VILCEK, J. (1980). Le interféron mRNA from human fibroblasts. *Proceedings of the National Academy of Sciences of the USA*, **77,** 5341–5.

RAJ, N. B. K., FERNIE, B. F. & PITHA, P. M. (1979). Correlation between the induction of mouse interferon and the amount of its mRNA. *European Journal of Biochemistry*, **98,** 215–21.

RAJ, N. B. K. & PITHA, P. M. (1977). Relationship between interferon production and interferon messenger RNA synthesis in human fibroblasts. *Proceedings of the National Academy of Sciences of the USA*, **74,** 1483–7.

RAJ, N. B. K. & PITHA, P. M. (1980). Synthesis of new proteins associated with the induction of interferon in human fibroblast cells. *Proceedings of the National Academy of Sciences of the USA*, **77,** 4918–22.

RAJ, N. B. K. & PITHA, P. M. (1981). Analysis of interferon mRNA in human fibroblast cells induced to produce interferon. *Proceedings of the National Academy of Sciences of the USA*, **78,** 7426–30.

REYNOLDS, F. H. JR., PREMKUMAR, E. & PITHA, P. M. (1975). Interferon activity produced by translation of human interferon messenger RNA in cell-free ribosomal systems and in *Xenopus* oocytes. *Proceedings of the National Academy of Sciences of the USA*, **72,** 4881–5.

RUBINSTEIN, M. (1982). Purification and structural analysis of interferon. *Philosophical Transactions of the Royal Society of London, Series B*, **299,** 39–50.

SAGAR, A. D., PICKERING, L. A., SUSSMAN-BERGER, P., STEWART, W. E. II & SEHGAL, P. B. (1981). Heterogeneity of interferon mRNA species from Sendai virus-induced human lymphoblastoid (Namalwa) cells and Newcastle diseases virus-induced murine fibroblastoid (L) cells. *Nucleic Acids Research*, **9,** 149–60.

SAITO, S., MATSUNO, T., SUDO, T., FURUYA, E. & KOHNO, S. (1976). Priming activity of mouse interferon: effect on messenger RNA synthesis. *Archives of Virology*, **52,** 159–63.

SECHER, D. S. (1981). An immunoradiometric assay for human leukocyte interferon using monoclonal antibody. *Nature*, **285,** 446–50.

SEHGAL, P. B. (1981). Regulation of stability of human β interferon mRNA in poly (I).poly (C) induced diploid fibroblasts: anchorage independence of the shut off mechanisms. *Virology*, **112,** 738–45.

SEHGAL, P. B., DOBBERSTEIN, B. & TAMM, I. (1977). Interferon messenger RNA content of human fibroblasts during induction, shut-off, and superinduction of interferon production. *Proceedings of the National Academy of Sciences of the USA*, **74,** 3409–12.

SEHGAL, P. B. & GUPTA, S. L. (1980). Regulation of the stability of poly(I).poly(C)-induced human fibroblast interferon mRNA: selective inactivation of interferon mRNA and lack of involvement of 2′5′-oligo (A) synthetase activation during the shut-off of interferon production. *Proceedings of the National Academy of Sciences of the USA*, **77,** 3849–93.

SEHGAL, P. B., LYLES, D. S. & TAMM, I. (1978). Superinduction of human fibroblast interferon: further evidence for the increased stability of interferon mRNA. *Virology*, **89,** 186–90.

SEHGAL, P. B. & SAGAR, A. D. (1980). Heterogeneity of poly(I).poly(C) induced human fibroblast mRNA species. *Nature*, **288,** 95–7.

SEHGAL, P. B. & TAMM, I. (1979). Two mechanisms contribute to the superinduction of poly(I).poly(C)-induced human fibroblast interferon production. *Virology*, **92,** 240–4.

SEHGAL, S. & KHOURY, S. (1979). Differentiation as a requirement for simian virus

40 gene expression in F-9 embryonal carcinoma cells. *Proceedings of the National Academy of Sciences of the USA*, **76,** 5611–15.

SEHGAL, S., LEVINE, A. J. & KHOURY, S. (1979). Evidence for non-spliced SV40 RNA in undifferentiated murine teratocarcinoma stem cells. *Nature*, **280,** 335–8.

SHUTTLEWORTH, J., MORSER, J. & BURKE, D. C. (1982a). Control of interferon mRNA levels and interferon yields in butyrate and 5′ bromodeoxyuridine-treated Namalwa cells. *Journal of General Virology*, **58,** 25–35.

SHUTTLEWORTH, J., MORSER, J. & BURKE, D. C. (1982b). Protein synthesis in human lymphoblastoid cells (Namalwa) after treatment with butyrate and 5′-bromodeoxyuridine. *Biochimica et Biophysica Acta*, **698,** 1–10.

SHUTTLEWORTH, J., MORSER, J. & BURKE, D. C. (1983). Coordinate expression of interferon-α and interferon-β mRNA in human lymphoblastoid (Namalwa) cells *European Journal of Biochemistry* (in press).

SLATE, D. L. & RUDDLE, F. H. (1979). Fibroblast interferon is coded by two loci on separate chromosomes. *Cell*, **16,** 171–80.

SLOCOMBE, P., EASTON, A., BOSELEY, P & BURKE, D. C. (1982). High level expression of an interferon α 2 gene cloned in phage M13 mp7 and subsequent purification with a monoclonal antibody. *Proceedings of the National Academy of Sciences of the USA*, **79,** 5455–9.

SNOEK, G. T., VOORMA, H. O. & VAN VIJK, R. (1981). A post-transcriptional site of induction of tyrosine aminotransferase by dexamethasone in Reuber H35 hepatoma cells. *FEBS Letters*, **125,** 266–70.

SOREQ, H., SAGAR, A. D. & SEHGAL, P. B. (1981). Translational activity and functional stability of human fibroblast β_1 and β_2 interferon mRNAs lacking 3′-terminal RNA sequences. *Proceedings of the National Academy of Sciences of the USA*, **78,** 1741–5.

STEWART, W. E. II. (1979 or 1981). *The Interferon System*. Wien, New York: Springer-Verlag.

STEWART, C. L., STUHLMAN, H., JAHNER, D. & JAENISH, R. (1982). *De novo* methylation, expression, and infectivity of retroviral genomes introduced into embryonal carcinoma cells. *Proceedings of the National Academy of Sciences of the USA*, **79,** 4098–102.

TAN, Y. H. & BERTHOLD, W. (1977). A mechanism for the induction and regulation of human interferon genetic expression. *Journal of General Virology*, **34,** 401–12.

TOMKINS, G. H., THOMPSON, E. B., HAYASHI, S., GELEHRTER, T. & PETERKOFSKY, D. (1966). Tyrosine transaminase induction in mammalian cells in tissue culture. *Cold Spring Harbor Symposium on Quantitative Biology*, **31,** 349–60.

TOVEY, M. G., BEGON-LOURS, J., GRESSER, I. & MORRIS, A. G. (1977). Marked enhancement of interferon production in 5-bromodeoxy-uridine treated human lymphoblastoid cells. *Nature*, **267,** 455–7.

TUITE, M. F., DOUBSON, M. J., ROBERTS, N. A., KING, R. M., BURKE, D. C., KINGSMAN, S. M. & KINGSMAN, A. J. (1982). Regulated high efficiency expression of human interferon-alpha in *Saccharomyces cerevisiae. The EMBO Journal*, **1,** 603–8.

VEOMETT, W. P., HEKMAN, A. C. P. & VEOMETT, G. E. (1980). Characteristics of the interferon system of an embryonal carcinoma cell are recessive in intraspecific seromatic cell hybrids. *Somatic Cell Genetics*, **6,** 325–32.

VONK, W. P., HEKMAN, A. C. P. & TRAPMAN, J. (1981). Properties of poly(I). poly C-induced L cell interferon. *Virology,* **113,** 388–91.

WEISSENBACH, J., CHERNAJOUSKY, Y., ZEEVI, M., SHULMAN, L., SOREQ, H., NIR, U., WALLACH, D., PERRICAUDET, M., TIOLLAIS, P. & REVEL, M. (1980). Two interferon mRNAs in human fibroblasts: *in vitro* translation and *Escherichia coli* cloning studies. *Proceedings of the National Academy of Sciences of the USA*, **77,** 7152–6.

WEISSMANN, C., NAGATA, S., BOLL, W., FOUNTOULAKIS, M., FUJISAWA, A., FUJISAWA, J.-I., HAYNES, J., HENCO, K., MANTEI, N., RAGG, H., SCHEIN, C., SCHMID, J., SHAW, G., STREULI, M., TAIRO, H., KODKORO, K. & WEIDLE, U. (1982). Structure and expression of human IFN-α genes. *Philosophical Transactions of the Royal Society of London, Series B*, **299,** 7–28.

WILSON, V., JEFFREYS, A. J., BARRIE, P. A., BOSELEY, P. G., SLOCOMBE, P. M., EASTON, A. & BURKE, D. C. (1983). A comparison of vertebrate interferon gene families detected by hybridization with human interferon DNA. *Journal of Molecular Biology* (in press).

YAMAMOTO, Y. (1981). Antigenicity of mouse interferons: two distinct molecular species common to interferons of various sources. *Virology*, **111,** 312–19.

THE BIOCHEMISTRY OF THE ANTIVIRAL STATE

MARTIN McMAHON AND IAN M. KERR

Imperial Cancer Research Fund Laboratories, P.O. Box 123, Lincoln's Inn Fields, London WC2A 3PX, UK

The interferons are a family of proteins, some of which are glycosylated. All of the interferons which have been tested individually possess both antiviral and cell growth regulatory activities. Moreover, although it is now clear that there is no single antiviral mechanism of action, this diversity cannot simply be attributed to the diversity of interferon molecules: each appears capable of inducing a similarly complex response in appropriate cells. On the other hand the individual interferons are differentially active on different cell types. It may be that the major importance of the molecular diversity of the interferons will prove to be in conferring (or avoiding) tissue specificity in the response. The effects of interferons on cell growth are described in the accompanying chapter by Joyce Taylor-Papadimitriou: here we will be concerned only with their antiviral activity. We will first discuss briefly the different interferons and what is known concerning their cell surface receptors. The sections that follow will be concerned with the changes induced in cells on treatment with interferon, the development of the antiviral state and with the various levels at which the interferons are thought to affect virus replication. In relation to the last of these, emphasis will be placed on 2–5A (ppp(A2′p)$_n$A; $n = 2$ to $\geqslant 4$) system as this is the best understood of the mechanisms thought to be involved.

The human interferons are classified according to their origin and antigenic specificity: interferon-α (IFN-α; leucocyte) and interferon-β (IFN-β; fibroblast), sometimes referred to as Type I interferons, and interferon-γ (IFN-γ; immune) or Type II interferon. Recent studies using molecular cloning techniques have shown that there are at least 12 IFN-α genes located on chromosome 9 (Brack, Nagata, Mantei & Weissmann, 1981; of necessity only selected references are given throughout). The number of IFN-β genes remains unclear. There may be only one gene, again on chromosome 9 (Owerbach *et al.*, 1981), but multiple IFN-β messenger RNA (mRNA) species have been reported (Sehgal & Sagar, 1980). None of the identified IFN-α or IFN-β genes have introns. There appears to be a single IFN-γ gene containing three introns on chromosome 12 (Gray *et al.*, 1982). Early work suggested a common mode of action but more recently it has become apparent that IFN-γ is different from IFN-α and β, at least in the early stages of action (Dianzani, Zucca, Scupham & Georgiades, 1980). IFN-γ may also be a stronger antiproliferative than antiviral agent (Rubin & Gupta, 1980; Verhaegen-Lewalle *et al.*, 1982). On the other hand it may protect cells more efficiently against some viruses than IFNs-α or β and vice versa, depending on which virus and cells are involved (Rubin & Gupta, 1980). It is possible that these differences are important in the response to different stimuli especially if one considers that IFNs-α and β are induced by virus infection, whereas IFN-γ is induced in response to immune stimulation.

In addition to the major antigenic heterogeneity there is considerable heterogeneity in activity within the different subspecies of IFNs-α. Small differences in protein structure are important in determining cell or tissue specificity. Studies using cloned purified human IFN-α subspecies indicate that they are differentially active by a factor of up to 10^5 against encephalomyocarditis virus (EMC) and vesicular stomatitis virus (VSV) when tested on the same cells (Weck, Apperson, May & Stebbing, 1981). Results of this type cannot easily be reconciled with a single IFN-α receptor triggering a unique biochemical response. They suggest that, on some cells at least, different subspecies of IFN-α may be capable of inducing significantly different responses, either through differing effects on a common receptor or through subtly different receptors. The identification and characterization of the receptors for the interferons will be crucial to the understanding of the mechanisms involved. Meanwhile it is clear that the antiviral activity of interferon can be expressed in a number of ways. The particular mechanism(s) observed depends on the specific virus-cell system involved. It is possible, for example, for a cell to be in an antiviral state against one virus but not another (Nilsen, Wood & Baglioni, 1980).

RECEPTORS AND INTERFERON-INDUCED RNA AND PROTEIN SYNTHESIS

Interferon is active in extremely small amounts. Only one or a few molecules per cell are required to induce the many interferon-related effects. The interferon receptors on the cell surface have not yet been purified. It appears, however, that human IFNs-α and β share a common receptor(s) which is coded by chromosome 21 (Epstein, McManus & Epstein, 1982). IFN-γ has a separate receptor (Branca & Baglioni, 1981). Little is known of the events immediately following receptor binding but the antiviral state takes several hours to develop during which time both RNA and protein synthesis are required (Taylor, 1964; Lockhart, 1964). The simplest model is that events at the cell membrane trigger signals to the nucleus which lead to transcription of mRNA coding for the proteins responsible for mediating the effects of interferon. With IFN-γ however, the early stages of induction appear more complex, possibly involving the increased or *de novo* translation of pre-existing mRNA to yield a protein that is involved in further mediating the development of the antiviral state (Dianzani *et al.*, 1980).

Interferon-induced mRNA and protein synthesis

A number of changes in mRNA and protein composition and in enzyme levels have been reported in response to interferon. For example, using pure human IFN-α or β on human fibroblasts, proteins of 120 000, 88 000, 80 000, 67 000 and 56 000 daltons have been reported to be induced (Gupta, Rubin & Holmes, 1979). Both the proteins induced and the extent of their induction show some difference when the effects of IFN-γ are compared with those of IFNs-α and β. On the other hand, the two major enzymes known to be induced by interferon, the 2–5A synthetase and protein kinase, are both induced by all three types of interferon (Hovanessian *et al.*, 1980). The isolation and translation of mRNA from interferon-treated and control cells has shown an induction of mRNAs for proteins of 60 000 to 70 000 daltons (Colonno, 1981) and, in the case of the 2–5A synthetase, the increase in enzyme level has been shown to be preceded by an increase in its mRNA (Shulman & Revel, 1980). One of the clearest examples of interferon-induced mRNAs is seen in human lymphoblastoid cells. There is an increase in HLA-A, B and C mRNAs detectable within a few hours after interferon treatment followed by an increase in HLA-A, B, C antigens on the cell membrane (Fellous *et al.*, 1982; Burrone & Milstein, 1982). All of the above changes in mRNA and protein levels induced by interferon are blocked by actinomycin D, indicating that control is at least partly transcriptional.

THE EFFECTS OF INTERFERONS ON CELL MEMBRANES AND THE CYTOSKELETON

Interferon is not just an antiviral agent; it has potent antiproliferative activity against many cell lines (reviewed by Taylor-Papadimitriou, 1980). Part of both its antiviral and its antiproliferative activity may be expressed through its effects on the cell membrane and cytoskeleton. It has been shown that pure human IFN-β inhibits the growth of human diploid fibroblasts. It leads to an increase in actin fibres and the fibronectin network with a decrease in cell motility and ruffling (Pfeffer, Landsberger & Tamm, 1981). It also induces changes in the cell membrane including a decrease in membrane fluidity and in unsaturated fatty acids, an inhibition in cap formation and an increase in surface antigen expression, especially HLA-A, B and C and β2 microglobulin (Pfeffer, Wang & Tamm, 1980; Apostolov & Barker, 1981; and reviewed by Friedman, 1979, and Taylor-Papadimitriou, 1980). There is also a report of a 16 000 dalton cell-surface protein detected by iodination, induced in human lymphoblastoid cells treated with pure IFN-α (Burrone & Milstein, 1982). Changes in membrane structure and composition may be of critical importance in the development of the antiviral state. Inhibitors of two enzymes, fatty acid cyclooxygenase and superoxide dismutase which are involved in the synthesis of membrane components, can inhibit the development of the antiviral state (Pottathil, Chandrabose, Cuatrecasas & Lang, 1980, 1981).

INTERFERON AND CYCLIC NUCLEOTIDES

The initial intracellular events following the binding of interferon to its cell surface receptor(s) are not known. Considerable effort has gone into studying interferon-induced changes in the concentrations of cAMP and cGMP and the activities of the adenyl and guanyl cyclases. There appears to be a rapid calcium-dependent increase in the level of cGMP occurring within minutes, followed much later by an increase in cAMP (Tovey & Rochette-Egly, 1981; Meldolesi, Friedman & Kohn, 1977). A rapid increase in guanyl cyclase has also been observed (Rochette-Egly & Tovey, 1982). Changes in the activity of adenyl cyclase have not, however, been detected. The significance of the increase in cGMP remains uncertain as, although depletion of calcium ions and cyclooxygenase inhibitors both prevent the increase in cGMP, only the latter has any effect on the development of the antiviral state (Tovey & Rochette-Egly, 1981). Recent work has once again invoked a role for cAMP in the interferon response. Treatment of mouse macrophages with IFN-β leads to an increase in this nucleotide followed by an increase in phagocytosis, growth inhibition and the development of the antiviral state. Macrophage lines deficient in adenyl cyclase or cAMP-dependent protein kinase show no increase in phagocytosis in response to IFN-β and a reduced sensitivity to the antiproliferative activity of the interferon. They retain, however, the ability to develop an antiviral state (Schneck, Rager-Zisman, Rosen & Bloom, 1982). For macrophages at least, therefore, cAMP may play a role in modulating the antiproliferative and immunomodulatory but not the antiviral actions of interferon.

MULTIPLE SITES OF EXPRESSION OF THE ANTIVIRAL STATE

The interferons suppress the growth of a wide range of DNA and RNA viruses. They have been reported to affect virus adsorption, entry or uncoating, viral RNA synthesis and methylation, protein synthesis and post-translational modification and virus assembly and release. Some of these activities are more important against certain viruses than others. Interferon does not necessarily protect the cell. The interferon-treated vaccinia virus-infected L-cell dies but no virus is produced (Joklik & Merigan, 1966). On the other hand the interferon-treated SV40-infected cell survives and is cured of virus infection. Not only is the virus important in determining the outcome but under laboratory conditions at least, this can vary with the interferon concentration and multiplicity of infection. For example, with picornaviruses high interferon levels protect the cell against low multiplicities of virus, whereas at lower interferon concentrations and higher multiplicities cell death ensues (Munoz & Carrasco, 1981). Cells in monolayers appear to survive better than those in spinner culture. The effect of interferon is not always, therefore, exclusively upon viral functions: in many interferon-treated, virus-infected cells both host and viral protein synthesis are inhibited. Interferon induces a potential antiviral state. How this is expressed depends on the virus. The ultimate outcome can vary with the state of the cell and the exact experimental conditions employed.

Effects on virus adsorption, entry and uncoating

Early experiments with infectious RNA suggested that interferon exerts its effect at points after the entry and uncoating of the virus. Despite this, in view of the profound changes induced by interferon in the cell membrane and cytoskeleton, it would be almost surprising if there were no effects on adsorption and entry, especially with the enveloped viruses. In the case of the RNA tumour viruses it is generally agreed that interferon has an effect early in replication at some point prior to integration of the viral DNA, but the exact site is unknown. It remains possible that the effect is upon entry or uncoating in part at least. Infections carried out with SV40 DNA rather than with whole virus, escape the interferon-mediated apparent inhibition of early viral transcription (Yamamoto, Yamaguchi & Oda, 1975). This inhibition may, therefore, result from an incomplete or delayed uncoating of the virus DNA rather than be a true effect on transcription. An alternative explanation for these results, however, is that the relatively large numbers of DNA molecules involved in successful DNA infections may simply swamp out a more subtle effect of interferon.

Effects on RNA transcription and methylation

In interferon-treated SV40 or VSV-infected cells there is an apparent inhibition of early viral transcription (Oxman & Levin, 1971; Marcus, Engelhardt, Hung & Sekellick, 1971). There is a decrease in the incorporation of radioactive precursors into early viral mRNA in such cells. On the face of it this would involve the inhibition of a host DNA-dependent RNA polymerase for SV40 and a viral RNA-dependent RNA polymerase for VSV. In the case of SV40 the inhibition is seen with nuclear as well as mature cytoplasmic mRNAs (Metz, Levin & Oxman, 1976). This reduces the possibility that the decrease in labelling reflects an increased breakdown or an inhibition of processing rather than decreased synthesis. Alternatively, as indicated above, the apparent inhibition of transcription could be a consequence of delayed or defective uncoating rather than of a true inhibition of enzyme activity. The major difficulty in pursuing the basis for these apparent effects on transcription has been the failure to reproduce them in cell-free systems.

Interferon can affect the methylation patterns of both viral and host mRNAs. Inhibitions of cap methylation have been observed in reovirus and VSV-infected cells and there is an increase in the abundance of internal 6-methyladenosine residues in mRNA late in the infectious cycle of SV40 virus in interferon-treated cells (Sen *et al.*, 1977; Shaila *et al.*, 1977; de Ferra & Baglioni, 1981; Kahana, Yakobson, Revel & Groner, 1981). There is an unstable inhibitor of cap methylation in extracts of interferon-treated Ehrlich ascites tumour and HeLa cells (Sen *et al.*, 1977) while a decrease in the ratio of *S*-adenosylmethionine to *S*-adenosylhomocysteine has been observed in a variety of interferon-treated cells (F. de Ferra & C. Baglioni, personal communication). The problem here is that no consistent pattern emerges; even the inhibitions of cap methylation observed differ in detail in the different systems.

Effects on protein synthesis and post-translation modification

There is a dramatic inhibition of protein synthesis in some interferon-treated, virus-infected cells and cell-free systems isolated from them. In 1966 Joklik & Merigan showed that in the interferon-treated, vaccinia virus-infected L-cell, early vaccinia messenger RNA is made but not translated. There is a very rapid inhibition of both host and viral protein synthesis in such cells (Metz & Esteban, 1972). A similar inhibition of both host and viral protein synthesis is observed under appropriate conditions in interferon-treated picornavirus-infected cells. It was the investigation of these systems which led to the suggested involvement of viral double-stranded RNA (dsRNA) in the inhibitions observed (Kerr, Brown & Ball, 1974), and subsequently to the discovery of the interferon and dsRNA-mediated 2–5A and protein kinase systems discussed in detail below. A more subtle and apparently selective inhibition of virus protein synthesis occurs in SV40 virus-infected cells treated with interferon up to 24 h after infection (Yakobson *et al.*, 1977) and in picornavirus-infected cells subjected to high doses of interferon and low multiplicities of infection (Vaquero, Aujean-Rigaud, Sanceau & Falcoff, 1981; Munoz & Carrasco, 1981). To date such selectivity has not been reproduced in cell-free systems. The basis for it is not, therefore, known. A role for the 2–5A and kinase systems cannot, however, be excluded (see below).

Whether interferon affects the post-translational processing or modification of viral proteins is less clear. The evidence for this comes from the investigation of its effects on infection by the RNA tumour viruses and VSV where it has been suggested that the inhibition of the maturation and release of the virus may reflect either a defect in the processing or glycosylation of viral precursor polypeptides. Maheshwari *et al.* (1980) reported that extracts of interferon-treated cells show a reduced ability to catalyse the first step in protein glycosylation but stressed the uncertainty of the physiological significance of this observation. Further investigation of the effects on glycosylation observed in the intact cell have suggested that in some systems at least they may be more apparent than real (Olden *et al.*, 1982).

Effects on virus maturation and release

Interferon inhibits the maturation and release of RNA tumour viruses and VSV (reviewed by Friedman, 1979). Typically, virus particles accumulate at the cell membrane and those that are released show a greatly reduced infectivity. Which of these two effects predominates varies with the different virus-cell systems employed. Their molecular basis is not known. It may be no more than that the effects of interferon on the cell membrane alter the position and orientation of the viral precursor polypeptides in such a way as to affect adversely the interactions essential to particle assembly and release. Whatever the mechanism the inhibition of maturation and release is clearly a major effect in the inhibition of RNA tumour virus replication by interferon. Its effects on early events in the replication of these viruses is much less dramatic. The reverse, or something approaching it, may be true, however, in the case of other groups of enveloped viruses. With these, the major effect of interferon may be on earlier events in the growth cycle, these early inhibitions being augmented by a late inhibition of maturation and release.

THE 2–5A AND PROTEIN KINASE SYSTEMS

Protein synthesis in cell-free extracts from interferon-treated cells is extremely sensitive to inhibition by double-stranded RNA (dsRNA) (Kerr *et al.*, 1974). This appears to be due to the activity of two interferon-induced, dsRNA-activated enzymes; a cAMP-independent protein kinase and the 2–5A synthetase. The kinase and 2–5A systems have been extensively reviewed (Baglioni, 1979;

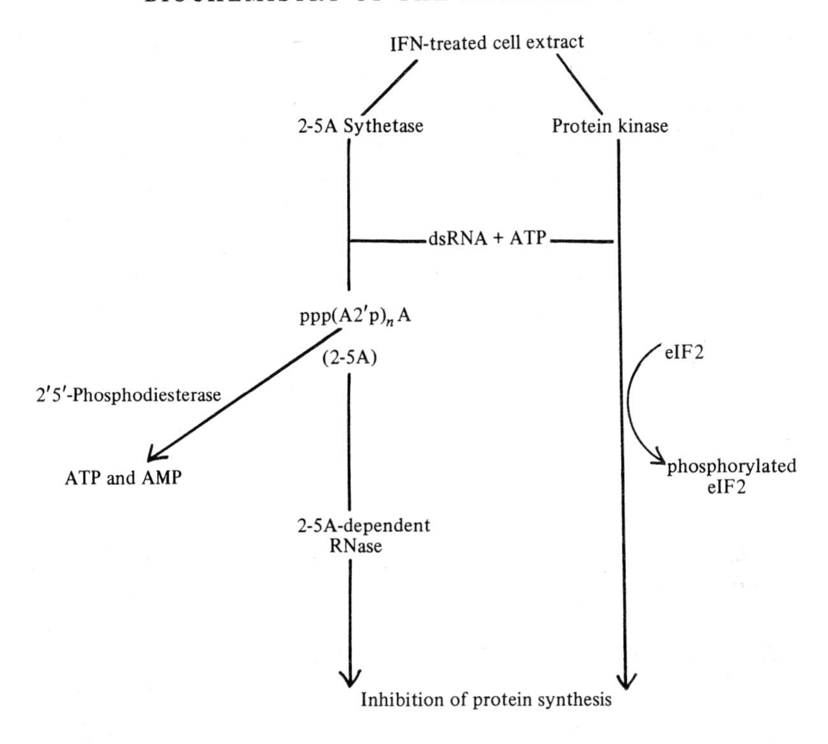

Revel, 1979; Lengyel, 1981, 1982) and are thought to operate as shown in Fig. 1 (above).

The protein kinase system

The dsRNA-activated protein kinase which appears analogous to that first described in rabbit reticulocytes (Farrell, Balkow, Hunt & Jackson, 1977), has been observed by a number of groups in a variety of different systems. It phosphorylates an endogenous 67 000 dalton protein. This may be an autophosphorylation serving to regulate its own activity. It has been suggested recently, however, that the kinase may also be activated by polyamines to phosphorylate ornithine decarboxylase (Sekar, Atmar, Krim & Keuhn, 1982). The dsRNA activated kinase also phosphorylates the 34 000 dalton α subunit of eukaryotic protein synthesis initiation factor 2 (eIF2α). This results indirectly in an inhibition of protein synthesis by preventing the repeated formation of methionyl tRNA-40S ribosome subunit initiation complexes (e.g. Farrell *et al.*, 1977; Siekierka, Mauser & Ochoa, 1982). In addition to activating the kinase the dsRNA may also augment its effect by inhibiting an eIF2α protein phosphatase (Epstein, Torrence & Friedman, 1980). All of these studies have been carried out with cell-free systems. The kinase has recently been shown to be active in interferon-treated, reovirus-infected cells (Gupta, Holmes & Mehra, 1982) but it has not been possible as yet to establish its general significance in the antiviral action of interferon (Discussion and Conclusions).

The 2–5A system

The system is composed of three enzymes (Fig. 1). (1) the 2–5A synthetase which synthesizes 2–5A; (2) a 2′,5′-phosphodiesterase which degrades 2–5A and (3) the 2–5A-dependent RNase (RNase L or RNase F). 2–5A is not a single molecule but an oligomeric series ppp(A2′p)$_n$*A*(n = 2 to $\geq$4) (Kerr & Brown, 1978). Oligomers up to the dodecamer at least, have been reported to be synthesized enzymatically in progressively decreasing amounts. The trimer, tetramer and pentamer are the predominant products. They are equally biologically active at nanomolar concentrations in cell-free systems. All three have been found to occur naturally in interferon-treated, EMC-infected cells (Williams *et al.*, 1979a; Knight *et al.*, 1980). When introduced into intact cells they inhibit protein and DNA synthesis and, hardly surprisingly therefore, virus growth (Williams, Golgher & Kerr, 1979b; Hovanessian & Wood, 1980). Although the non-phosphorylated 'core' (A2′p)$_n$A is inactive in cell extracts it has been found in intact interferon-treated cells (Knight *et al.*, 1980) and treatment of mitogen-stimulated lymphocytes or serum-stimulated 3T3 cells with 'core' can inhibit the mitogenic response (Kimchi, Shure & Revel, 1981b; Kimchi *et al.*, 1981a).

The 2–5A synthetase

The synthetase has been purified to homogeneity from both mouse and human cells where it is reported to have molecular weights of 105 000 and 100 000 respectively (Yang *et al.*, 1981). The basal levels and extent of induction of the synthetase by interferon vary considerably depending on the cell type and their growth and hormone status (Stark *et al.*, 1979; Krishnan & Baglioni, 1980). For example, HeLa cells have very high basal levels of the enzyme and show a 3- to 100-fold induction on interferon treatment (Silverman *et al.*, 1982a). Chick cells on the other hand have a considerably lower basal level of the enzyme but show a 10 000-fold induction in response to interferon (Ball & White, 1979).

The 2′-5′-phosphodiesterase

2–5A is unstable both in the intact cell and in the cell-free system (Williams *et al.*, 1978). There is a 2′,5′-phosphodiesterase activity which degrades it to AMP and ATP (Schmidt *et al.*, 1979, and Fig. 1). Both interferon treatment and mitogenic stimulation have been reported to increase the levels of this enzyme several-fold. It may, therefore, play a part in the regulation of 2–5A levels in the cell (Schmidt *et al.*, 1979). Although the partially purified enzyme shows a preference for 2′-5′ it will also cleave 3′-5′ linkages. In addition it has been reported to remove the CCA terminus from tRNA (Schmidt *et al.*, 1979). The significance of this is not known.

The 2–5A-dependent RNase and the inhibition of protein synthesis

The 2–5A-dependent nuclease is usually either not induced or induced only slightly in response to interferon (e.g. Silverman *et al.*, 1982b). When activated by 2–5A it cleaves both cellular and viral RNA on the 3′ side of UN doublets with a preference for UU or UA, to yield UpNp terminated products (Wreschner, McCauley, Skehel & Kerr, 1981; Floyd-Smith, Slattery & Lengyel, 1981). The significance of this specificity is not known. For activity the enzyme requires the continued presence of 2–5A. Since 2–5A is unstable in cells and cell-free systems, the activation of the RNase is transient in the absence of a 2–5A regenerating system (Williams *et al.*, 1978). The mechanism of activation is not known. Cell extracts contain a protein which binds 2–5A with high affinity and specificity. The ribonuclease and binding protein do not separate through several purification steps (Slattery, Ghosh, Samanta & Lengyel, 1979; Wreschner *et al.*, 1982) and it is generally assumed that the two activities are associated with the same molecule. Accordingly it can be affinity-labelled with a radioactive analogue of 2–5A. Labelled in this way the enzyme from rabbit, mouse, and human cell extracts has an apparent molecular weight of about 85 000 when analysed by electrophoresis on SDS polyacrylamide gels (Wreschner *et al.*, 1982).

The 2–5A system in interferon-treated, virus-infected cells

2–5A is present and the 2–5A-dependent ribonuclease is active in interferon-treated, EMC virus-infected mouse-L and human HeLa cells and in interferon-treated, reovirus-infected HeLa cells (Knight *et al.*, 1980; Wreschner, James, Silverman & Kerr, 1981a; Nilsen, Maroney & Baglioni, 1982). 2–5A *per se* has been detected in TCA extracts from such cells by a combination of HPLC analysis with both radiobinding and biological assays (Knight *et al.*, 1980). Evidence for the activation of the ribonuclease has been provided by the detection of a characteristic pattern of 2–5A-mediated rRNA cleavages in such cells (Wreschner, James, Silverman & Kerr, 1981a). Moreover, EMC virus growth is resistant to interferon in NIH-3T3 clone 1 cells which are thought to be deficient in the 2–5A-dependent ribonuclease (Epstein *et al.*, 1981). It must be remembered, however, that it is not known what other function(s) may also be deficient in these cells. Less-extensive studies have similarly established the activity of the 2–5A system in interferon-treated, Semliki Forest virus-infected cells (P. J. Cayley & I.M.K. unpublished). The observation of rRNA breakdown late in SV40 virus-infected cells treated with interferon 18 h post infection, also suggests the possible involvement of 2–5A (Revel *et al.*, 1979). In accord with this the natural occurrence of 2–5A has recently been demonstrated in this system (C. L. Hersh, G. R. Stark & I.M.K., unpublished).

Control of the 2–5A system at the level of the 2–5A-dependent RNase

HeLa cells have an unusually high basal level of synthetase and a functional 2–5A-dependent RNase. 2–5A-mediated rRNA cleavages are observed in response to EMC virus-infection in both interferon-treated and untreated cells. Despite this, EMC virus grows well in untreated but not in interferon-treated cells. A possible explanation for this apparent paradox is that in the absence of interferon treatment the 2–5A-dependent RNase is either lost or inactivated in response to virus infection. It is, therefore, only transiently activated, producing limited rRNA cleavage. After this the virus presumably takes over and grows normally. Interferon pretreatment prevents the inactivation of the 2–5A-dependent RNase. In this system, therefore, it seems that the crucial aspect of interferon action is the prevention of the virus-mediated inhibition of the RNase rather than the induction of the 2–5A synthetase (Silverman *et al.*, 1982a). The loss of RNase activity is not unique to the EMC virus HeLa cell system since it is also observed in the EMC virus-infected mouse-L or EAT cells (Cayley, Knight & Kerr, 1982). The mechanism of inactivation of the RNase is not known. It is not a direct consequence of the virus-mediated shut-off of host protein synthesis or of a generalized degradation or inactivation of host proteins; nor does it reflect the presence of a low molecular weight inhibitor or inhibitory metabolite of 2–5A *per se*.

The 2–5A system and the selective inhibition of virus protein synthesis

One model for the operation of the 2–5A and protein kinase systems in the interferon-treated, EMC virus-infected cell is that on virus infection the production of viral dsRNA activates both the 2–5A synthetase and the protein kinase. This leads to a shut down of both host and viral protein synthesis, followed by cell death which prevents the growth and release of progeny virus. Cell death, however, does not always occur. For example cells treated with low concentrations of interferon and high multiplicities of Mengo or EMC virus die, but cells treated with high concentrations of interferon and low multiplicities of virus survive (Munoz & Carrasco, 1981; Vaquero *et al.*, 1981; Cayley *et al.*, 1982). Under the latter set of conditions viral, but not host, protein synthesis is inhibited. By the criterion of 2–5A-mediated rRNA cleavages, however, the 2–5A system operates under both sets of conditions (Cayley *et al.*, 1982). It also remains possible that the 2–5A system may be active in the selective inhibition of late viral protein synthesis which occurs in SV40-infected cells treated with interferon after infection (Revel *et al.*, 1979). One possible model for how such a selective inhibition might operate involves a localized activation of the 2–5A system in the neighbourhood of partially double-stranded replicating viral RNA. Evidence suggesting that such localized activation can indeed occur in extracts from interferon-treated cells, has been provided by Nilsen & Baglioni (1979). Irrespective of the mechanism, the rRNA cleavage data show that a selective inhibition of protein synthesis can occur under conditions in which the 2–5A system is active. Whether or not 2–5A or rRNA cleavage is directly involved remains to be established.

Alternative products and alternative modes of action

The 2–5A synthetase will add AMP in 2′-5′ linkage to a number of interesting metabolites, including NAD, ADP, ribose, A5′$p_4$5′A and tRNA (Ball & White, 1979; Ferbus, Justesen, Besançon & Thang, 1981). To date none of these novel products has been detected in the intact cell (Cayley & Kerr, 1982). On the other hand, 2–5A has recently been shown to occur naturally in normal mouse tissues (L. Lawrence, J. Marti, H. Cailla and I.M.K., unpublished). This emphasizes that the 2–5A system may be more diverse than is currently recognized. Its wider possible significance in the control of cell growth, differentiation, or development, lies outside the scope of this particular review and has been discussed elsewhere (e.g. Stark *et al.*, 1979; Kimchi *et al.*, 1981b; Nilsen, Wood & Baglioni, 1982).

DISCUSSION AND CONCLUSIONS

With respect to interferon action in general, multisite models and analogies with hormone action are currently favoured. It is thought that the different interferons have fundamental mechanisms of action in common. It also appears, however, that against this common background subtle to very significant differences may be displayed. These are likely to be most dramatic between the Type I (α and β) and Type II (γ) interferons, but may also turn out to be important even amongst the different IFN-αs.

The 2–5A system operates in interferon-treated EMC-, Reo- and SFV-infected cells and the effect of interferon on virus maturation and release is clearly significant in the inhibition of RNA tumour virus growth. The general importance of these relative to the protein kinase and other possible mechanisms, however, remains to be firmly established. It remains possible that effects on uncoating, RNA transcription and methylation and as yet unrecognized alternative mechanisms, will prove to be of equal importance in other cell-virus systems. The emphasis laid on the 2–5A system here, therefore, is a reflection of the fact that it is by far the best understood of the systems involved. It should not be taken as an index of its relative importance. For example, from what we know of the requirements for the activation of the synthetase and kinase in cell extracts, conditions which activate the 2–5A system will also trigger the kinase. Accordingly, evidence for the operation of the 2–5A system suggests that the kinase is also likely to be active. In more general terms it may be quite normal for more than one of the possible antiviral mechanisms to operate in any given cell-virus situation. Certainly the establishment of a role for one does not necessarily exclude others.

The most important developments of the last few years have been the recognition of the multiplicity of the interferons and their increasing availability in purified form. This has permitted a start to be made on the identification and characterization of the interferon receptors. It should, therefore, be only a matter of time until we know whether the varied effects of the different human interferons reflect a spectrum of similar but subtly different receptors or the ability of different interferons to evoke different responses from only two major receptors (those for α and β and those for γ).

The area in which our ignorance remains most profound is that concerning the nature of the second message(s) responsible for the induction of the interferon-induced proteins and how that induction is controlled. Once again, however, it can be only a matter of time until libraries of cDNAs are available which will allow one to monitor accurately the synthesis and breakdown of the interferon-induced messages, and to probe the structure of the interferon-inducible genes. Indeed, for the interferon-inducible HLA-A, B and C antigens such cDNAs are available and the corresponding gene analysis is well under way. It remains a long step to the definition of second messages but equally we have come a long way and the 'black box' between the interaction of interferon with the cell membrane and the multiplicity of its effects on cell and virus growth does not look quite as daunting as it did only a few years ago.

REFERENCES

Apostolov, K. & Barker, W. (1981). The effects of interferon on the fatty acids in uninfected cells. *FEBS Letters*, **126,** 261–4.

Baglioni, C. (1979). Interferon induced enzyme activities and their role in the antiviral state. *Cell*, **17,** 255–64.

Ball, L. A. & White, C. N. (1979). Induction, purification and properties of 2′5′-oligoadenylate synthetase. In *Regulation of Macromolecular Synthesis*, ed. G. Koch & D. Richter, pp. 303–17. New York: Academic Press.

Brack, C., Nagata, S., Mantei, N. & Weissman, C. (1981). Molecular analysis of the human interferon-α family. *Gene*, **15,** 379–94.

Branca, A. A. & Baglioni, C. (1981). Evidence that Type I and Type II interferons have different receptors. *Nature*, **294,** 768–70.

BURRONE, O. R. & MILSTEIN, C. (1982). Control of HLA-A, B, C synthesis and expression in interferon treated cells. *EMBO Journal*, **1,** 345–9.

CAYLEY, P. J. & KERR, I. M. (1982). Synthesis, characteristion and biological significance of (2′-5′)oligoadenylate derivatives of NAD^+, ADP-ribose and adenosine (5′)tetraphospho(5′)adenosine. *European Journal of Biochemistry*, **122,** 601–8.

CAYLEY, P. J., KNIGHT, M. & KERR, I. M. (1982). Virus mediated inhibition of the $ppp(A2'p)_nA$ system and its prevention by interferon. *Biochemical and Biophysical Research Communications*, **104,** 376–82.

CAYLEY, P. J., SILVERMAN, R. H., BALKWILL, F. R., MCMAHON, M., KNIGHT, M. & KERR, I. M. (1982). The 2–5A system and interferon action. In *Vol. XXV UCLA Symposia of Molecular and Cellular Biology*, ed. R. M. Friedman & T. C. Merigan. New York: Academic Press (in press).

COLONNO, R. J. (1981). Accumulation of newly synthesized mRNAs in response to human fibroblast (β) interferon. *Proceedings of the National Academy of Sciences of the USA*, **78,** 4763–6.

DE FERRA, F. & BAGLIONI, C. (1981). Viral messenger RNA unmethylated in the 5′ terminal guanosine in interferon-treated HeLa cells infected with vesicular stomatitis virus. *Virology*, **112,** 426–5.

DIANZANI, F., ZUCCA, M., SCUPHAM, A. & GEORGIADES, J. A. (1980). Immune and virus induced interferons may activate cells by different de-repressional mechanisms. *Nature*, **283,** 400–2.

EPSTEIN, D. A., CZARNIECKI, C. W., JACOBSON, H., FRIEDMAN, R. M. & PANET, A. (1981). A mouse cell line which is unprotected by interferon against lytic virus infection lacks ribonuclease F activity. *European Journal of Biochemistry*, **118,** 9–15.

EPSTEIN, C. J., MCMANUS, N. H. & EPSTEIN, L. B. (1982). Direct evidence that the gene product of human chromosome 21 locus IFRC is the interferon α receptor. *Biochemical and Biophysical Research Communications*, **107,** 1060–6.

EPSTEIN, D. A., TORRENCE, P. F. & FRIEDMAN, R. M. (1980). Double stranded RNA inhibits a phosphoprotein phosphatase present in interferon-treated cells. *Proceedings of the National Academy of Sciences of the USA*, **77,** 107–11.

FARRELL, P. J., BALKOW, K., HUNT, T. & JACKSON, R. (1977). Phosphorylation of initiation factor eIF-2 and the control of reticulocyte protein synthesis. *Cell*, **11,** 187–200.

FELLOUS, M., NIR, U., WALLACH, D., MERLIN, G., RUBINSTEIN, M. & REVEL, M. (1982). Interferon dependent induction of mRNA for the major histocompatibility antigens in human fibroblasts and lymphoblastoid cells. *Proceedings of the National Academy of Sciences of the USA*, **79,** 3082–6.

FERBUS, D., JUSTESEN, J., BESANÇON, F. & THANG, M. N. (1981). The 2′5′ oligoadenylate synthetase has a multifunctional 2′5′ nucleotidyl-transferase activity. *Biochemical and Biophysical Research Communications*, **100,** 847–56.

FLOYD-SMITH, G., SLATTERY, E. & LENGYEL, P. (1981). Interferon action: RNA cleavage pattern of a (2′-5′)oligoadenylate-dependent endonuclease. *Science*, **212,** 1030–1.

FRIEDMAN, R. M. (1979). Interferons: interactions with cell surfaces. In *Interferon 1*, ed. I. Gresser, pp. 53–74. New York: Academic Press.

GRAY, P. W. & GOEDDEL, D. V. (1982). Structure of the human immune interferon gene. *Nature*, **298,** 859–63.

GUPTA, S. L., HOLMES, S. L. & MEHRA, L. L. (1982). Interferon action against reovirus: activation of interferon-induced protein kinase in mouse L 929 cells upon reovirus infection. *Virology*, **120,** 495–9.

GUPTA, S. L., RUBIN, B. Y. & HOLMES, S. L. (1979). Interferon action: induction of specific proteins in mouse and human cells by homologous interferons. *Proceedings of the National Academy of Sciences of the USA*, **76**, 4817–21.

HOVANESSIAN, A. G., MEURS, E., AUJEAN, O., VACQUERO, C., STEFANOS, S. & FALCOFF, E. (1980). Antiviral response and induction of specific proteins in cells treated with immune T (Type II) interferon analogous to that from viral interferon (Type I)-treated cells. *Virology*, **104**, 195–204.

HOVANESSIAN, A. G. & WOOD, J. N. (1980). Anticellular and antiviral effect of pppA(2′p5′A)$_n$. *Virology*, **101**, 81–90.

JOKLIK, W. K. & MERIGAN, T. C. (1966). Concerning the mechanism of action of interferon. *Proceedings of the National Academy of Sciences of the USA*, **56**, 558–65.

KAHANA, C., YAKOBSON, E., REVEL, M. & GRONER, Y. (1981). Increased methylation of RNA in SV40-infected interferon treated cells. *Virology*, **112**, 109–18.

KERR, I. M., BROWN, R. A. & BALL, L. A. (1974). Increased sensitivity of cell free protein synthesis to double stranded RNA after interferon treatment. *Nature*, **250**, 57–9.

KERR, I. M. & BROWN, R. E. (1978). pppA2′p5′A2′p5′A: An inhibitor of protein synthesis synthesized with an enzyme fraction from interferon-treated cells. *Proceedings of the National Academy of Sciences of the USA*, **75**, 256–60.

KIMCHI, A., SHURE, H., LAPIDOT, Y., RAPOPORT, S., PANET, A. & REVEL, M. (1981a). Antimitogenic effects of interferon and (2′-5′)-oligoadenylate in synchronized 3T3 fibroblasts. *FEBS Letters*, **134**, 212–16.

KIMCHI, A., SHURE, H. & REVEL, M. (1981b). Anti-mitogenic function of interferon-induced (2′-5′)oligo(adenylate) and growth-related variations in enzymes that synthesize and degrade this oligonucleotide. *European Journal of Biochemistry*, **114**, 5–10.

KNIGHT, M., CAYLEY, P. J., SILVERMAN, R. H., WRESCHNER, D. H., GILBERT, C. S., BROWN, R. E. & KERR, I. M. (1980). Radioimmune, radiobinding and HPLC analysis of 2–5A and related oligonucleotides from intact cells. *Nature*, **288**, 189–92.

KRISHNAN, I. & BAGLIONI, C. (1980). Increased levels of (2′-5′)oligo(A) polymerase activity in human lymphoblastoid cells treated with glucocorticoids. *Proceedings of the National Academy of Sciences of the USA*, **77**, 6506–10.

LENGYEL, P. (1981). Mechanism of interferon action: The (2′-5′)(A)$_n$ synthetase-RNase L pathway. In *Interferon 3* (ed. I. Gresser), pp. 78–99. New York: Academic Press.

LENGYEL, P. (1982). Biochemistry of interferons and their actions. *Annual Review of Biochemistry*, **51**, 251–82.

LOCKHART, R. Z., JR. (1964). The necessity for cellular RNA and protein synthesis for viral inhibition resulting from interferon. *Biochemical and Biophysical Research Communications*, **15**, 513–18.

MAHESHWARI, R. K., BANERJEE, D. K., WAECHTER, C. J., OLDEN, K. & FRIEDMAN, R. M. (1980). Interferon treatment inhibits glycosylation of a viral protein. *Nature*, **287**, 454–6.

MARCUS, P. I., ENGELHARDT, D. L., HUNG, J. M. & SEKELLICK, M. J. (1971). Interferon action: inhibition of VSV RNA synthesis induced by virion-bound polymerase. *Science*, **174**, 593–8.

MELDOLESI, M. F., FRIEDMAN, R. M. & KOHN, L. D. (1977). An interferon-induced increase in cyclic AMP levels precedes the establishment of the antiviral state. *Biochemical and Biophysical Research Communications*, **79**, 239–46.

METZ, D. H. & ESTEBAN, M. (1972). Interferon inhibits viral protein synthesis in L-cells infected with vaccinia virus. *Nature*, **238,** 385–8.

METZ, D. H., LEVIN, M. J. & OXMAN, M. N. (1976). Mechanism of interferon action: further evidence for transcription as the primary site of action of Simian virus 40 infection. *Journal of General Virology*, **32,** 227–40.

MUNOZ, A. & CARRASCO, L. (1981). Protein synthesis and membrane integrity in interferon treated HeLa cells infected with encephalomyocarditis virus. *Journal of General Virology*, **56,** 153–62.

NILSEN, T. W. & BAGLIONI, C. (1979). Mechanism for discrimination between viral and host mRNA in interferon-treated cells. *Proceedings of the National Academy of Sciences of the USA*, **76,** 2600–4.

NILSEN, T. W., MARONEY, P. A. & BAGLIONI, C. (1982). Synthesis of (2′-5′)-oligoadenylate in interferon treated HeLa cells infected with reovirus. *Journal of Virology*, **42,** 1039–45.

NILSEN, T. W., WOOD, D. L. & BAGLIONI, C. (1980). Virus specific effects of interferon in embryonal carcinoma cells. *Nature*, **286,** 178–80.

NILSEN, T. W., WOOD, D. L. & BAGLIONI, C. (1982). Presence of 2′5′-oligo(A) and of enzymes that synthesize, bind and degrade 2′5′-oligo(A) in HeLa cell nuclei. *Journal of Biological Chemistry*, **257,** 1602–5.

OLDEN, K., BERNARD, B. A. & WHITE, S. L. (1982). Interferon is not an inhibitor of protein glycosylation. *Journal of Cell Biology*, **91,** 397.

OWERBACH, D., RUTTER, W. J., SHOWS, T. B., GRAY, P., GOEDDEL, D. V. & LANVIN, R. M. (1981). Leukocyte and fibroblast interferon genes are located on human chromosome 9. *Proceedings of the National Academy of Sciences of the USA*, **78,** 3123–7.

OXMAN, M. N. & LEVIN, M. J. (1971). Interferon and transcription of early virus specific RNA in cells infected with Simian virus 40. *Proceedings of the National Academy of Sciences of the USA*, **68,** 299–302.

PFEFFER, L. M., LANDSBERGER, F. R. & TAMM, I. (1981). Interferon induced time dependent changes in the plasma membrane lipid bilayer of cultured cells. *Journal of Interferon Research*, **1,** 613–20.

PFEFFER, L. M., WANG, E. & TAMM, I. (1980). Interferon effects on microfilament organisation, cellular fibronectin distribution and cell motility in human fibroblasts. *Journal of Cell Biology*, **85,** 9–17.

POTTATHIL, R., CHANDRABOSE, K. A., CUATRECASAS, P. & LANG, D. J. (1980). Establishment of the interferon-mediated antiviral state: role of fatty acid cyclooxygenase. *Proceedings of the National Academy of Sciences of the USA*, **77,** 5437–40.

POTTATHIL, R., CHANDRABOSE, K. A., CUATRECASAS, P. & LANG, D. J. (1981). Establishment of the interferon-mediated antiviral state: possible role of superoxide dismutase. *Proceedings of the National Academy of Sciences of the USA*, **78,** 3343–7.

REVEL, M. (1979). Molecular mechanisms involved in the antiviral effect of interferon. In *Interferon 1*, ed. I. Gresser, pp. 101–63. New York: Academic Press.

REVEL, M., KIMCHI, A., SCHMIDT, A., SHULMAN, L., CHERNAJOVSKY, Y., RAPOPORT, S. & LAPIDOT, Y. (1979). Studies on interferon action: synthesis, degradation and biological activity of (2′-5′)isoadenylate. In *Regulation of Macromolecular Synthesis*, ed. G. Koch & D. Richter, pp. 341–59. New York: Academic Press.

ROCHETTE-EGLY, C. & TOVEY, M. G. (1982). Interferon enhances guanylate cyclase activity in human lymphoma cells. *Biochemical and Biophysical Research Communications*, **107,** 150–6.

RUBIN, B. Y. & GUPTA, S. L. (1980). Differential efficacies of human Type I and

Type II interferons as antiviral and antiproliferative agents. *Proceedings of the National Academy of Sciences of the USA*, **77,** 5928–32.
SCHMIDT, A., CHERNAJOVSKY, Y., SHULMAN, L., FEDERMAN, P., BERISSI, H. & REVEL, M. (1979). An interferon-induced phosphodiesterase degrading (2′-5′)oligoadenylate and the CCA terminus of tRNA. *Proceedings of the National Academy of Sciences of the USA*, **76,** 4788–92.
SCHNECK, J., RAGER-ZISMAN, B., ROSEN, O. M. & BLOOM, B. R. (1982). Genetic analysis of the role of cAMP in mediating effects of interferon. *Proceedings of the National Academy of Sciences of the USA*, **79,** 1879–83.
SEHGAL, P. B. & SAGAR, A. D. (1980). Heterogeneity of poly(I).poly(C) induced human fibroblast mRNA species. *Nature*, **288,** 95–7.
SEKAR, V., ATMAR, V. J., KRIM, M. & KUEHN, G. D. (1982). Interferon induction of a polyamine-dependent protein kinase activity in Ehrlich ascites tumor cells. *Biochemical and Biophysical Research Communications*, **106,** 305–11.
SEN, G., SHAILA, S., LEBLEU, B., BROWN, G. E., DESROSIERS, R. C. & LENGYEL, P. (1977). Impairment of reovirus mRNA methylation in extract of interferon-treated Ehrlich ascites tumour cells. *Journal of Virology*, **21,** 69–83.
SHAILA, S., LEBLEU, B., BROWN, G. E., SEN, G. C. & LENGYEL, P. (1977). Characteristic of extracts from interferon treated HeLa cells: presence of a protein kinase and endoribonuclease activated by double stranded RNA and of an inhibitor of mRNA methylation. *Journal of General Virology*, **37,** 535–46.
SHULMAN, L. S. & REVEL, M. (1980). Interferon-dependent induction of mRNA activity for (2′-5′)oligoisoadenylate synthetase. *Nature*, **288,** 98–100.
SIEKIERKA, J., MAUSER, L. & OCHOA, A. (1982). Mechanism of polypeptide chain initiation in eukaryotes and its control by phosphorylation of the α subunit of eIF2. *Proceedings of the National Academy of Sciences of the USA*, **79,** 2537–40.
SILVERMAN, R. H., CAYLEY, P. J., KNIGHT, M., GILBERT, C. S. & KERR, I. M. (1982a). Control of the ppp(A2′p)$_n$A system in HeLa cells. Effects of interferon and virus infection. *European Journal of Biochemistry*, **124,** 131–8.
SILVERMAN, R. H., WATLING, D., BALKWILL, F. R., TROWSDALE, J. & KERR, I. M. (1982b). The ppp(A2′p)$_n$A and protein kinase systems in wild type and interferon resistant Daudi cells. *European Journal of Biochemistry*, **126,** 333–41.
SLATTERY, E., GHOSH, N., SAMANTA, H. & LENGYEL, P. (1979). Interferon, double stranded RNA and RNA degradation: activation of an endonuclease by (2′-5′)A$_n$. *Proceedings of the National Academy of Sciences of the USA*, **76,** 4778–82.
STARK, G. R., DOWER, W. J., SHIMKE, R. T., BROWN, R. E. & KERR, I. M. (1979). 2–5A synthetase: assay, distribution and variation with growth or hormone status. *Nature*, **278,** 471–3.
TAYLOR, J. (1964). Inhibition of interferon action by actinomycin. *Biochemical and Biophysical Research Communications*, **14,** 447–51.
TAYLOR-PAPADIMITRIOU, J. (1980). Effects of interferon on cell growth and function. In *Interferon 2*, ed. I. Gresser, pp. 13–46. New York: Academic Press.
TOVEY, M. G. & ROCHETTE-EGLY, C. (1981). Rapid increase in guanosine 3′,5′ cyclic monophosphate in interferon treated mouse leukemia L1210 cells: relationship to the development of the antiviral state and inhibition of cell multiplication. *Virology*, **115,** 272–81.
VAQUERO, C., AUJEAN-RIGAUD, O., SANCEAU, J. & FALCOFF, R. (1981). Effect of interferon on transient shut-off of cellular RNA and protein synthesis induced by mengovirus infection. *Antiviral Research*, **1,** 123–34.
VERHAEGEN-LEWALLE, M., KUWATA, T., ZHANG, Z.-X., DE CLERQ, E., CANTELL, K. & CONTENT, J. (1982). 2–5A synthetase activity induced by interferon α, β and γ in human cell lines differing in their sensitivity to the anticellular and antiviral activities of these interferons. *Virology*, **117,** 425–34.

WECK, P. K., APPERSON, S., MAY, L. & STEBBING, N. (1981). Comparison of the antiviral activities of various cloned human interferons in mammalian cell culture. *Journal of General Virology*, **57,** 233–7.

WILLIAMS, B. R. G., GOLGHER, R. R., BROWN, R. E., GILBERT, C. S. & KERR, I. M. (1979a). Natural occurrence of 2–5A in interferon-treated EMC virus-infected L cells. *Nature*, **282,** 582–6.

WILLIAMS, B. R. G., GOLGHER, R. R. & KERR, I. M. (1979b). Activation of a nuclease by pppA2′p5′A2′p5′A in intact cells. *FEBS Letters*, **105,** 47–52.

WILLIAMS, B. R. G., KERR, I. M., GILBERT, C. S., WHITE, C. N. & BALL, L. A. (1978). Synthesis and breakdown of pppA2′p5′A2′p5′A and transient inhibition of protein synthesis in extracts from interferon-treated and control cells. *European Journal of Biochemistry*, **92,** 455–62.

WRESCHNER, D. H., JAMES, T. C., SILVERMAN, R. H. & KERR, I. M. (1981a). Ribosomal RNA cleavage, nuclease activation and 2–5A in interferon-treated cells. *Nucleic Acid Research*, **9,** 1571–81.

WRESCHNER, D. H., McCAULEY, J. W., SKEHEL, J. J. & KERR, I. M. (1981b). Interferon action: sequence specificity of the ppp$(A2'p)_nA$-dependent ribonuclease. *Nature*, **289,** 414–17.

WRESCHNER, D. H., SILVERMAN, R. H., JAMES, T. C., GILBERT, C. S. & KERR, I. M. (1982). Affinity labelling and characterisation of the ppp$(A2'p)_nA$-dependent endoribonuclease from different mammalian sources. *European Journal of Biochemistry*, **124,** 261–8.

YAKOBSON, E., PRIVES, C. L., HARTMAN, Y., WINOCOUR, E. & REVEL, M. (1977). Selective inhibition of viral protein synthesis in monkey cells treated with interferon late after SV40 infection. *Cell*, **12,** 73–81.

YAMAMOTO, K., YAMAGUCHI, N. & ODA, K. (1975). Mechanism of interferon induced inhibition of early Simian virus 40 (SV40) functions. *Virology*, **68,** 58–70.

YANG, K., SAMANTA, H., DOUGHERTY, J., JAYARAM, B., BROEZE, R. & LENGYEL, P. (1981). Interferons, double stranded RNA, and RNA degradation. *Journal of Biological Chemistry*, **256,** 9324–8.

THE EFFECTS OF INTERFERON ON THE GROWTH AND FUNCTION OF NORMAL AND MALIGNANT CELLS

JOYCE TAYLOR-PAPADIMITRIOU

Imperial Cancer Research Fund, P.O. Box 123, Lincoln's Inn Fields, London WC2A 3PX, UK

INTRODUCTION

As well as being antiviral agents, the interferons represent an important group of naturally occurring regulatory molecules, indeed the first to be well characterized, which can inhibit cell growth and affect differentiation. These effects of interferons are seen in a variety of cell types which display a wide range of differentiated functions, and whose growth-regulatory mechanisms may be diverse. Because of this, and because the phenomena of growth and differentiation are not well defined at the molecular level, the study of interferons as regulatory agents is complex. Moreover, in addition to the complexity of the biological phenomena under study, there is now the diversity of the interferon molecules themselves to consider. In the case of human interferons, more than eight different α interferon genes as well as a HuIFN-β, and HuIFN-γ gene have been cloned, and expressed in micro-organisms. We have yet to determine the profile of activity on cell growth and function for each interferon species, and how widely these profiles differ. What studies have been done suggest that differences do exist, and moreover, synergism or antagonism between the different species may be observed.

Because of the complex nature of the systems, the study of interferons as regulatory agents has been, until recently, mainly at the descriptive level. Effects on cell growth have been largely studied using tumour cell lines, although the proliferation of normal cells can also be inhibited. On the other hand, effects on differentiated function have been studied intensively using the various effector cells of the immune system. This to some extent reflects the availability of experimental systems for studying regulation of cell proliferation and differentiated function. Studies on molecular mechanisms underlying the regulatory effects of interferons are just beginning and are already giving results which not only relate to interferon action but also focus on cellular events important in the control of cell growth and function. Conversely it is to be expected that definition of mechanisms involved in regulation of growth and differentiation in different cell types, will stimulate progress in understanding the action of interferons on these processes.

This review will deal with the growth-inhibitory effects of interferons, their effects on differentiated functions, mainly in cells other than those of the haemopoietic system, and will discuss interferon-induced changes in cell membranes which may be related to these two general effects. Much of the work relates to experiments done in cell culture systems, mainly because the effects of interferons in multicellular organisms, where cell interactions obscure a primary response, are difficult to interpret. However, it should be borne in mind that effects on cell growth and function *in vivo* may contribute not only to interferons' action in eliminating virally infected or malignant cells, but may also lead to the toxic side effects which have already been seen, even in patients receiving purified interferons (Priestman, 1980; Tyrrell, Scott, Secher & Cantell, 1981) or a single molecular species produced in bacteria (Gutterman *et al.*, 1981).

EFFECTS OF INTERFERONS ON CELL GROWTH

(i) *General*

The first report describing an effect of interferons on cells distinct from its antiviral action appeared in 1962, when Kurt Paucker and his colleagues reported that the growth of mouse fibroblasts was inhibited by preparations of mouse interferon (Paucker, Cantell & Henle, 1962). These early observations of growth inhibition were treated with some scepticism, mainly because of the impurity of the interferon preparations which contained many other factors. However, now that purified preparations of human and mouse interferons have been shown to inhibit cell growth (Knight, 1976; Gresser, De Maeyer-Guignard, Tovey & De Maeyer, 1979; Evinger, Rubinstein & Pestka, 1981a; Evinger, Maeda & Pestka, 1981b) it can be stated unequivocally that the interferon molecules themselves inhibit cell proliferation in a wide range of cell types.

In an adult multicellular organism the balance of cell growth is very different for different cell types, some of which proliferate extensively and continuously to replenish shed cells (as in skin or the intestine), while others grow only periodically in response to a specific stimulus (for example, lymphocytes or the mammary gland at puberty and pregnancy). Thus while there are almost certainly features about growth regulation common to all cells, there are bound to be differences, and in considering interferons as inhibitors of cell growth, it should be kept in mind that we may be talking about a collection of phenomena.

Interferons have been found in many culture systems to be cytostatic rather than cytotoxic (Gresser, 1977). These studies usually relate to short-term experiments in fibroblast systems. However, if interferons are present for several weeks (Strander & Einhorn, 1977) or if the more slowly growing cell shows an enhanced tendency to differentiate terminally (see below), then cytotoxicity can be observed (Greenberg & Mosney, 1977; Van't Hull, Schellekens, Lowenberg & de Vries, 1978; Shibata & Taylor-Papadimitriou, 1981). This is an important question in relation to using interferons as anti-tumour agents, where direct cytotoxic effects on tumour cells are desirable.

(ii) *Sensitivity of different cell types*

Activities of interferons have been defined in terms of antiviral units, and the amount of a given interferon required to affect the growth of cultured cells is between 0.2 and 10 000 units/ml depending on the cell type. With interferons these numbers can be defined in terms of mg of protein, and with human interferon from Namalwa cells with a specific activity of 3×10^8/mg (Fantes & Allen, 1981) the effective dose range is roughly 2 pg – 50 ng/ml.

Normal cells

There are not a great deal of data available on growth inhibition of cells from normal tissues, much of the work having been done with cell lines. However, human diploid fibroblasts (Pfeffer, Murphy & Tamm, 1979), human mammary epithelium (Balkwill, Watling & Taylor-Papadimitriou, 1978) and cells from the haemopoietic system (Lindahl-Magnusson, Leary & Gresser, 1972; Greenberg & Mosney, 1977; van't Hull *et al.*, 1978) all show reduced proliferation after treatment with interferons. Comparisons have not been made of the relative sensitivity of normal cells from different tissues in the same individual or animal to the growth-inhibitory effect of the same interferon preparation. However, there are some indications that there may be differences in sensitivity of cells from different normal tissues. For example human mammary epithelial cells cultured from milk are more sensitive to HuIFN-α from Namalwa cells (HuIFN-αN) than are human embryo fibroblasts or fibroblasts from the human mammary gland (Taylor-Papadimitriou, Shearer, Balkwill & Fantes, 1982). Other investigators have reported that human fibroblast growth can be inhibited by only 40 units of HuIFN-β (Pfeffer *et al.*, 1979). Direct comparisons, however, are required to draw conclusions on the relative sensitivity of human fibroblasts to HuIFN-α and HuIFN-β.

Tumour cells

Inhibition of the growth of tumour cells by interferons has been studied widely using cell lines, and in some instances, primary cultures of human tumours (Balkwill & Oliver, 1977; Bradley & Ruscetti, 1981). While this method would seem to be a good way of testing the sensitivity of individual tumours to the growth-inhibitory effect of interferons, there are serious limitations to its application. Only some tumours yield proliferating cells, which may or may not be representative of the original *in vivo* stem cells. Moreover, a direct inhibition of cell growth may be only partially responsible for interferons' anti-tumour effect (see below).

If a range of cell lines deriving from tumours of the same type are considered, a wide range of sensitivity is noted (Strander & Einhorn, 1977; Adams, Strander & Cantell, 1975; Shibata & Taylor-Papadimitriou, 1981; Creasey, Bartholomew & Merigan, 1980). Thus while the lymphoblastoid cell line Daudi (derived from a Burkitt lymphoma) and the BT20 cell line (derived from a primary breast cancer – Lasfargues & Ozello, 1958), can be significantly inhibited by less than 1 unit or approximately 3 pg (per ml) of interferon, cell lines from ostensibly similar tumours may require 1000 times this amount to show an effect on growth. It has been suggested that tumour cells are more sensitive to interferons' growth-inhibitory effect than normal cells (Strander & Einhorn, 1977). While the most sensitive cells have so far been found to be tumour cell lines, there are also resistant lines from similar tumours which are less sensitive than the normal cell from which they derive. Thus, while *in vivo*, tumour cells may be more sensitive because of the indirect effects of the immune system, it is not possible to generalize with cultured cells.

The differences in sensitivity of the various cell lines and strains may be due to a variety of factors. Cells may lack the membrane receptors for interferon as for the resistant L1210 cell line (Aguet, 1980) or they may produce a growth factor (Bourne & Rozengurt, 1976; De Larco & Todaro, 1978; Todaro, Fryling & De Larco, 1980) antagonistic to the inhibitory effect of interferon (Taylor-Papadimitriou, Shearer & Rozengurt, 1981; Taylor-Papadimitriou & Rozengurt, 1982). The metabolic profile of the cell will certainly influence its reaction to interferon treatment, and if it is capable of differentiation, it may be more sensitive. It is perhaps relevant to point out that while a marked sensitivity to the anti-growth effect of interferons may correlate well with sensitivity to their antiviral effects in some series of cell lines, this is not always the case, and may depend on the virus used.

Heterologous cells

Although originally, the interferons were thought to be species specific, it is now clear that human interferons, particularly those of the α type can act very effectively as antiviral agents on heterologous cells, and that each subspecies has a different range of activity in cells from a variety of mammalian species (Stewart *et al.*, 1980; Goeddell *et al.*, 1980; Berg *et al.*, 1982). Bovine cells are particularly sensitive to the antiviral action of most of the HuIFN-α species; indeed they are protected more effectively against virus infection than the homologous cells by far example, HuIFN-α1 (or HuIFN-D). However, the effect on virus growth is not always paralleled by an effect on cell growth (Taylor-Papadimitriou *et al.*, 1982). Fig. 1 shows the effect of HuIFN-α from Namalwa cells (HuIFN-αN) on virus growth and on cell growth in four types of bovine cells; the same result is seen with HuIFN-α2 (Taylor-Papadimitriou *et al.*, 1982). The lack of effect on cell growth is most obvious with bovine embryo kidney cells, which can be protected by less than one unit of HuIFN-α2 or HuIFN-αN against EMC infection, but whose growth is not inhibited by 1000 units of the same interferons. The lack of sensitivity to the growth-inhibitory effect is not an intrinsic property of BEK cells, since inhibition of their growth can be effected by either bovine interferon or HuIFM-α1. This result will be discussed further in the section on interferons' effects on membranes since it implies that the interaction of two different interferons with the same membrane receptor can lead to different functional changes in the cell.

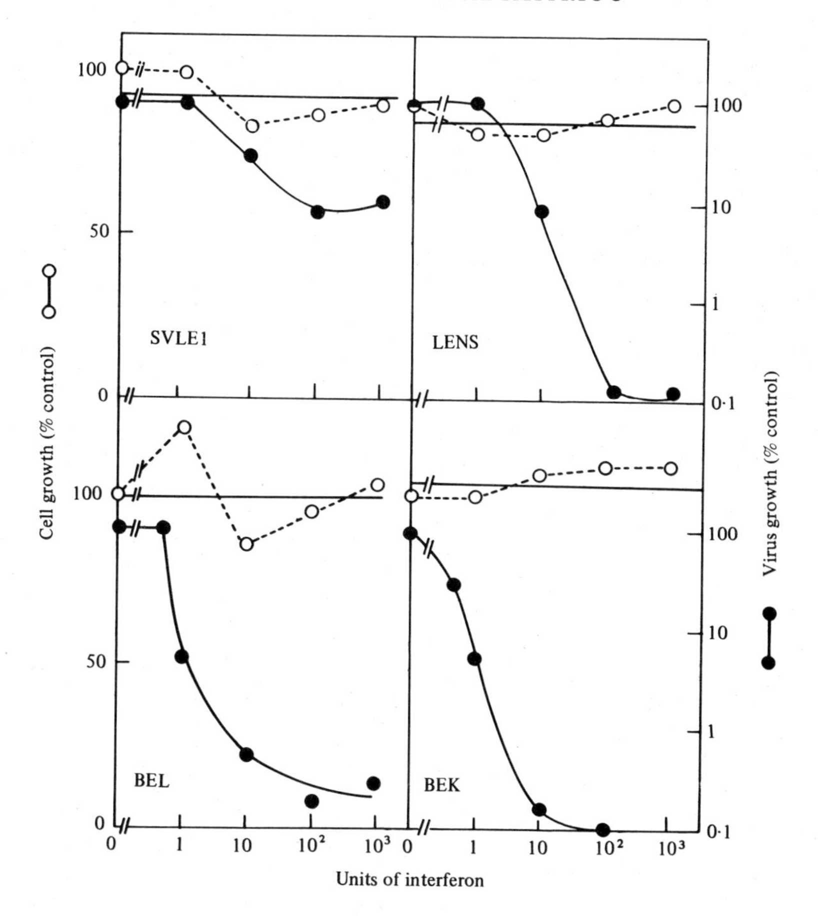

Fig. 1. Dose/response curves for the HuIFN-αN inhibition of growth of EMC virus (●—●) and for inhibition of cell growth (○—○), on bovine cells. SVLE1 is a cell line derived by SV40 transformation of bovine lens. BEL and BEK are diploid cell strains derived from bovine embryo lung and bovine embryo kidney respectively.

(iii) *Efficacy of different interferons*

In both mouse and man, which are the organisms where most is known about the interferon genes, there appear to be three major classes, α, β and γ, and in the human system several distinct molecular species are found within a class (for review see Sehgal, 1982). Single molecular species of interferons have been produced in bacteria from cloned genes, and α interferons have also been fractionated from the mixed preparations produced by leucocytes and lymphoblastoid cells. These single species have usually been tested for growth-inhibitory ability in the very sensitive Daudi cell line (Adams *et al.*, 1975; Masucci *et al.*, 1980; Evinger *et al.*, 1981b). In this system all those tested show some anti-proliferative activity. However, as yet there are few data comparing the individual interferons with each other and with the mixed preparations of HuIFN-α produced in leucocytes and lymphoblastoid lines, for their growth inhibitory ability. The different but immunologically related α interferons not only show a different relative antiviral activity on mammalian cells from different species but also against different viruses in the same cell type (Weck, Apperson, May & Stebbing, 1981). It becomes difficult therefore to define equivalent amounts in terms of a biological activity and is more meaningful to compare the activity of equivalent amounts of protein, which necessitates, initially at least, using pure preparations. A comparison of pure HuIFN-α2 and HuIFN-αN has shown equivalent amounts of protein to be equally active in inhibiting the growth of cultured breast cancer cell lines; α2 was less effective than HuIFN-αN however in inhibiting the growth of a breast cancer grown as a xenograft in the nude mouse (Taylor-Papadimitriou *et al.*, 1982). Until such controlled comparisons have been made, it is difficult to interpret for example the observation that equivalent numbers of antiviral units of γ interferons are more effective in inhibiting cell growth than the same number of antiviral units of β interferons (Rubin & Gupta, 1980). γ interferons may be poor inhibitors of virus growth, and on a mg protein basis, only equivalent in efficacy in inhibiting cell growth to the α and β interferons. The observation that α and β interferons may synergize with γ interferons in inhibiting cell or tumour growth is however of great potential interest (Fleischman, Kleyn & Bron, 1980); γ interferon in the mouse and human systems does not interact with the receptor used by α and β interferons (Ankel *et al.*, 1980; Branca & Baglioni, 1981) and may induce host reactions different from those induced by the α and β interferons.

(iv) *Interferons and the cell cycle*

Studies using both asynchronously and synchronously growing cells, have indicated that all phases of the cycle can be affected by interferon treatment. That the cell cycle is extended was shown by Collyn d'Hooghe, Brouty-Boye, Malaise & Gresser (1977) who measured the interval between mitosis by time-lapse in L cells cultured with mouse interferon. Pfeffer and colleagues (1979) noted a similar effect on cell cycle time with HuIFN-β and human diploid fibroblast. These workers also measured cellular macromolecular synthesis and found that while cell division was inhibited, the rates of RNA, DNA and protein synthesis were only marginally affected over a period of three days. An increase in the number of binucleate cells and in cell volume was also seen, confirming the suggestion that cell division was inhibited. The observation that the rates of macromolecular synthesis are unaffected and cell volume is increased by interferon treatment, has been made in other cell types including mouse Erlich ascites cells (Panniers & Clemens, 1981) and human breast cancer cells (Shibata & Taylor-Papadimitriou, 1981).

Other studies using cytofluorometric methods and asynchronously growing mouse and human tumour cell lines have shown interferon treatment to induce elongation of all phases of the cell cycle, G_1 and G_2 being usually extended more than S (Killander *et al.*, 1976; Matarese and Rossi, 1977; Balkwill *et al.*, 1978; Creasey *et al.*, 1980) but not in all cases (Lundblad & Lundgren, 1981; Panniers & Clemens, 1981). Studies using quiescent cells stimulated to grow by serum or purified mitogenic factors have indicated that both G_1 and G_2 may be markedly extended (Balkwill & Taylor-Papadimitriou, 1978; Creasey *et al.*, 1980), or that the rate of entry of cells into S from G_1 is reduced (Sokawa, Watanabe, Watanabe & Kawade, 1977).

The cell types showing a selective block in G_1 are fibroblasts and some unusual tumour cell lines which may be arrested in G_1 by removal of serum. Virally-transformed and malignant cells and even some normal epithelial cells do not usually arrest in G_1, upon removal of mitogenic factors. This may be due to the autonomous production of growth factors which keep the cells cycling, or to a constitutive production of some proteins crucial to proliferation, and normally induced by mitogens. The tendency for some cells to be blocked in G_1 by interferons probably reflects the growth behaviour of these cell types under conditions of reduced growth, and is not a general response of all cells, particularly of most tumour cells.

Although G_1 arrest may not be a feature of all tumour cells, there is a great advantage in using systems which can be arrested in G_1 for studying the effects of interferon on cell growth, since when the quiescent cells are stimulated to divide by mitogens, the whole population moves through G_1 synchronously and events can be studied at the molecular level (see below).

(v) *Interactions of interferons with factors which stimulate growth*

Cells proliferate in response to a mitogenic stimulus, which in tissue culture has usually been provided by serum. The main growth factor in serum is platelet-derived growth factor (PDGF) although insulin, somatomedins, steroid hormones and epidermal growth factor (EGF) are also present. Most of the past work following the effect of interferons on cell proliferation has been done using cells cultured in medium containing serum. Perhaps the closest *in vivo* analogy to this is the wounding response where PDGF is released and stimulates growth. Tissues and tumours, however, do not normally see PDGF *in vivo* and are stimulated to proliferate by the synergistic action of other hormones or growth factors (e.g. oestrogens, EGF), some of which may be produced by the tumours themselves (Todaro *et al.*, 1980). It is now possible to use purified growth factors to stimulate cell growth, and these are particularly useful when applied to the quiescent fibroblast system, since their synergistic actions can be analysed unambiguously (Rozengurt, 1980).

In analysing the effect of interferons on the growth-stimulatory effects of the various growth factors using synchronized cells arrested in G_1, it is found that the intensity of the inhibitory effect of interferon on cell growth is inversely related to the intensity of the mitogenic stimulus. For example, mouse β interferon added to quiescent 3T3 cells stimulated by three growth factors, markedly inhibits DNA synthesis but has relatively little effect on DNA synthesis stimulated by five growth factors (Taylor-Papadimitriou *et al.*, 1981). It is not easy to work out which growth factors are most effective in reversing interferon's growth-inhibitory effect because several are required to obtain synergism. However, PDGF, a potent mitogen in the absence of other growth factors, is reported to reverse interferon's growth-inhibitory effect (Oleszak & Inglot, 1980). Now that the single molecular species of interferons are available it will be important to test them to see if they synergize in inhibiting cell growth in the way that the growth factors do in stimulating it.

The growth factors which have been shown to be produced by most virally transformed cells (Bourne & Rozengurt, 1976; De Larco & Todaro, 1978) and some human tumours (Todaro *et al.*, 1980) are particularly important to consider as possible antagonists to interferon's growth-inhibitory effect since they may actually be active *in vivo*. The 'transforming growth factors' appear to be effective in inducing cells which are normally anchorage-dependent to grow in agar. This brings us to another problem in measuring interferon's effect on cell growth, namely that certain growth conditions are more stringent than others and require more growth factors. Tumour cells which grow in agar in 10% serum (and often do so because they are producing their own transforming factor) will grow with negligible amounts of serum on plastic (O'Neill, Riddle & Jordon, 1979). Conversely, fibroblasts which grow on plastic in 5% serum, require high serum concentrations for growing in agar. Thus, it is perhaps not surprising that interferons are able to inhibit the anchorage-independent growth of cells better than their growth on plastic (Shibata & Taylor-Papadimitriou, 1981) since the relative mitogenic stimulus is lower.

The fact that interferons can inhibit the mitogenic stimulus, which in turn if intensified can reverse the growth inhibitory effect of interferon, creates a problem in quantitating growth inhibition by interferons. It suggests however that interferons and growth factors may be acting in opposite ways on the same cellular events crucial to cell proliferation. These may involve changes in levels of certain enzymes, such as ornithine decarboxylase whose activity is increased during proliferation and decreased by interferon treatment (see below). Interferons and mitogens can also affect the cytoskeleton and these changes may also be related to the effects of these agents on cell growth.

(vi) *Changes in the cytoskeleton*

Some features of the cytoskeleton, including the actin filaments and tubulin network, are common to all cells. However, the proteins making up the intermediate filaments are different in different cell types, as are some of the major components of the extracellular matrix. It should also be kept in mind that the functional role played by various cytoskeletal elements in the physiology and growth of cells is not well understood, so that in interpreting the effects of interferon on the expression of cytoskeletal elements and vice versa, some caution must be exercised.

Some experiments investigating the relation between interferon action and the cytoskeleton followed the effect of substances which disrupt or stabilize elements of the cytoskeleton on interferon action (mainly antiviral). Thus butyrate which is thought to stabilize the tubulin filaments was reported to enhance interferon action (Bourgeade & Chany, 1979), while agents which disrupt these filaments (vinblastine and colchicine), inhibited interferon action (Bourgeade & Chany, 1976). These indirect experiments are suggestive of a role of microtubules in interferon action as an antiviral agent, and possibly as an inhibitor of cell growth but are subject to the criticism that the agents used to alter the cytoskeleton may be having other effects on cells.

There are a few reports of studies showing a direct effect of interferon on certain elements of the cytoskeleton (Bourgeade, Rousset, Paulin & Chany, 1981). In mouse and human fibroblast and in HeLa cells, treatment with interferon results in an increase in the actin cables seen at the cell membrane (Pfeffer, Wang & Tamm, 1980a; Brouty-Boye, Cheng & Chen, 1981; Wang, Pfeffer & Tamm, 1981b). Increased expression of fibronectin is also seen, but no real increase in the number of intermediate filaments or microtubules (above that to be expected in keeping with the increase in cell volume) was noted. The actin fibres are probably involved in cell motility and the increase in their number and rigidity seen in interferon-treated cells may be responsible for the decrease in cell motility (Brouty-Boye & Zetter, 1980; Pfeffer *et al.*, 1980a) and in membrane receptor mobility (Pfeffer, Wang & Tamm, 1980b; Matsuyama, 1979) noted in such cells.

The cytoskeletal changes seen in interferon-treated macrophages are quite extensive. Interferon treatment of thioglycolate-elicited macrophages resulted in an aggregation of microtubules and 10 nm (intermediate) filaments in the perinuclear region while the peripheral cytoplasm lost these structures. These changes induced by interferon may be causally related to the decreased pinocytosis seen in the interferon-treated macrophages, while their increased phagocytic activity appears to be related to an increased rate of association of actin filaments with the phagocytic cup (Wang *et al.*, 1981a).

(vii) *Possible molecular mechanisms involved in growth inhibition*

Studies on ornithine decarboxylase

The polyamines are implicated in the control of many cell functions and are found in higher levels in proliferating cells (Raine & Janne, 1975). The enzyme which catalyses the rate-limiting step in the synthesis of polyamines and largely regulates the level of these compounds is ornithine decarboxylase (ODC) and an increase in the activity of this enzyme is invariably seen as an accompaniment to the proliferative response (Janne, Poso & Raine, 1978). In the early embryo, regenerating liver, tumour cells, and many normal cells and tissues, increased activity of ODC has been found to be one of the earliest events to follow the mitogenic stimulus, and inhibition of this increase results in an inhibition of embryogenesis (Fozard *et al.*, 1980) or of cell growth (Mamont *et al.*, 1976; Mamont, Duchesne, Grove & Bey, 1978).

The kinetics of the induction of ODC activity can be followed accurately in 3T3 cells stimulated from quiescence to growth by serum or other mitogenic factors (Clarke, 1974; Sreevalsan, Taylor-Papadimitriou & Rozengurt, 1979). In this system, when interferon is added with the mitogenic stimulus a dramatic inhibition of induction of ODC activity is seen, whether serum or combinations of pure mitogenic factors are used to stimulate growth (Sreevalsan *et al.*, 1979; Sreevalsan, Rozengurt, Taylor-Papadimitrion & Burchell, 1980). The inhibition is specific, since overall protein synthesis is not affected (Sreevalsan *et al.* 1979; Taylor-Papadimitriou, 1980) and neither are the very early increases in the transport of ions, nucleotides and sugars which occur within minutes of addition of mitogens (Rozengurt, 1980). Studies with metabolic inhibitors suggest that the increase in ODC activity involves regulation of transcription as well as a reduction in the half life of the enzyme (Clarke, 1974). Interferon action in inhibiting ODC induction may also involve transcription (Sreevalsan *et al.*, 1980). However, the regulation of ODC activity is notoriously complex, and without the measurement of levels of message and of protein, it is not possible to say unequivocally how ODC activity is stimulated by mitogens, and inhibited by interferon. The system does however suggest itself as a suitable model for the study of interferon action both on cell growth and on induced protein synthesis. ODC is certainly a crucial enzyme in growth regulation in almost all cells, and the inhibition of the induction of its activity by mitogens is likely to be an important factor in interferon-induced inhibition of cell growth.

A recent observation of Sekar, Atmar, Krim & Kuehn (1982) has focused attention again on ODC in the context of interferon action. These investigators showed that the protein kinase which is induced by interferon, and which is activated by dsRNA, can also be activated by polyamines. They suggest that by analogy with another system – the slime mould *Physarum polycephalum*, the 70K protein which is phosphorylated by the kinase is ornithine decarboxylase. If this were so, then interferons could be mobilizing a natural feedback mechanism to suppress ODC activity and consequently growth.

Cyclic nucleotides and interferon action

The possible involvement of cyclic nucleotides in growth control and in interferon action is controversial. Cyclic AMP (cAMP) has been implicated as an inhibitor of cell growth for some time, but recent developments indicate that while this may be true for some types, such as certain lymphoma cell lines (Hochman *et al.*, 1975), it is not a general rule. On the contrary, cAMP may act as a mitogenic signal in many cells including keratinocytes (Green, 1978), mammary epithelial cells (Taylor-Papadimitriou *et al.*, 1980), Schwann cells (Raffe, Abney, Brocker & Hornby-Smith, 1978) and 3T3 cells (Rozengurt, Legg, Strang & Courtenay-Luck, 1981). These different and opposing effects of cAMP on cell growth may depend on the metabolic profile of the cells. For example, cAMP inhibition of the growth of S49 lymphoma cells correlates with decreased ODC activity and the presence of cAMP-dependent protein kinase. Mutants of S49 lymphoma cells which are 'protein kinase-less' do not show a reduction in either growth or ODC activity upon treatment with analogues of cAMP (Insel & Fenno, 1978). In such cell types where cAMP and its analogues inhibit growth, it may be that interferon's antigrowth effect is mediated at least in part by cyclic nucleotides. This concept is supported by the work of Schneck, Rager-Isman, Rosen & Blum (1982) who have analysed the effect of interferon on a series of variants derived from a cloned macrophage-like cell line J774.2. These workers found that while all the cell lines were equally protected against virus infection, only those with a functional cAMP-dependent protein kinase were sensitive to the growth-inhibitory effect of interferon and to its effect on differentiated function.

Where cAMP stimulates growth, it is unlikely to be involved directly in interferon-induced growth inhibition. Conceivably in these cell types other protein kinases, like the one activated by polyamines, are recruited. As indicated earlier, different cell types may use different signals in growth regulation and this diversity will almost certainly be reflected in the mechanisms of growth inhibition by interferon.

Interferon-induced proteins involved in antiviral action

The last few years have seen significant advances in our understanding at the molecular level of some features of the antiviral action of interferon. To induce an antiviral state in cells, interferons require host RNA and protein synthesis (Taylor, 1964; Friedman & Sonnabend, 1964) and several new proteins are synthesized (Knight & Korant, 1978). It is difficult to design an experiment to ask whether the requirement for host macromolecular synthesis is also necessary to the growth-inhibitory effect of interferon. It is possible, however, to ask whether any of the enzymes which are induced in the interferon-treated cell and are involved in the antiviral effect, play any role in growth inhibition by interferon. The two most studied enzymes are the 2′5′-oligoadenylate synthetase (Kerr & Brown, 1978) and the protein kinase (Lebleu *et al.*, 1976; Zilberstein, Federman, Shulman & Revel, 1976; Roberts *et al.*, 1976) which phosphorylates a 70K protein; both enzymes are activated by dsRNA, and it seems likely that the protein kinase is also activated by polyamines (Sekar *et al.*, 1982). The protein kinase has not been studied in detail in relation to interferon's growth-inhibitory effect. However, the fact that polyamines can activate the kinase gets rid of the problem of how it would be activated in an uninfected cell, which would contain little if any dsRNA. The possibility of ODC being the protein phosphorylated and thus inactivated by the kinase makes this an area of great potential interest.

There is some evidence that the 2′5′-oligoadenylate synthetase system may be related to inhibition of mitogen-stimulated DNA synthesis in lymphocytes (Kimchi, Shure & Revel, 1979). These cells normally show a high level of synthetase when not proliferating (Shimizu & Sokawa, 1979) and the enzyme activity is reduced in stimulated lymphocytes, and increased after interferon treatment (Kimchi et al., 1979). However, there are several examples of human cell lines resistant to interferon's growth-inhibitory effect which still make levels of 2′5′A synthetase comparable to those seen in the wild type sensitive cells (Vandenbussche *et al.*, 1981; Content & Verhaegen-Lewalle, 1981; Silverman *et al.*, 1982). It should be remembered that there are several components to the 2′5′A oligoadenylate synthetase system, in particular the enzyme which degrades the oligomers of 2′5′A and the nuclease which is activated by the oligomers. Levels of 2′5′ adenylate oligomers will depend on the relative rates of degradation and synthesis and whether they are active will depend on the level of ribonuclease F, the enzyme which is activated by the nucleotides (Clemens & Williams, 1978; Schmidt *et al.*, 1978; Farrell *et al.*, 1978). In a line of NIH 3T3 cells, absence of ribonuclease F activity has been found to be associated with a reduced antiviral activity of interferon against EMC and VSV viruses, and also with a reduced activity on cell proliferation (Epstein *et al.*, 1981; Czarniecki, Sreevalsan, Friedman & Panet, 1981). While these observations might suggest that 2′5′A can play a role in interferon-induced growth inhibition, in the one case where actual levels of nucleotide have been measured, no correlation could be found. Thus, Silverman *et al.* could not detect any 2′5′A oligonucleotides in either the highly sensitive Daudi cell line, or an interferon-resistant variant. Moreover, no 2′5′A-mediated RNA cleavage could be detected inside the cell. A cautious interpretation of the relevant data available so far would be that the 2′5′A system may be involved in growth inhibition by interferon in some cell types (e.g. lymphocytes), but not necessarily in others. With the 2′5′A system there remains the problem of whether there is enough dsRNA in the uninfected cell to activate the synthetase sufficiently.

(viii) *Growth inhibition and the anti-tumour effects of interferon*

In model systems, interferons have been shown to inhibit the growth of chemically induced, transplantable and spontaneously-arising tumours, as well as those of known viral etiology (Gresser & Tovey, 1978). Since interferons can have an anti-tumour effect *in vivo* on cells which are resistant to its growth-inhibitory effect *in vitro* (Gresser, Maury & Brouty-Boye, 1972), it is relevant to ask whether a direct inhibition of cell growth plays any role at all in interferons observed effects on tumour development. Although this is an important question, since it relates to the possibility of using *in vitro* assays for inhibition of tumour cell growth as predictive of clinical effectiveness, little information is available. There are two reports which suggest that direct inhibition of tumour cell growth may be important in interferons antitumour effect. Firstly, Gresser & Bourali-Maury (1973) reported that inhibition of the immune response (by anti-lymphocytic serum or X-irradiation) did not abrogate the inhibitory effect of interferon on transplantable tumours in the mouse. Secondly, Balkwill, Taylor-Papadimitriou, Fantes & Sebesteny (1980) found that HuIFN-αN could inhibit the development of human breast tumours grown as xenografts in the *nude* (or athymic) mouse. It is quite likely that different tumour types vary both in their sensitivity to interferon's growth-inhibitory effect, and in the importance of this effect *in vivo* in eliminating malignant cells.

(ix) *Stimulation of cell growth by interferon*

In examining the effects of interferon on the growth of various cells in culture, some investigators have occasionally noted a small stimulation of growth (Bradley & Ruscetti, 1981). Some caution should be exercised in interpreting these observations where the interferon preparations used were not pure and were possibly contaminated by mitogenic factors. However, using purified HuIFN-α2 or HuIFN-αN, we have found a small but consistent stimulation of the growth of bovine cells. The stimulation is more marked in low serum, and, as shown in Fig. 2, was observed with four different bovine cell types which are sensitive to the antiviral action of the interferons. As mentioned earlier, no inhibition of growth of these heterologous cells is observed even with high concentrations of HuIFN-α2 and HuIFN-αN (Taylor-Papadimitriou *et al.*, 1982). With some human cell lines whose growth is inhibited only by high concentrations of interferon, stimulation is seen with low concentrations. These observations may reflect a real phenomenon which could occur in cancer patients receiving exogeneously administered interferon, and therefore certainly warrant attention. It is tempting to speculate that interferon may act as a growth factor by stimulating intracellular levels of cAMP in cells where cAMP is a mitogenic signal.

EFFECTS ON DIFFERENTIATED FUNCTION

(i) *General*

As with growth regulation, it has to be kept in mind that differentiation in different cell types may involve several mechanisms for the control of differentiated function. However, for clarity of presentation, it is useful to consider the many effects which interferons have on cell functions as examples of either enhancement of an already operating differentiated function, or inhibition of the acquisition of such a function. These stimulatory and inhibitory effects of interferons are well illustrated in lymphocytes, where the evidence for a

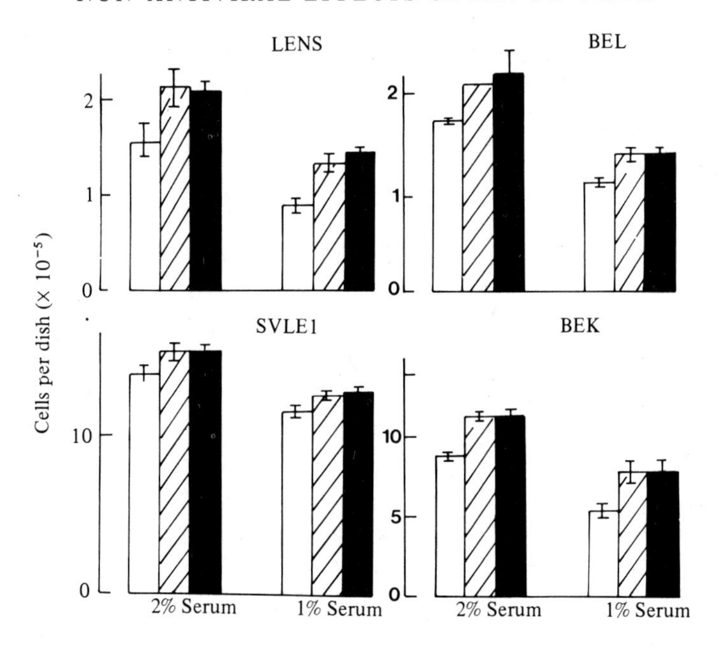

Fig. 2. Stimulation of growth of bovine cells in low serum by HuIFN-αN (hatched columns) and HuIFN-α2 (solid columns).

natural regulatory role for interferons is strongest. However, interferons also affect function in cells other than those of the immune system. In what follows, we will discuss the inhibitory effects of interferon separately from those effects resulting in enhancement of cell function, bearing in mind that until actions underlying these mechanisms are better understood, the unification under the two headings is a pragmatic one.

(ii) *Inhibition of cell function*

Apart from its effect on cell division and those functions which are probably altered via changes in the cytoskeleton (cell movement, pinocytosis), interferon's inhibitory effects (Table 1) can be characterized as inhibition of the induced synthesis of certain enzymes or proteins. This is seen clearly in the inhibition of steroid-induced enzyme synthesis, in the inhibition of DMSO-induced haemoglobin synthesis in Friend cells, and in the inhibition of induction of thymidine kinase by mitogens. Ornithine decarboxylase is an enzyme which is invariably induced by mitogens, growth factors and tumour promoters, and its inhibition by interferons suggests that in

Table 1. *Cell functions inhibited by interferons*

Function	References
Growth	
Cell proliferation	Paucker *et al.* (1962); Lindahl Magnusson, Leary & Gresser (1972); Frayssinet, Gresser, Tovey & Lindahl (1973); van't Hull *et al.* (1978); Balkwill *et al.* (1978)
Tumourigenicity	Gresser *et al.* (1972); Gresser & Tovey (1978)
Inducible activities or proteins	
Thymidine uptake	Brouty-Boye & Tovey (1978); Gewert, Shah & Clemens (1981)
Steroid inducible glycerol-3-phosphatase	Illinger, Coupin, Richards & Poindron, 1976
Steroid inducible tyrosine aminotransferase in hepatoma cells	Beck *et al.* (1974)
Glutamine synthetase in chick neural retina	Matsuno & Shirasawa (1978)
Induction of ornithine decarboxylase activity in fibroblasts by serum growth factors and tumour promoters	Sreevalsan *et al.* (1979); Sreevalsan *et al.* (1980)
Serum induced release of plasminogen activator	Schroder, Chou, Jaken & Black (1978)
Dimethyl sulfoxide induced production of haemoglobin in Friend cells	Rossi *et al.* (1977a,b)
Cellular differentiation	
Formation of antibody-producing cells	For reviews see Gresser (1977); Balkwill (1979)
Delayed hypersensitivity reaction	De Maeyer & De Maeyer-Guignard (1977)
Insulin induced differentiation of 3T3 cells to adipocytes	Keay & Grossberg (1980)
Maturation of human monocytes to macrophages	Lee & Epstein (1980)

inhibiting cell growth, interferons may be inhibiting the synthesis of a group of proteins and/or enzymes, induced by growth factors and crucial to cell growth. Some of the more complex phenomena may be the result of an effect of interferon on both cell growth and induced protein synthesis, as for example, in the inhibition of the production of antibody-forming cells in response to a specific stimulus, and in the inhibition of the maturation of macrophages. Differentiation may involve cell division, the production of specific factors (e.g. proteins), and of surface receptors for these factors (glycoproteins), and may require the synthesis of specific proteins by the differentiating cell in response to the factors. All of these stages could be affected by interferon, and where inhibition of differentiation is observed, an effect on any one or all of them could be involved.

The selective inhibition by interferons of the synthesis of certain host proteins is clearly an interesting problem which may be approached at the molecular level. In several instances, investigators have asked whether the inhibition is at the transcriptional level. In glutamine synthetase induction by steroids and DMSO-induced haemoglobin synthesis in Friend cells, an effect on the production of the respective message has been reported although an effect on translation of haemoglobin mRNA is also observed (Shirasawa & Matsuno, 1979; Rossi *et al.*, 1977a). In the inhibition of ODC, indirect experiments with antimetabolites again suggest a major effect on transcription although some effect on translation is also indicated (Sreevalsan *et al.*, 1980). It has been suggested that a selective effect of interferon on protein synthesis may occur at the translational level by a non-specific inhibition of elongation of the polypeptide, which would lead to a stronger inhibition of synthesis of those messages where elongation is rate limiting (Yau *et al.*, 1978). However, this does not appear to be the explanation for the selective inhibition of the induction of ornithine decarboxylase activity by mitogens (Taylor-Papadimitriou, 1980; Sreevalsan *et al.*, 1980).

(iii) *Enhancement of cell functions*

An examination of the functions which can be stimulated by interferon and which are listed in Table 2, shows that many are specific functions performed by highly differentiated cells. There are few satisfactory model systems for studying the control mechanisms involved in the acquisition of a differentiated function, particularly using cell types from solid tissues. Studies in those systems which are available (keratinocytes, lymphocytes, muscle cells) suggest that differentiation is accompanied by a reduction in the proliferative capacity of the cell. It is possible therefore that the enhancement of differentiated function seen in interferon-treated

Table 2. *Cell functions enhanced by interferons*

Function	References
Phagocytosis by macrophages	Donahoe & Huang (1976)
Cytotoxicity of sensitized lymphocytes	Lindahl, Leary & Gresser (1972)
Activity of natural killer cells	Trimchieri & Santoli (1978)
Production of antibody	Gisler, Lindahl & Gresser (1974)
IgE-mediated histamine release by basophils	Ida, Hooks, Siraganian & Notkins, (1977)
Synthesis of prostaglandins	Yaron *et al.* (1977)
Synthesis of ketosteroids in adrenal cells	Chany *et al.* (1980)
Beat frequency of myocardial cells	Blalock & Stanton (1980)
Excitability of cultured neurones	Calvet & Gresser (1979)
Uptake of iodide by thyroid cells	Friedman *et al.* (1981)
Expression of Fc receptors	Fridman *et al.* (1980)
Induction of differentiation of mouse myeloid leukemic cells	Lotem & Sachs (1978); Tomida, Yamamoto & Hozumi (1980)
Synthesis of t-RNA methylase	Rozee, Katz & McFarlane (1969)
Expression of histocompatibility antigens in mouse and human cells	Lindahl, Leary & Gresser (1973, 1974); Dolei, Ameglio, Capobianchi & Tosi, (1981); Heron, Berg & Cantell (1976); Fellous, Kamoun, Gresser & Bond (1979); Imai, Ng, Glassy & Ferrone (1981); Liao, Kwong, Khosravi & Dent (1982)
Expression of carcino-embryonic antigen	Attallah, Needy, Noguchi & Elisberg (1979)
Induction of aryl hydrocarbon hydroxylase	Nebert & Friedman (1973)
Production of interferon (Priming)	Stewart, Gosser & Lockhart (1971)
Cytotoxicity of dsRNA	Stewart *et al.* (1972)

cells is merely a result of an inhibition of cell growth. Conversely, the effect of interferon on cell growth may be more apparent in a cell which has the capacity to differentiate (Balkwill & Oliver, 1977; Shibata & Taylor-Papadimitriou, 1981; Taylor-Papadimitriou *et al.*, 1982). Whether there is a causal relationship between the two effects is not easy to determine. However, the data of Schneck *et al.* (1982) with the mouse macrophage-like cell line J774.2 and its derivatives are perhaps relevant to this point. The variants of J774.2 which were sensitive to growth inhibition by mouse interferon also showed increased Fc-mediated phagocytosis of opsonized sheep red blood cells after interferon treatment. Conversely, variants whose proliferation was not affected by interferon did not show enhanced differentiated function (phagocytosis) after interferon treatment. It may be that interferons will prove to be useful in investigating the control mechanisms (so far little understood) involved in moving cells from a proliferative to a differentiated state.

Related to the effects of interferons on differentiated function is the increased expression of surface antigens and secreted proteins. Increased expression of histocompatibility antigens has been noted in mouse thymocytes. Many tumour cells express HLA antigens, and their increased expression could conceivably result in an enhancement of the immune rejection mechanism. Certainly, the enhanced expression of HLA on the normal cells of the haemopoietic system is likely to enhance the cell interactions involved in the immune response. The recent observation that increased levels of β-microglobulin are found in the serum of patients treated with HuIFN-α suggests that the phenomenon can operate *in vivo* (Lucero *et al.*, 1982).

The increased level of aryl hydrocarbon hydroxylase which can be induced in interferon-treated cells again suggests a general role for interferons in protection of the host against foreign substances – in this case, potential carcinogens, while the effects on histamine production and prostaglandins, normally involved in the inflammatory response, would indicate an enhanced reaction to infection.

Two phenomena which are affected by interferon perhaps deserve separate attention, namely, sensitivity to dsRNA, and the production of interferon itself (priming). The sensitivity to dsRNA almost certainly reflects the increased level of dsRNA-sensitive enzymes which inhibit protein synthesis, and which are induced in interferon-treated cells. The priming phenomenon is less easily explained. Mouse L cells treated with small amounts of interferon and then stimulated with NDV, produce more interferon earlier than unprimed cells (Stewart, Gosser & Lockharde, 1971; Stewart *et al.*, 1972). Primed cells can also produce interferon in response to poly I:C in the absence of DEAE-dextran which is required for interferon induction by dsRNA in unprimed cells (Stewart *et al.*, 1972; Fujita & Kohno, 1981). The mechanism of priming appears to be different in viral and dsRNA induction. In viral induction, priming appears to increase the efficiency of translation of the interferon messenger RNA (Abreu, Bancroft & Stewart, 1979; Content *et al.*, 1980), while with poly I:C induction, pretreatment with interferon increased the level of mRNA (Fujita, Saito & Kohno, 1979). All of the work on priming has been done with mouse β interferon, which is effective even when purified to homogeneity (De Maeyer-Guigard, Cachard & De Maeyer, 1980). An important observation, recently reported by Fujita & Kohno (1981) is that cellular protein synthesis is required for interferon to exert its priming effect on cells. This is the only effect of interferons, other than their antiviral action, which has been shown to depend on host protein synthesis. If priming is a general phenomenon, not restricted to L cells, it may play an important role in amplifying the initial defence of the host tissues against virus infection.

(iv) In vivo *effects*

The enhancement of the differentiated function of the effector cells of the immune system (macrophages, NK cells, T and B lymphocytes) almost certainly plays a crucial part in the *in vivo* action of interferons as antiviral and possibly antitumour agents. However, it may be that effects on other cells (such as myocardial cells) could lead to the toxic symptoms (for example, tachycardia) which are seen in patients receiving interferon parenterally. Recently, effects of interferon on levels of hormones and serum proteins have been seen in normal patients receiving low doses of HuIFN-α from leucocytes. Thus Kaupilla *et al.* (1982) have noted a decrease in levels of oestrogen and progestin in the serum level of normal patients treated with low doses of HuIFN-α from leucocytes. This could be an important effect in interferon treatment of hormone-sensitive tumours. An effect on plasma lipoproteins and the activity of post-heparin plasma ligases has also been seen (Ehnholm *et al.*, 1982). These observations serve to remind us of the complexity of the interferon system when operating *in vivo* in man.

EFFECTS ON CELL MEMBRANES

(i) *General*

The first step in interferon's interaction with the cell is to bind to high-affinity binding sites – receptors, in the cell membrane. As a result of this interaction, a number of physical, chemical and functional changes are induced in the membrane, and these are

Table 3. *Changes in cell membranes induced by interferons*

Change	References
Increased expression of surface antigens	Lindahl *et al.* (1973); Lucero *et al.* (1982)
Inhibition of movement of membrane receptors	Pfeffer *et al.* (1980b); Matsuyama (1979)
Increased binding of lectins	Huet, Gresser, Bandu & Lindahl (1974)
Increased net negative charge	Knight & Korant (1977)
Increase in relative proportion of saturated acyl side chains in membrane phospholipids	Chandrabose *et al.* (1981)
Increase in intramembrane particles	Change *et al.* (1978)
Increase in membrane rigidity	Pfeffer *et al.* (1981)
Increased association of actin with the cell membrane	Pfeffer *et al.* (1979); Wang *et al.* (1981b); Bourgeaude *et al.* (1981)
Inhibition of binding of cholera toxin and TSH	Kohn *et al.* (1976)
Decreased exposure of oligosaccharide moieties of some gangliosides	Grollman *et al.* (1978)
Stimulation of adenyl cyclase activity	Friedman & Pastan (1969)
Changes in ion transport in membrane vesicles	Grollman *et al.* (1978)
Effects on release of mouse Leukemia virus particles from infected cells	Billiau, Somis & De Somer (1973); Friedman & Ramseur (1974); Chang, Mims, Triche & Friedman (1977); Pitha, Rowe & Oxman (1976)
Effects on incorporation of glycoprotein (G) and membrane protein (M) into Vesicular Stomatitis Virus	Maheshwari, Demsey, Mohanty & Friedman (1980)

listed in Table 3. Presumably some of the changes are crucial steps in the transmission of the extracellular signal to the inside of the cell, resulting in the synthesis of new proteins, inhibition of viral and cell growth and possibly in the enhancement of cell function. At the moment, understandably, the mechanisms involved in the transmission of the extracellular signal are poorly understood. However, with the availability of pure single species of interferon, and monoclonal antibodies directed to them, the interaction with the receptor can be analysed. Before discussing these data, some discussion of the membrane changes is warranted.

(ii) *Changes effected in membranes treated with interferons*

Interferon treatment has been reported to affect the charge on the cell membrane of L cells, making them more electronegative, and to alter the buoyant density of AKRC cells. These changes could be related to the changes in the chemical composition which have also been noted. Chandrabose, Cuatrecasas & Pottathill (1981) have made the interesting observation that the phospholipids in the membranes of S-180 mouse sarcoma cells show an increase in the relative proportion of saturated acyl side chains, while Chang, Jay & Friedman (1978) found an increase in the concentration of some membrane glycoproteins and the number of intramembraneous particles in interferon-treated AKRC cell membranes. The increased levels of histocompatibility antigens discussed previously may represent a similar phenomenon, i.e. increased expression of membrane glycoproteins. Recent work suggests that the action of interferon on the expression of HLA proteins appears to be at the gene level, and increased levels of the relevant mRNAs can be detected in lymphoblastoid cells as early as two hours after interferon treatment (Revel *et al.*, 1982). Whether this is true for other membrane glycoproteins, particularly those, like β-microglobulin, which can be secreted, remains to be seen.

The increase in membrane rigidity, as determined by spin label electron spin resonance (Pfeffer, Landsberger & Tamm, 1981) which is seen in interferon-treated fibroblasts, may be related to the reduced mobility of surface receptors for Con A, and to the increased association of actin with the cell membrane. The importance of membrane rigidity in growth control is not clear, but it is possible that these effects of interferons are important for its action in inhibiting cell growth; they are almost certainly related to the inhibitory effect on cell motility, and may also relate to the decreased exposure of the oligosaccharide moieties of certain gangliosides seen in membranes of interferon-treated cells (Grollman *et al.*, 1978).

Apart from the effects on hormone binding (Kohn, Friedman, Holmes & Lee, 1976; Grollman *et al.*, 1978) which may be considered functional changes, interferons appear to affect at least three membrane functions; they increase membrane adenyl cyclase, affect the transport of small ions and molecules (in membrane vesicles) and affect the budding of enveloped viruses. The effect on adenyl cyclase may be a non-specific effect resulting from the interferon–

membrane interaction, like the transient early change in membrane rigidity noted in interferon-treated fibroblasts (Pfeffer *et al.*, 1981). The increased transport of ions and small molecules seen in membrane vesicles was not seen in interferon-treated quiescent 3T3 cells, where the increased transport of uridine, phosphate, deoxy-glucose, and Rb^+, seen early after mitogenic stimulation, was not inhibited by interferon (Rozengurt & Taylor-Papadimitriou, unpublished observations). The effect on release or infectivity of mouse leukaemia virus and on the incorporation of a membrane protein into VSV virus represent antiviral effects of interferons which appear to be mediated via an effect on the host cell membrane.

(iii) *Receptor interactions*

Only in the last year or two has it been possible to study the interaction of interferons with the cell surface using radioactively labelled ligands (Aguet, 1980; Aguet & Blanchard, 1981; Branca & Baglioni, 1981). However indirect evidence indicated that interferons bound to specific receptors in the membrane. Since they could be shown to bind to gangliosides (Besancon & Ankel, 1974a; Vengris, Reynolds, Hollenberg & Pitha, 1976) and to compete with other hormones, cholera toxin, and lectins, for sites on the cell membrane (Besancon & Ankel, 1974b; Grollman *et al.*, 1978), it was deduced that a specific membrane receptor existed, and that membrane gangliosides were involved in the binding. A role for protein in receptor binding was directly demonstrated by Grollman *et al.* (1978), and was inferred from the studies of human–mouse hybrids containing only human chromosome 21. There is fairly strong evidence that sensitivity to the antiviral effect of HuIFN-α and HuIFN-β is directed by chromosome 21 (Tan, Tischfield & Ruddle, 1973; Chany *et al.*, 1973, 1975). There is little direct evidence to prove that chromosome 21 codes for a membrane receptor. Two reports suggest, however, that this might be the case. Firstly, there is the reported inhibition of interferon action by an antibody to a membrane component coded for by chromosome 21 (Revel, Bash & Ruddle, 1976). Secondly, a mouse–human hybrid carrying only chromosome 21 is sensitive to HuIFN-β and the phosphorylated 67K protein implicated in interferon's antiviral action is of mouse origin (Slater *et al.*, 1978). Most of the work on membrane interactions of interferon relates to its antiviral effect. The interactions which lead to inhibition of cell growth or to effects on cell function are relatively unexplored. It is relevant to note, however, that while chromosome 21 appears to code for sensitivity to interferon's growth-inhibitory effect (Tan, 1976) and for its effect on monocytes (Epstein, Lee & Epstein, 1980), priming and increased sensitivity to dsRNA are not increased in cells trisomic for chromosome 21 (De Clercq, Edy & Cassiman, 1975).

Work following the direct binding of ^{125}I-labelled interferons has been done by Aguet and colleagues in the mouse system and by Branca and Baglioni in human cells. The data indicate that while there appears to be one high-affinity binding site for α and β interferons, a separate site is involved in binding γ interferon. These studies confirm earlier data of Ankel *et al.* (1980) indicating that type 1 (α and β) and type II (γ) interferons use separate receptors.

It would appear therefore that all the HuIFN-α species (probably eight in all) and HuIFN-β interact with the same receptor. Yet each of these molecules shows a specific profile of activity against different viruses and against cell function. For example, HuIFN-αF protects Vero cells (a simian cell line) 1000 times better against EMC virus infection than against VSV infection, while HuIFN-αA (HuIFN-α2) is 10 times less effective in protecting these cells against EMC than against VSV. Moreover, although HuIFN-α2 and HuIFN-α1 are both effective antiviral agents in bovine cells, the one molecule stimulates the growth of cells of this species, while the other appears to inhibit. This is illustrated in Fig. 3 which shows the effect of both interferons on the growth of bovine embryo kidney cells. It appears therefore, that while only one high-affinity binding site is involved in binding several interferons, the effect on the cell of binding the different interferons is not always the same. Since little is known about the way in which the membrane interaction is translated into an effect on the cell metabolism, it is difficult to speculate how the different effects are produced from an ostensibly similar interaction. Some clues may be obtained, however, by a consideration of the structural differences of the proteins themselves. Consider the two interferons HuIFN-α1 and HuIFN-α2 which show differences in the extent of their reaction with heterologous species. HuIFN-α1 protects bovine cells better than human cells, and even acts effectively on mouse cells, while HuIFN-α2 does not protect mouse cells and although effective on bovine cells, is less so than on homologous cells; also as mentioned above HuIFN-α2 does not inhibit the growth of bovine cells, while HuIFN-α1 does. Streuli *et al.* (1981) have made hybrid DNAs from the α1 and α2

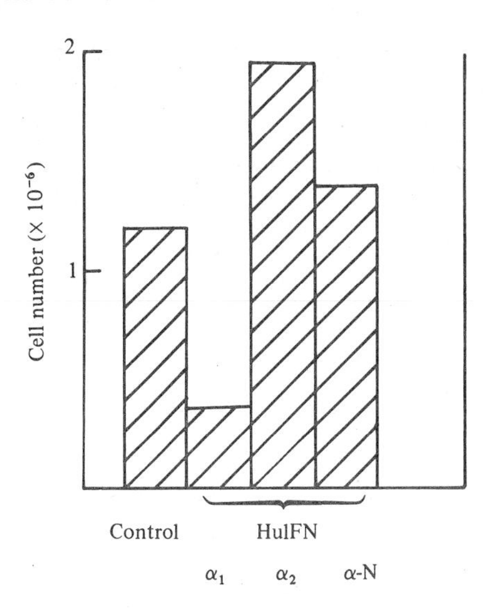

Fig. 3. Effect of different species of HuIFN-α (1000 units per ml) on the growth of BEK cells. Units here defined by inhibition of growth of Semliki forest virus in WISH cells.

genes and produced hybrid interferons with (a) the NH_2-proximal half of $\alpha1$, and (b) with the NH_2-proximal half of $\alpha2$, and followed the spectrum of activity of the hybrids on different species. Their data are consistent with there being two binding sites in each IFN molecule (or one site consisting of two idiotypes), one in the NH_2-proximal region, and the other in the COOH-proximal; triggering efficiency would then be some function of the two interactions, which could vary independently of each other in different cell lines.

Another way to approach the problem of receptor interactions is to use monoclonal antibodies to interferons, to characterize those parts of the molecule which are important for binding and to compare these domains in the different interferon molecules. There are already some preliminary indications that this line of attack may be useful. Shand and Ivanyi and colleagues have obtained a monoclonal antibody (HC46) to HuIFN-αN which in an antiviral assay neutralizes only six of the eight species of α found in preparations of HuIFN-αN, suggesting that a different determinant is involved in binding the other two species to cell membranes. When tested against HuIFN-$\alpha1$ and HuIFN-$\alpha2$, HC46 is found to neutralize HuIFN-$\alpha2$ but not $\alpha1$, suggesting that the different patterns of reaction of these two interferons on the functions of homologous and heterologous cells, may be attributable to differences in the structural properties of the binding site(s) of the two interferons. It is interesting to note that as early as 1975, Paucker and colleagues working with a mixture of HuIFN-α produced by leucocytes, was able to show that the component or determinant involved in the interaction with heterologous cells (rabbit) was not neutralized by an anti-HuIFN-α serum (produced in the rabbit) which neutralized the antiviral action on homologous cells (Paucker, Dalton, Ogburn & Torma, 1975). It may be that the dominant antibodies in his preparation were of the HC-46 type, which does not neutralize α1 (or αD) – the HuIFN-α with the strongest effect on rabbit cells (Weck *et al.*, 1981).

We have emphasized the use of comparing activities of the various human interferons on cells of different species. It will be of great interest to know whether the various HuIFNs also show differences in specificity with regard to tissue and to the cell functions they effect. Although there is apparently only one receptor for interferons of the α and β type, there is not enough information available yet to know whether the K_m for a single IFN species is the same in all human cells. The same glycoprotein placed differently in the membrane could present an altered binding site to the same interferon. Alternatively, as we have indicated above, interferons with structural differences will have different interactions with the same receptor which could lead to variation in cell responses. Again it may be possible to obtain monoclonal antibodies which may neutralize one function but not another, and this approach should be pursued.

DISCUSSION

The study of interferons as cell regulators is clearly entering a new phase. Now that purified interferons are available to more investigators, and the possibility of artefacts arising due to contaminating materials is minimized, progress in studying the regulatory effects of interferons should be more rapid. Furthermore, although the complexity of the system has been amplified by the discovery that interferons are coded for by a family of genes, the application of the tools of recombinant DNA technology and monoclonal antibodies to the investigation of these molecules will most certainly help in the study of the mechanisms involved in their action. For example, in studying the antigrowth effects of interferon attempts have been made in the past to compare metabolic changes in two or more cell lines with different sensitivities to this action of interferons. Now, it may be possible to do the complementary type of experiment, which may be more meaningful, namely to look at changes in the same cell treated with different purified bacterially-produced interferons which affect or do not affect a specific function like growth, and whose actions may or may not be modified by various monoclonal antibodies.

Advances in determining whether and how interferons are going to be useful in the treatment of human disease should be helped by progress in the basic biology of the interferon system, although predictions from *in vitro* studies are not always feasible. However, the cautionary years of interferon research, where clinical trials were restricted by the availability of human interferon, have led to a clinical approach which attempts to get as much basic information on the *in vivo* actions of interferon as possible. It is hoped that with the advent of larger supplies this approach will be maintained, and the information coming from patients will complement the work in *in vitro* systems in studying the interferons as regulators of cell growth.

REFERENCES

Abreu, S. L., Bancroft, F. C. & Stewart, W. E. II. (1979). Interferon priming: effects on interferon messenger RNA. *Journal of Biological Chemistry*, **254,** 4114–18.

Adams, A., Strander, H. & Cantell, K. (1975). Sensitivity of Epstein-Barr virus transformed human lymphoid cell lines to interferon. *Journal of General Virology*, **28,** 207–17.

Aguet, M. (1980). High-affinity binding of ^{125}I-labelled mouse interferon to a specific cell surface receptor. *Nature*, **284,** 459–61.

Aguet, M. & Blanchard, B. (1981). High affinity binding of ^{125}I-labelled mouse interferon to a specific cell surface receptor. *Virology*, **115,** 249–61.

Ankel, H., Krishnamurti, C., Besancon, F., Stefanos, S. & Falcoff, E. (1980). Mouse fibroblast (type I) and immune (type II) interferons: pronounced differences in affinity for gangliosides and in antiviral and antigrowth effects on mouse leukemia L-1210R cells. *Proceedings of the National Academy of Sciences of the USA*, **77,** 2528–32.

Attallah, A. M. & Strong, D. M. (1979). Differential effects of interferon on the MHC expression of human lymphocytes: enhanced expression of HLA without effect on Ia. *International Archives of Allergy and Applied Immunology*, **60,** 101–7.

Attallah, A. M., Needy, C. F., Noguchi, P. D. & Elisberg, B. L. (1979).

Enhancement of carcinoembryonic antigen expression by interferon. *International Journal of Cancer*, **24,** 49–52.

BALKWILL, F. R. (1979). Interferons as cell regulatory molecules. *Cancer Immunology and Immunotherapy*, **7,** 7–14.

BALKWILL, F. & OLIVER, R. T. D. (1977). Growth inhibitory effects of interferons on normal and malignant human haemopoietic cells. *International Journal of Cancer*, **20,** 500–5.

BALKWILL, F. & TAYLOR-PAPADIMITRIOU, J. (1978). Interferon affects both G_1 and $S + G_2$ in cells stimulated from quiescence to growth. *Nature*, **274,** 798–800.

BALKWILL, F., TAYLOR-PAPADIMITRIOU, J., FANTES, K. H. & SEBESTENY, A. (1980). Human lymphoblastoid interferon can inhibit the growth of human breast cancer xenografts in athymic (nude) mice. *European Journal of Cancer*, **16,** 569–73.

BALKWILL, F., WATLING, D. & TAYLOR-PAPADIMITRIOU, J. (1978). Inhibition by lymphoblastoid interferon of growth of cells derived from the human breast. *International Journal of Cancer*, **22,** 258–65.

BECK, G., POINDRON, P., ILLINGER, D., BECK, J. P., EBEL, J. P. & FALCOFF, R. (1974). Inhibition of steroid inducible tyrosine aminotransferase by mouse and rat interferon in hepatoma tissue culture cells. *FEBS Letters*, **48,** 297–300.

BERG, K., HOKLAND, M. & HEREN, I. (1982). Biological activities of pure HuIFN-α. In *Interferon Properties, Mode of Action, Production and Clinical Application*. Basle: Karger Verlag (in press).

BESANCON, F. & ANKEL, H. (1974a). Binding of interferon to gangliosides. *Nature*, **252,** 478–80.

BESANCON, F. & ANKEL, H. (1974b). Inhibition of interferon action by plant lectins. *Nature*, **250,** 784–6.

BILLIAU, A., SOBIS, H. & DE SOMER, P. (1973). Influence of interferon on virus particle formation in different oncornavirus carrier cell lines. *International Journal of Cancer*, **12,** 646–53.

BLALOCK, J. E. & STANTON, J. D. (1980). Common pathways of interferon and hormonal action. *Nature*, **283,** 406–8.

BOURGEADE, M. F. & CHANY, C. (1976). Inhibition of interferon action by cytochalasin B, colchicine, and vinblastine. *Proceedings of the Society of Experimental Biology and Medicine*, **153,** 501–4.

BOURGEADE, M. F. & CHANY, C. (1979). Effect of sodium butyrate on the antiviral and anti-cellular action of interferon on normal and MSV-transformed cells. *International Journal of Cancer*, **24,** 314–18.

BOURGEADE, M. F., ROUSSET, S., PAULIN, D. & CHANY, C. (1981). Reorganization of the cytoskeleton by interferon in MSV-transformed cells. *Journal of Interferon Research*, **1,** 323–7.

BOURNE, H. R. & ROZENGURT, E. (1976). An 18,000 molecular weight polypeptide induces early events and stimulates DNA synthesis in cultured cells. *Proceedings of the National Academy of Sciences of the USA*, **73,** 4555–9.

BRADLEY, E. C. & RUSCETTI, F. W. (1981). Effect of fibroblast, lymphoid and myeloid interferons on human tumour colony formation *in vitro*. *Cancer Research*, **41,** 244–9.

BRANCA, A. A. & BAGLIONI, C. (1981). Evidence that types I and II interferons have different receptors. *Nature*, **294,** 768–70.

BROUTY-BOYE, D., CHENG, Y. S. E. & CHEN, L. B. (1981). Association of phenotypic reversion of transformed cells induced by interferon with morphological and biochemical changes in the cytoskeleton. *Cancer Research*, **41,** 4174–84.

BROUTY-BOYE, D. & TOVEY, M. G. (1978). Inhibition by interferon of thymidine uptake in chemostat cultures of L1210 cells. *Interviology*, **9,** 243–52.

BROUTY-BOYE, D. & ZETTER, B. R. (1980). Inhibition of cell motility by interferon. *Science*, **208,** 516–18.

CALVET, M.-C., & GRESSER, I. (1979). Interferon enhances the excitability of cultured neurones. *Nature*, **278,** 558–60.

CHANDRABOSE, K., CUATRECASAS, P. & POTTATHIL, R. (1981). Changes in fatty acyl chains of phospholipids induced by interferon in mouse sarcoma S-180 cells. *Biochemical and Biophysical Research Communications*, **98,** 661–8.

CHANG, E. H., MIMS, S. J., TRICHE, T. J. & FRIEDMAN, R. M. (1977). Interferon inhibits mouse leukaemia virus release: an electron microscope study. *Journal of General Virology*, **34,** 363–7.

CHANG, E. H., JAY, F. T. & FRIEDMAN, R. M. (1978). Physical, morphological and biochemical alterations in the membrane of AKR mouse cells after interferon treatment. *Proceedings of the National Academy of Sciences of the USA*, **75,** 1859–63.

CHANY, C., GREGOIRE, A., VIGNAL, M., LEMAITRE-MONCUIT, J., BROWN, P., BESANCON, F., SUAREZ, H. & CASSINGENA, R. (1973). Mechanism of interferon uptake in parental and somatic monkey-mouse hybrid cells. *Proceedings of the National Academy of Sciences of the USA*, **70,** 557–61.

CHANY, C., MATHIEU, D. & GREGOIRE, A. (1980). Induction of delta (4)3 ketosteroid synthesis by interferon in some adrenal tumor cell cultures. *Journal of General Virology*, **50,** 447–50.

CHANY, C., VIGNAL, M., COUILLIN, P., CONG, N. V., BOUE, J. & BOUE, A. (1975). Chromosomal localization of human genes governing the interferon-induced antiviral state. *Proceedings of the National Academy of Sciences of the USA*, **72,** 3129–33.

CLARKE, J. L. (1974). Specific induction of ornithine decarboxylase in 3T3 mouse fibroblasts by putuitary growth factors. Cell density dependent biphasic response and alteration of half life. *Biochemistry*, **13,** 4668–74.

CLEMENS, M. J. & WILLIAMS, B. R. G. (1978). Inhibition of cell-free protein synthesis by pppA2′A2′p5′A: a novel oligonucleotide synthesized by interferon-treated cell extracts. *Cell*, **13,** 565–72.

COLLYN D'HOOGHE, M. C., BROUTY-BOYE, D., MALAISE, E. P. & GRESSER, I. (1977). Interferon and cell division. XII. Prolongation by interferon of the intermitotic time of mouse mammary tumour cells *in vitro*. Microcinematographic analysis. *Experimental Cell Research*, **105,** 73–5.

CONTENT, J., JOHNSTON, M. I., DE WIT, L., DE MAEYER-GUIGNARD, J. & DE CLERQ, E. (1980). Kinetics and distribution of interferon mRNA in interferon-primed and unprimed mouse L-929 cells. *Biochemical and Biophysical Research Communications*, **96,** 415–24.

CONTENT, J. & VERHAEGEN-LEWALLE, M. (1981). Molecular basis of interferon action. In *Intracellular Signals, Control of Growth and Differentiation*, ed. J. E. Dumont, pp. 275–98. London: Plenum Press.

CREASEY, A. A., BARTHOLOMEW, J. C. & MERIGAN, T. C. (1980). Role of G0–G1 arrest in the inhibition of tumour cell growth by interferon. *Proceedings of the National Academy of Sciences of the USA*, **77,** 1471–5.

CZARNIECKI, W., SREEVALSAN, T., FRIEDMAN, R. M. & PANET, A. (1981). Dissociation of interferon effects on murine leukemia virus and EMC virus replication in mouse cells. *Journal of Virology*, **37,** 827–31.

DE CLERCQ, E., EDY, V. G. & CASSIMAN, J. J. (1975). Non-antiviral activities of interferon are not controlled by chromosome 21. *Nature*, **256,** 132–4.

DE LARCO, J. E. & TODARO, G. J. (1978). Growth factors for murine sarcoma virus-transformed cells. *Proceedings of the National Academy of Sciences of the USA*, **75,** 4001–5.

De Maeyer, E. & De Maeyer-Guignard, J. (1977). Effect of interferon on cell-mediated immunity. The interferon system: a current review to 1978. *Texas Reports on Biology and Medicine*, **35,** 370–4.

De Maeyer-Guignard, J., Cachard, A. & De Maeyer, E. (1980). Electrophoretically pure mouse interferon has priming but no blocking activity in poly(IC) induced cells. *Virology*, **102,** 222–5.

Dicker, P., Pohjanpelto, P., Pettican, P. & Rozengurt, E. (1981). Similarities between fibroblast-derived growth factors and platelet derived growth factors. *Experimental Cell Research*, **135,** 221–7.

Dolei, A., Ameglio, F., Capobianchi, M. R. & Tosi, R. (1981). Human β-type interferon enhances the expression and shedding of Ia-like antigens. Comparison to HLA-A,B,C and β2-microglobulin. *Antiviral Research* **1,** 367–81.

Donahoe, R. M. & Huang, K.-Y. (1976). Interferon preparations enhance phagocytosis *in vivo. Infection and Immunology*, **13,** 1250–6.

Edy, V. G., Billiau, A. & De Somer, P. (1978). Non-appearance of injected fibroblast interferon in the circulation. *Lancet*, **i,** 451–2.

Ehnholm, C., Aho, K., Huttumen, J., Kostiainen, E., Mattila, K., Pikkarainen, J. & Cantell, K. (1982). Effect of interferon on plasma lipoproteins and on the activity of postheparin plasma lipases. *Atherosclerosis* (in press).

Epstein, D. A., Czarniecki, C. W., Jacobsen, H., Friedman, R. M. & Panet, A. (1981). A mouse cell line, which is unprotected by interferon against lytic virus infection, lacks ribonuclease F activity. *European Journal of Biochemistry*, **118,** 9–15.

Epstein, L B., Lee, S. H. S. & Epstein, C. J. (1980). Enhanced sensitivity of Trisomy 21 monocytes to the maturation-inhibiting effect of interferon. *Cell Immunology*, **50,** 191–4.

Evinger, M., Maeda, S. & Pestka, S. (1981b). Recombinant human leucocyte interferon produced in bacteria has antiproliferative activity. *Journal of Biology and Chemistry*, **256,** 2113–14.

Evinger, M., Rubinstein, M. & Pestka, S. (1981a). Antiproliferative and antiviral activities of human leucocyte interferon. *Archives of Biochemistry and Biophysics*, **210,** 319–29.

Fantes, K. H. & Allen, G. (1981). Specific activity of pure human interferons and a non-biological method for estimating the purity of highly purified interferon preparations. *Journal of Interferon Research*, **1,** 465–74.

Farrell, P. J., Sen, G. C., Dubois, M. F., Ratner, L., Slattery, E. & Lengyel, P. (1978). Interferon action: two distinct pathways for inhibition of protein synthesis by double-stranded RNA. *Proceedings of the National Academy of Sciences of the USA*, **75,** 5893–7.

Fellous, M., Kamoun, M., Gresser, I. & Bond, R. (1979). Enhanced expression of HLA antigens and β2-microglobulin on interferon-treated human lymphoid cells. *European Journal of Immunology*, **9,** 446–9.

Fleischman, W. R., Kleyn, K. M. & Bron, S. (1980). Potentiation of antitumour effect of virus induced interferon by mouse immune interferon. *Journal of the National Cancer Institute*, **65,** 963–6.

Fozard, J. R., Part, M.-L., Prakash, N. J., Grove, J., Schechter, P. J., Sjoerdamsa, A. & Koch-Weser, J. (1980). L-Ornithine decarboxylase: an essential role in early mammalian embryogenesis. *Science*, **208,** 505–8.

Frayssinet, C., Gresser, I., Tovey, M. & Lindahl, P. (1973). Inhibitory effect of potent interferon preparations on the regeneration of mouse liver after partial hepatectomy. *Nature*, **245,** 146–7.

Fridman, W. H., Gresser, I., Bandu, M. T., Aguet, M. & Neaupor Sautes, C. (1980). Interferon enhances the expression of Fcγ receptors. *Journal of Immunology*, **124,** 2436–43.

FRIEDMAN, R. M., KOHN, L. D., LEE, G., EPSTEIN, D. & JACOBSEN, H. (1981). Stimulatory activity by mouse interferon on the uptake of iodide by rat thyroid cells. In *The Biology of the Interferon System*, ed. De Maeyer, G. Galasso, & H. Schellekens, pp. 109–14. London, Amsterdam: Elsevier/North-Holland Biomedical Press.

FRIEDMAN, R. M. & PASTAN, I. (1969). Monophosphate: potentiation of antiviral activity. *Biochemical and Biophysical Research Communications*, **36,** 735–40.

FRIEDMAN, R. M. & RAMSEUR, J. M. (1974). Inhibition of murine leukemia virus production in chronically infected AKR cells: A novel effect of interferon. *Proceedings of the National Academy of Sciences of the USA*, **71,** 3542–4.

FRIEDMAN, R. M. & SONNABEND, J. A. (1964). Inhibition of interferon action by p-fluoro-phenylalanine. *Nature*, **203,** 366–8.

FUJITA, T., SAITO, S. & KOHNO, S. (1979). Priming increases the amount of interferon mRNA in poly(rI) poly(rc)-treated L cells. *Journal of General Virology*, **45,** 301–6.

FUJITA, T. & KOHNO, S. (1981). Studies on interferon priming: cellular responses to viral and non-viral inducers and requirements of protein synthesis. *Virology*, **112,** 62–9.

GEWERT, D. R., SHAH, S. & CLEMENS, M. J. (1981). Inhibition of cell division by interferons. Changes in the transport and intracellular metabolism of thymidine in human lymphoblastoid (Daudi) cells. *European Journal of Biochemistry*, **116,** 487–92.

GISLER, R. H., LINDAHL, P. & GRESSER, I. (1974). Effects of interferons on antibody synthesis *in vitro*. *Journal of Immunology*, **113,** 438–44.

GOEDDEL, D. V., YELVERTON, E., ULLRICH A., HEYNEKER, H. L., MIOZZARI, E., HOLMES, W., SEEBURG, P. H., DULL, T., MAY, L., STEBBING, N., CREA, R., MAEDA, S., MCCANDISS, R., SLOMA, A., TABOR, J. M., GROSS, M., FAMILLETTI, P. C. & PESTKA, S. (1980). Human leucocyte interferon produced by *E. coli* is biologically active. *Nature*, **290,** 20–6.

GREEN, H. (1978). Cyclic AMP in relation to proliferation. *Cell*, **15,** 801–11.

GREENBERG, P. & MOSNEY, S. (1977). Cytotoxic effects of interferon *in vitro* on granulocytic progenitor cells. *Cancer Research*, **37,** 1794–9.

GRESSER, I. (1977). On the varied biological effects of interferon. *Cell Immunology*, **34,** 406–15.

GRESSER, I. & BOURALI-MAURY, C. (1973). The antitumour effect of interferon in lymphocyte and macrophage-depressed mice. *Proceedings of the Society of Experimental Biology and Medicine*, **44,** 896–900.

GRESSER, I., DE MAEYER-GUIGNARD, J., TOVEY, M. G. & DE MAEYER, E. (1979). Electrophoretically pure mouse interferon exerts multiple biologic effects. *Proceedings of the National Academy of Sciences of the USA*, **76,** 5308–12.

GRESSER, I., MAURY, C. & BROUTY-BOYE, D. (1972). Mechanism of antitumour effect of interferon in mice. *Nature*, **239,** 167–8.

GRESSER, I. & TOVEY, M. G. (1978). Antitumour effects of interferon. *Biochimica et Biophysica Acta*, **516,** 231–47.

GROLLMAN, E. F., LEE, G., RAMOS, S., LAZO, P. S., KABACK, R., FRIEDMAN, R. & KOHN, L. D. (1978). Relationships of the structure and function of the interferon receptor to hormone receptors and establishment of the antiviral state. *Cancer Research*, **38,** 4172–85.

GUTTERMAN, J. U., FEIN, S., QUESADA, J., HORNING, S. J., LEVINE, J. L., ALEXANIAN, R., BERNHARDT, L., KRAMER, M., SPEIGEL, H., COLBURN, W., TROWN, P., MERIGAN, T. & DZIEWANOWSKA, Z. (1981). Recombinant human leucocyte interferon (IFLrA): a clinical study of pharmacokinetics, single dose tolerance and biologic effects in cancer patients. *Annals of Internal Medicine*, Dec. 1981.

HERON, I., BERG, K. & CANTELL, K. (1976). Regulatory effect of interferon on T cells *in vitro. Journal of Immunology*, **117**, 1370–3.

HOCHMAN, J., INSEL, P. A., BOURNE, H. R., COFFINO, P. & TOMKINS, G. M. (1975). A structural gene mutation affecting the regulatory subunit of cyclic AMP-dependent protein kinase in mouse lymphoma cells. *Proceedings of the National Academy of Sciences of the USA*, **72**, 5051–5.

HUET, C., GRESSER, I., BANDU, M. T. & LINDAHL, T. (1974). Increased binding of concanavalin A to interferon-treated murine leukemia L1210 cells. *Proceedings of the Society of Experimental Biology and Medicine*, **147**, 52–7.

IDA, S., HOOKS, J. J., SIRAGANIAN, R. P. & NOTKINS, A. B. (1977). Enhancement of IgE-mediated histamine release from human basophils by viruses: role of interferon. *Journal of Experimental Medicine*, **145**, 892–906.

ILLINGER, D., COUPIN, G., RICHARDS, M. & POINDRON, P. (1976). Rat interferon inhibits steroid-inducible glycerol 3-phosphate dehydrogenase in a rat glial cell line. *FEBS Letters*, **64**, 391–5.

IMAI, K., NG, A.-K., GLASSY, M. C. & FERRONE, S. (1981). Differential effect of interferon on the expression of tumor-associated antigens and histocompatibility antigens on human melanoma cells: relationship to susceptibility to immune lysis mediated by monoclonal antibodies. *Journal of Immunology*, **127**, 505–9.

INSEL, P. A. & FENNO, J. (1978). Cyclic AMP-dependent protein kinase mediates a cyclic AMP-stimulated decrease in ornithine and S-adenosylmethionine decarboxylase activities. *Proceedings of the National Academy of Sciences of the USA*, **75**, 862–5.

JANNE, J., POSO, H. & RAINE, A. (1978). Polyamines in rapid growth. *Biochimica et Biophysica Acta*, **473**, 241–93.

KAUPILLA, A. C., ANTELL, K., JANNE, O., KOKKO, E. & VIHKO, R. (1982). Serum sex steroid and peptide hormone concentrations, and endometrial oestrogen and progestin receptor levels during administration of human leukocyte interferon. *International Journal of Cancer* (in press).

KEAY, S. & GROSSBERG, S. E. (1980). Interferon inhibits the conversion of 3T3-L1 mouse fibroblasts into adipocytes. *Proceedings of the National Academy of Sciences of the USA*, **77**, 4099–103.

KERR, I. M. & BROWN, R. E. (1978). pppA2′p5′A2′p5′A: an inhibitor of protein synthesis synthesized with an enzyme fraction from interferon-treated cells. *Proceedings of the National Academy of Sciences of the USA*, **75**, 256–60.

KILLANDER, D., LINDHAL, P., LUNDIN, L., LEARY, P. & GRESSER, I. (1976). Relationship between the enhanced expression of histocompatibility L1210 antigens on interferon-treated L1210 cells and their position in the cell cycle. *European Journal of Immunology*, **6**, 56–9.

KIMCHI, A., SHURE, H. & REVEL, M. (1979). Regulation of lymphocyte mitogenesis by (2′–5′) oligo-isoadenylate. *Nature*, **282**, 849–51.

KNIGHT, E. (1976). Antiviral and cell growth inhibitory activities reside in the same glycoprotein of human fibroblast interferon. *Nature*, **262**, 302–3.

KNIGHT, E. & KORANT, B. D. (1977). A cell surface alteration in mouse L cells induced by interferon. *Biochemical and Biophysical Research Communications*, **74**, 707–13.

KNIGHT, E. & KORANT, B. D. (1978). Fibroblast interferon induces synthesis of four proteins in human fibroblast cells. *Proceedings of the National Academy of Sciences of the USA*, **76**, 1824–7.

KOHN, L. D., FRIEDMAN, R. M., HOLMES, J. M. & LEE, G. (1976). Use of thyotropin and cholera toxin to probe the mechanism by which interferon initiates its antiviral activity. *Proceedings of the National Academy of Sciences of the USA*, **73**, 3695–9.

LASFARGUES, E. Y. & OZELLO, L. (1958). Cultivation of human breast carcinomas. *Journal of the National Cancer Institute*, **21**, 1131–47.

LEBLEU, B., SEN, G. C., SHAILA, S., CABRER, B. & LENGYEL, P. (1976). Interferon, double-stranded RNA, and protein phosphorylation. *Proceedings of the National Academy of Sciences of the USA*, **73**, 3107–11.

LEE, S. H. S. & EPSTEIN, L. B. (1980). Reversible inhibition by interferon of the maturation of human peripheral blood monocytes to macrophages. *Cell Immunology*, **50**, 177–90.

LIAO, S.-K., KWONG, P. C., KHOSRAVI, M. & DENT, D. P. (1982) Enhanced expression of melanoma-associated antigens and B2-microglobulin on cultured human melanoma cells by interferon. *Journal of the National Cancer Institute*, **68**, 19–25.

LINDAHL, P., LEARY, P. & GRESSER, I. (1972). Enhancement by interferon of the specific cytotoxicity of sensitized lymphocytes. *Proceedings of the National Academy of Sciences of the USA*, **69**, 721–5.

LINDAHL, P., LEARY, P. & GRESSER, I. (1973). Enhancement by interferon of the expression of surface antigens on murine leukemia L1210 cells. *Proceedings of the National Academy of Sciences of the USA*, **70**, 2785–8.

LINDAHL, P., LEARY, P. & GRESSER, I. (1974). Enhancement of the expression of histocompatibility antigens of mouse lymphoid cells by interferon *in vitro*. *European Journal of Immunology*, **4**, 779–84.

LINDAHL-MAGNUSSON, P., LEARY, P. & GRESSER, I. (1972). Interferon inhibits DNA synthesis induced in mouse lymphocyte suspensions by phytohaemagglutinin or by allogeneic cells. *Nature New Biology*, **237**, 120–5.

LOTEM, J. & SACHS, L. (1978). Genetic dissociation of different cellular effects of interferon on myeloid leukemic cells. *International Journal of Cancer*, **22**, 214–20.

LUCERO, M. A., MAGDELENAT, H., FRIDMAN, W. H., POUILLART, P., BILLARDON, C., BILLIAU, A., CANTELL, K. & FALCOFF, E. (1982). Pharmacological properties of human alpha and beta interferons. *European Journal of Cancer* (in press).

LUNDBLAD, D. & LUNDGREN, E. (1981). Block of a glioma cell line in S by interferon. *International Journal of Cancer*, **27**, 749–54.

MAHESHWARI, R. K., DEMSEY, A. E., MOHANTY, S. B. & FRIEDMAN, R. M. (1980). Interferon-treated cells release vesicular stomatitis virus particles lacking glycoprotein spikes: correlation with biochemical data. *Proceedings of the National Academy of Sciences of the USA*, **44**, 2284–7.

MAMONT, P. S., BOHLEN, P., MCCANN, P. P., BEY, P., SCHUBER, F. & TARDIF, C. (1976). α-Methyl ornithine, a potent competitive inhibitor of ornithine decarboxylase, blocks proliferation of rat hepatoma. *Proceedings of the National Academy of Sciences of the USA*, **73**, 1626–30.

MAMONT, P. S., DUCHESNE, M.-C., GROVE, J. & BEY, P. (1978). Antiproliferative properties of DL-α-difluoromethyl ornithine in cultured cells. A consequence of the irreversible inhibition of ornithine decarboxylase. *Biochemical and Biophysical Research Communications*, **81**, 58–66.

MASUCCI, M., SZIGETI, R., KLEIN, E., KLEIN, G., GRUEST, J., MONTAGNIER, L., TAIRA, H., HALL, H., NAGATA, S. & WEISSMANN, C. (1980). Effect of interferon-αI from *E. coli* on some cell functions. *Science*, **209**, 1431–8.

MATARESE, G. P. & ROSSI, G. B. (1977). Effect of interferon on growth and division cycle of Friend erythroleukaemic murine cells *in vitro*. *Journal of Cell Biology*, **75**, 344–54.

MATSUNO, G. & SHIRASAWA, N. (1978). Interferon suppresses steroid-inducible glutamine synthetase biosynthesis in embryonic chick neural retina. *Biochimica et Biophysica Acta*, **538**, 188–94.

MATSUYAMA, M. (1979). Action of interferon on cell membrane of mouse lymphocytes. *Experimental Cell Research*, **124**, 253–9.

NEBERT, D. W. & FRIEDMAN, R. M. (1973). Stimulation of aryl hydrocarbon hydroxylase induction in cell cultures by interferon. *Journal of Virology*, **11,** 193–7.

OLESZAK, E. & INGLOT, A. D. (1980). Platelet derived growth factors (PDGF) inhibits antiviral and anticellular action of interferon in synchronised mouse or human cells. *Journal of Interferon Research*, **1,** 37–48.

O'NEILL, C. H., RIDDLE, P. N. & JORDON, P. W. (1979). The relationship between surface area and anchorage dependence of growth in hamster and mouse fibroblasts. *Cell*, **16,** 909–18.

PANNIERS, L. R. V. & CLEMENS, M. J. (1981). Inhibition of cell division by interferon: changes in cell cycle characteristics and in morphology of Erlich ascites tumour cells in culture. *Journal of Cell Science*, **48,** 259–79.

PAUCKER, K., CANTELL, K. & HENLE, W. (1962). Quantitative studies on viral interference in suspended L-cells. III. Effect of interfering viruses and interferon on the growth rate of cells. *Virology*, **17,** 324–34.

PAUCKER, K., DALTON, B. J., OGBURN, C. A. & TORMA, E. (1975). Multiple active sites on human interferons. *Proceedings of the National Academy of Sciences of the USA*, **72,** 4587–91.

PFEFFER, L. M., LANDSBERGER, F. R. & TAMM, I. (1981). Beta interferon induced time-dependent changes in the plasma membrane lipid-bilayer of cultured cells. *Journal of Interferon Research*, **1,** 613–19.

PFEFFER, L. M., MURPHY, J. S. & TAMM, I. (1979). Interferon effects on the growth and division of human fibroblasts. *Experimental Cell Research*, **121,** 111–20.

PFEFFER, L. M., WANG, E. & TAMM, I. (1980a). Interferon effects on microfilament organisation, cellular fibronectin distribution and cell motility in human fibroblasts. *Journal of Cell Biology*, **85,** 9–17.

PFEFFER, L. M., WANG, E. & TAMM, I. (1980b). Interferon inhibits the redistribution of cell surface components. *Journal of Experimental Medicine*, **152,** 469–74.

PITHA, P. M., ROWE, W. P. & OXMAN, M. N. (1976). Effect of interferon on exogenous, endogenous and chronic murine leukemia virus infection. *Virology*, **70,** 324–38.

PRIESTMAN, T. J. (1980). Initial evaluation of human lymphoblastoid interferon in patients with advanced malignant disease. *Lancet*, **ii,** 112–18.

RAFFE, M. C., ABNEY, E., BROCKER, J. P. & HORNBY-SMITH, A. (1978). Schwann cell growth factors. *Cell*, **15,** 813–22.

RAINE, A. & JANNE, J. (1975). Physiology of the natural polyamines putrescine, spermidine and spermine. *Medical Biology*, **53,** 123–47.

REVEL, M., BASH, D. & RUDDLE, F. H. (1976). Antibodies to cell-surface component coded by human chromosome 21 inhibit action of interferon. *Nature*, **260,** 139–41.

REVEL, M., FELLOUS, M., NIR, U., WALLACH, D., WOLF, D., MERLIN, G. & KIMCHI, A. (1982). Synthesis and action of interferon related to the control of cell functions. *Journal of Cell Biochemistry*, Suppl. 6, Abstract, p. 83.

ROBERTS, W. K., HOVANESSIAN, A., BROWN, R. E., CLEMENS, M. J. & KERR, I. M. (1976). Interferon-mediated protein kinase and low-molecular-weight inhibitor of protein synthesis. *Nature*, **264,** 477–80.

ROSSI, G. B., DOLEI, A., CIOE, L., BENEDETTO, A., MATARESE, G. P. & BELARDELLI, F. (1977a). Inhibition of transcription and translation of globin messenger RNA in dimethyl sulfoxide-stimulated Friend erythroleukemic cells treated with interferon. *Proceedings of the National Academy of Sciences of the USA*, **74,** 2036–40.

ROSSI, G. B., MATARESE, G. P., GRAPELLI, C., BELARDELLI, F. & BENEDETTO, A. (1977b). Interferon inhibits dimethyl sulphoxide-induced erythroid differentiation of Friend leukaemia cells. *Nature*, **267**, 50–2.

ROZEE, K. R., KATZ, L. J. & MCFARLANE, E. S. (1969). *Canadian Journal of Microbiology*, **15**, 969–74.

ROZENGURT, E. (1980). Stimulation of DNA synthesis in quiescent cultured cells: exogeneous agents, internal signals and early events. *Current Topics in Cellular Regulation*, **17**, 59–88.

ROZENGURT, E., LEGG, A., STRANG, G. & COURTENAY-LUCK, N. (1981). Cyclic AMP: a mitogenic signal for Swiss 3T3 cells. *Proceedings of the National Academy of Sciences of the USA*, **78**, 4392–6.

RUBIN, B. Y. & GUPTA, S. L. (1980). Differential efficacies of human type I and type II interferons as antiviral and antiproliferative agents. *Proceedings of the National Academy of Sciences of the USA*, **77**, 5928–32.

SCHMIDT, A., ZILBERSTEIN, A., SCHULMAN, L., FEDERMAN, P., BERISSI, H. & REVEL, M. (1978). Interferon action: isolation of nuclease F, a translational inhibitor activated by interferon-induced (2′–5′) oligo-isoadenylate. *FEBS Letters*, **95**, 257–64.

SCHNECK, J., RAGER-ZISMAN, B., ROSEN, O. M. & BLUM, B. R. (1982). Genetic analysis of the role of cAMP in mediating effects of interferon. *Proceedings of the National Academy of Sciences of the USA*, **79**, 1879–83.

SCHRODER, E. W., CHOU, I. N., JAKEN, S. & BLACK, P. H. (1978). Interferon inhibits the release of plasminogen activator from SV3T3 cells. *Nature*, **276**, 828–9.

SEHGAL, P. (1982). The interferon genes. *BBA*, **695**, 17–33.

SEKAR, V., ATMAR, V. J., KRIM, M. & KUEHN, G. D. (1982). Interferon induction of polyamine dependent protein kinase activity in Ehrlich ascites tumour cells. *Biochemical and Biophysical Research Communications*, **106**, 305–11.

SHIBATA, H. & TAYLOR-PAPADIMITRIOU, J. (1981). Effects of human lymphoblastoid interferon on cultured breast cancer cells. *International Journal of Cancer*, **38**, 447–53.

SHIRASAWA, N. & MATSUNO, T. (1979). Suppression of the accumulation of steroid-inducible glutamine synthetase mRNA on embryonic chick retinal polysomes by interferon preparation. *Biochimica et Biophysica Acta*, **562**, 271–80.

SHIMIZU, N. & SOKAWA, Y. (1979). 2′,5′-oligoadenylate synthetase activity in lymphocytes from normal mouse. *Journal of Biological Chemistry*, **254**, 12034–7.

SILVERMAN, R. H., WATLING, D., BALKWILL, F. R., TROWSDALE, J. & KERR, I. M. (1982). The ppp(A2′p)$_n$A and protein kinase systems in wild-type and interferon-resistant Daudi cells. *European Journal of Biochemistry*, **126**, 333–41.

SLATER, D. L., SHULMAN, L., LAWRENCE, J. B., REVEL, M. & RUDDLE, F. H. (1978). Presence of human chromosome 21 alone is sufficient for hybrid cell sensitivity to human interferon. *Journal of Virology*, **25**, 319–25.

SOKAWA, Y., WATANABE, Y., WATANABE, Y. & KAWADE, Y. (1977). Interferon suppresses the transition of quiescent 3T3 cells to a growing state. *Nature*, **268**, 236–8.

SREEVALSAN, T., ROZENGURT, E., TAYLOR-PAPADIMITRIOU, J. & BURCHELL, J. (1980). Differential effect of interferon on DNA synthesis, 2-deoxyglucose uptake and ornithine decarboxylase activity in 3T3 cells stimulated by polypeptide growth factors and tumour promoters. *Journal of Cellular Physiology*, **104**, 1–9.

SREEVALSAN, T., TAYLOR-PAPADIMITRIOU, J. & ROZENGURT, E. (1979). Selective inhibition by interferon of serum-stimulated biochemical events in 3T3 cells. *Biochemical and Biophysical Research Communications*, **87**, 679–85.

STEWART, II, W. E., DECLERQ, E., BILIAU, A., DESMYTER, J. & DESOMER, P. (1972). Increased susceptibility of cells treated with interferon to the toxicity of polyriboinosinic polyribocytidylic acid. *Proceedings of the National Academy of Sciences of the USA*, **69**, 1851–4.

STEWART, W. E., GOSSER, L. B., & LOCKHART, R. Z. (1971). Priming. A nonantiviral function of interferon. *Journal of Virology*, **7**, 792–801.

STEWART, II, W. E., SARKAR, F., TAIRA, H., HALL, A., NAGATA, S. & WEISSMANN, C. (1980). Comparisons of several biological and physicochemical properties of human leukocyte interferons produced by human leukocytes and by *E. coli*. *Gene*, **II**, 181–6.

STRANDER, H. & EINHORN, S. (1977). Effect of human leucocyte interferon on the growth of human osteosarcoma cells in tissue culture. *International Journal of Cancer*, **19**, 468–73.

STREULI, M., HALL, A., BOLL, W., STEWART, II, W., NAGATA, S. & WEISSMANN, C. (1981). Target cell specificity of two species of human interferon-α produced in *E. coli* and hybrid molecules derived from them. *Proceedings of the National Academy of Sciences of the USA*, **78**, 2848–52.

TAN, Y. H. (1976). Chromosome 21 and the cell growth inhibitory effect of human interferon preparations. *Nature*, **260**, 141–3.

TAN, Y H., TISCHFIELD, J. & RUDDLE, F. H. (1973). The linkage of genes for the human interferon-induced antiviral protein and indophenol oxidase-B traits to chromosome G-21. *Journal of Experimental Medicine*, **137**, 317–30.

TAYLOR, J. (1964). Inhibition of interferon action by actinomycin. *Biochemical and Biophysical Research Communications*, **14**, 447–51.

TAYLOR-PAPADIMITRIOU, J. (1980). Effects of interferon on cell growth and function. In *Interferon 2, 1980*, ed. I. Gresser, pp. 13–46. London: Academic Press.

TAYLOR-PAPADIMITRIOU, J., PURKIS, P. & FENTIMAN, I. (1980). Cholera toxin and analogs of cyclic AMP stimulate the growth of cultured human mammary epithelial cells. *Journal of Cell Physiology*, **102**, 317–22.

TAYLOR-PAPADIMITRIOU, J. & ROZENGURT, E. (1982). Modulation of interferons' inhibitory effect on cell growth by the mitogenic stimulus. *Texas Reports on Biology and Medicine, 'The Interferon System: a Review to 1982'* (in press).

TAYLOR-PAPADIMITRIOU, J., SHEARER, M., BALKWILL, F. R. & FANTES, K. H. (1982). Effects of HuIFN-α2 and HuIFN-α (Namalwa) on breast cancer cells grown in culture and as xenografts in the nude mouse. *Journal of Interferon Research* **2(4)**, 479–91.

TAYLOR-PAPADIMITRIOU, J., SHEARER, M. & ROZENGURT, E. R. (1981). Inhibitory effect of interferon on cellular DNA synthesis: modulation by pure mitogenic factors. *Journal of Interferon Research*, **1**(3), 401–9.

TODARO, G. J., FRYLING, C. & DE LARCO, J. E. (1980). Transforming growth factors produced by certain human tumour cells: polypeptides that interact with epidermal growth factor. *Proceedings of the National Academy of Sciences of the USA*, **77**, 5258–62.

TOMIDA, M., YAMAMOTO, Y. & HOZUMI, M. (1980). Stimulation by interferon of induction of differentiation of mouse myeloid leukemic cells. *Cancer Research*, **40**, 2919–24.

TRINCHIERI, G. & SANTOLI, D. (1978). Antiviral activity induced by culturing lymphocytes with tumor-derived or virus-transformed cells. Enhancement of human natural killer cell activity by interferon and antagonistic inhibition of susceptibility of target cells to lysis. *Journal of Experimental Medicine*, **147**, 1314–33.

TYRRELL, D., SCOTT, G., SECHER, D. & CANTELL, K. (1981). An interim report of the effects of highly purified interferon α from human leucocytes. *Abstract of the Second Annual International Interferon Conference.* San Francisco, USA.

VANDENBUSSCHE, P., DIVIZIA, M., VERHAEGEN-LEWALLE, M., FUSE, A., KUWATA, T., DE CLERCQ, E. & CONTENT, J. (1981). Enzymatic activities induced by interferon in human fibroblast cell lines differing in their sensitivity to the anticellular activity of interferon. *Virology*, **111,** 11–22.

VAN'T HULL, E. V., SCHELLEKENS, H., LOWENBERG, B. & DE VRIES, M. (1978). Influence of interferon preparations on the proliferative capacity of human and mouse bone marrow cells *in vitro*. *Cancer Research*, **38,** 911–14.

VENGRIS, V. E., REYNOLDS, F. H., HOLLENBERG, M. D. & PITHA, P. (1976). Interferon action: role of membrane gangliosides. *Virology,* **72,** 486–93.

WANG, E., MICHL, J., PFEFFER, L. M., SILVERSTEIN, S. C. & TAMM, I. (1981a). Interferon stimulates phagocytosis but suppresses pinocytosis in mouse macrophages; related changes in cytoskeletal organization. *Abstract of the Second International Congress for Interferon Research.* San Francisco, USA.

WANG, E., PFEFFER, L. M. & TAMM, I. (1981b). Interferon increases the abundance of submembraneous microfilaments in HeLa-S3 cells in suspension culture. *Proceedings of the National Academy of Sciences of the USA*, **78,** 6281–5.

WECK, P. K., APPERSON, S., MAY, L. & STEBBING, N. (1981). Comparison of the antiviral activities of various cloned human interferon-α subtypes in mammalian cell cultures. *Journal of General Virology*, **57,** 233–7.

YARON, M., YARON, I., GURANI-ROTMAN, D., REVEL, M., LINDER, H. R., & ZOR, U. (1977). Stimulation of prostaglandin E production in cultured human fibroblasts by poly(I).poly(C) and human interferon. *Nature*, **267,** 457–9.

YAU, P. M. P., GODEFROY-COLBURN, T., BIRGE, C. H., RAMABHADRAN, T. U. & THACH, R. E. (1978). Specificity of interferon action in protein synthesis. *Journal of Virology*, **27,** 648–58.

ZILBERSTEIN, A., FEDERMAN, P., SHULMAN, L. & REVEL, M. (1976). Specific phosphorylation *in vitro* of a protein associated with ribosomes of interferon-treated mouse L cells. *FEBS Letters*, **68,** 119–24.

INTERFERON AND THE IMMUNE SYSTEM 1: INDUCTION OF INTERFERON BY STIMULATION OF THE IMMUNE SYSTEM

M. WILKINSON AND A. G. MORRIS

Department of Biological Sciences, University of Warwick, Coventry CV4 7AL, UK

INTRODUCTION

Interferon (IFN) was first detected by virtue of its antiviral effect and produced experimentally by treatment of chick cells with influenza virus. It was originally described as 'the interferon', responsible for the phenomenon of viral interference (the prevention of replication of one virus by another). The research of the 25 years succeeding the publication of Isaacs & Lindenmann's original observations on 'the interferon' (1957) has demonstrated that IFN has a much wider significance than being simply 'the interferon'. Its antiviral effect, and of course, its production by virus infection, imply a role in resistance to virus disease. It is fairly clear that IFN is indeed important in resistance to virus disease; the best evidence for this comes from the demonstration that anti-IFN globulins given to a virus-infected animal exacerbate the course of the disease (Gresser *et al.*, 1979) and the observation that some children particularly susceptible to upper respiratory tract infections seem to have a defect in IFN production (Isaacs *et al.*, 1981). We now know that in addition to its direct antiviral effect IFN exerts other important biological effects, including inhibition of growth rate of at least some cell types, and effects on some immune responses. These facts imply that IFN has roles other than in virus diseases; the nature and variety of IFN's biological effects suggest it may play a central role in resistance to other diseases – bacterial, parasitic, malignant. There is, as yet, no direct evidence for this role, but the demonstration that there are several ways in which cells may be induced to produce IFN by mechanisms not requiring virus infection strengthens this suggestion. The mechanisms of IFN production that do not require virus infection involve stimulation of cells of the immune system, and may be divided into three broad classes: (1) stimulation by non-specific agents such as mitogens and *Corynebacterium parvum*; (2) stimulation by specific antigens (including virus antigens); (3) stimulation by tumour cells.

These three classes of 'immune induction' mechanisms, distinguished on an operational basis, may, in fact, be closely related in terms of molecular mechanism. The common denominator may be the interaction of the inducing agent (mitogen, antigen, tumour cell) with membrane receptors of cell types (lymphocytes, macrophages) which are programmed to thereupon produce IFN. Stimulation by mitogens and antigens may be by essentially the same mechanism; broadly similar cell types are involved, and in both cases IFN-gamma is the main IFN type which is produced. On the other hand, induction by tumour cells may be distinct, since clearly different cell types are involved and IFN-alpha appears to be the main product.

INDUCTION OF IFN BY NON-SPECIFIC STIMULATION OF THE IMMUNE SYSTEM

An immune response to an antigen results from specific recognition of that antigen by cells of the immune system. Recognition involves interaction of the antigen with clonally-distributed specific receptors on the surface of lymphoid cells: the type of response generated depends on both the cell types involved and the biochemical nature of the antigen, but is characterized by its specificity, e.g. the antibodies that are generated recognize that antigen but not unrelated antigens, and 'memory', i.e. a second exposure to the same antigen, results in a more vigorous and usually quicker, response.

Cells of the immune system may also be stimulated in a non-antigen-specific way by a variety of agents, of which the most important class is the mitogens. This type of response involves interaction of mitogens with cell surface receptors probably distinct from antigen-specific receptors. It is characterized by its antigen non-specific nature (e.g. antibodies recognizing a wide range of antigens may be generated) and its lack of memory, i.e. a second exposure to mitogen does not result in an increased level of response.

Both antigens and mitogens provoke similar metabolic changes in lymphocytes: the chief difference is the number of cells which respond. Mitogens will stimulate up to 80–90% of a given class of lymphocytes, whereas antigen stimulates only the small number of cells bearing specific receptors, usually of the order of 0.01–0.1% of the total. The cells respond by transforming into lymphoblasts: lymphocytes which were previously metabolically inactive start synthesizing RNA, proteins and DNA at a greatly increased rate; they enlarge and usually divide. This process may be conveniently followed by labelling with precursors of macromolecular synthesis, or simply by microscopic observation of the enlarged cells (lymphoblasts).

Mitogens in common use are listed in Table 1. They show varying degrees of specificity for different subpopulations of lymphocytes. Thus, monoclonal antibodies to T-lymphocytes are strictly specific for these cells; concanavalin A (Con A), and phytohaemagglutinin (PHA) stimulate more or less exclusively T-lymphocytes; LPS more or less exclusively B lymphocytes; whilst PWM activates both T- and B-lymphocytes (mostly B). All these mitogens induce IFN in cultures of fresh unfractionated mononuclear leucocytes.

A number of other agents, which are not primarily considered as mitogens, also may stimulate cells of the immune system to produce IFN. These agents, listed in Table 1, may be mitogenic for lymphocytes but have other important effects, for example, *Corynebacterium parvum* acts as a mitogen for human T-cells (Sugiyama & Epstein, 1978) but also activates macrophages.

Indeed, there is a very wide range of non-viral inducers of IFN which may act by stimulating the immune system. Thus, fungi, many different bacteria and protozoa have been reported as inducing IFN (see Stewart, 1979, for a list of references). However, it is unclear whether this induction is due to stimulation of the immune system by these agents. For example, fungi often harbour double-stranded RNA viruses which are potent inducers of interferon in a wide range of cells types. In addition the immune status of the experimental animals is often uncertain, particularly in the case of common pathogenic organisms.

In the following sections, we will discuss only those agents which appear to involve cells of the immune system in IFN induction, and will therefore not discuss, for example, protozoa, for which there is no clear evidence of such an involvement.

Induction of IFN by mitogens

Mitogen induction of IFN was first noted by Wheelock in 1965 who demonstrated the presence of an IFN-like virus inhibitor in supernatants from cultures of PHA-treated mouse spleen cells. Gamma

Table 1. *Agents inducing interferon by non-specific stimulation of the immune system*

Agent	Source	Main effects on immune system	Induction of IFN
Con A (concanavalin A)	*Canavalia ensiformis*	T-cell mitogen	Wallen *et al.*, 1973
PHA (phytohaemagglutinin)	*Phaseolus vulgaris*	T-cell mitogen	Wheelock, 1965
PWM (poke-weed mitogen)	*Phytolacca americana*	T- and B-cell	Friedman & Cooper, 1967
SEA (staphylococcal enterotoxin A)	*Staphylococcus aureus*	T(?)-cell mitogen	Johnson *et al.*, 1977
Staph A protein	*S. aureus*	T- and B-cell mitogen	McCool *et al.*, 1981
LPS (lipopolysaccharide 'endotoxin')	various bacteria	B-cell mitogen: activator of macrophages	Ho, 1964
Corynebacterium parvum		activator of macrophages	Sugiyama & Epstein, 1978
ALS (antilymphocyte serum)		T-cell mitogen	Falcoff *et al.*, 1972
Monoclonal T-cell antibodies		T-cell mitogen	Pang *et al.*, 1981

IFN is produced by treating cultures of leucocytes by any of the mitogens listed in Table 1. Titres from such cultures vary widely with the conditions of culture, the anatomical source of the leucocytes, the animal strain (Tyring & Lefkowitz, 1980) or individual human donor used, and the nature and concentration of the mitogen used. Certain tumour promoters (phorbol esters, teleocidins) which themselves induce small amounts of IFN can act synergistically with mitogens, increasing yields of IFN-gamma up to 30-fold (Yip, Pang, Urban & Vilček, 1981a; Yip *et al.*, 1981b; M. Wilkinson and A. G. Morris, unpublished). The thymic hormone, thymosin, has also been reported to increase IFN-gamma yields (Gray *et al.*, 1982; Huang, Kind, Jagoda & Goldstein, 1981; Shoham, Eshel, Aboud & Salzberg, 1980) from leucocytes treated with mitogens. In our laboratory, peak yields of IFN of about 10^4 units/ml are obtained three or four days after treating fresh, Ficoll-purified mononuclear leucocytes from human peripheral blood with a combination of SEA and the tumour-promoter mezerein. Similar results have been reported by other laboratories using similar induction procedures

(De Ley *et al.*, 1980; Yip *et al.*, 1981a). Cultures of fresh mouse spleen leucocytes treated with mitogens produce generally rather lower titres of IFN-gamma, and mouse peripheral blood leucocytes are no more productive than spleen leucocytes (C. Sutton and A. G. Morris, unpublished). In the murine system, tumour promoters seem not to increase IFN yields (C. Sutton and A. G. Morris, unpublished). However, there have been recent reports of techniques for obtaining high yields of IFN from murine leucocytes. Thus, Fuse, Heremans, Opdenakker & Billiau (1982) indicated that titres of the order of 10^4 could be obtained from spleen cell cultures treated with crude lysates of *Staphylococcus aureus* culture and Marcucci, Nowak, Krammer & Kirchner (1982) have shown that similar titres could be obtained by a second Con A stimulation of murine spleen cells that had been grown for a short period *in vitro* after a primary stimulation with Con A.

IFN-gamma is the major IFN-type produced when leucocyte cultures are treated with T-cell mitogens (Osborne, Georgiades & Johnson, 1980; Langford *et al.*, 1981a). However, other types of IFN may also be produced. It appears, for example, that at least in the mouse quite large amounts of IFN-alpha/beta are produced when spleen cells are treated with LPS (Kato *et al.*, 1980; Ascher, Apte & Pluznik, 1981; Maehara & Ho, 1977). There are reports that human leucocytes produce trace amounts of IFN-alpha in response to SEA, PHA or PWM (Wiranowska-Stewart, 1981) or IFN-beta in response to Con A (Van Damme *et al.*, 1982).

Recently we have obtained evidence for the production of an antigenically distinct type of IFN (IFN-delta?) produced by human peripheral blood mononuclear leucocytes treated with PHA and the tumour promoter, teleocidin. This IFN is not neutralized by monospecific antibodies which neutralize IFNs-alpha, -beta or -gamma and has physicochemical and biological properties quite different from those of IFN-alpha, -beta or -gamma (Wilkinson & Morris, 1982).

Cellular source of IFN produced in response to mitogens

Leucocytes are a complex mixture of many different cell types – polymorphonuclear leucocytes, monocytes/macrophages, T, B and null lymphocytes. Much effort has been expended in determining which particular subpopulations of leucocytes are involved in mitogen-induced IFN production. Two general techniques have been used to determine the cellular source of IFN. Different subpopulations of leucocytes have been obtained by the use of various fractionation techniques and the ability of each fraction to produce IFN determined. Alternatively, it is possible to grow and clone certain leucocytes *in vitro* – specifically, T-lymphocytes – and the production of IFN by these cloned cells can be determined and thus correlated with cell type. This latter technique has shown that T-lymphocytes can produce IFN-gamma, at least in response to Con A and PHA (Klein, Raulet, Pasternak & Bevan, 1982; Marcucci, Waller, Kirchner & Krammer, 1981; Matsuyama, Sugamura, Kawade & Hinuma, 1982); but, since at present, T-cells are the only type of non-tumour leucocyte to be propagated and cloned over extended periods *in vitro*, it gives no information on the involvement of other cell types. For this, experiments involving fractionation techniques are essential. Unfortunately, such techniques very rarely give clear-cut results, for a number of reasons including cross-contamination between fractions. Also, subpopulations of leucocytes may overlap; thus, in the human system, T-cells as defined by the presence of the receptor for sheep red blood cells do not exactly coincide with those defined by the 'pan-T' monoclonal antibody OKT-3, and the relationship of either of these with 'thymus processed cells' is not clear. In addition, subpopulations are often defined by negative selection: for example, T-cells may be removed from mononuclear leucocyte cultures by a number of techniques but the remaining non-T cells are by no means homogenous B-cells as has on occasion been assumed. Other, minor, cell types in a fraction of predominantly one cell type may be responsible for responses observed.

Nevertheless, cell fractionation studies have suggested that IFN production in response to mitogens usually requires co-operation between two cell types, lymphocytes and macrophages with no involvement of the other major class of leucocyte, the polymorphonuclear leucocytes. Several groups have shown that purified lymphocyte or macrophage cultures produce little or no IFN when treated with mitogens, whereas recombination of the two cell types results in IFN production (Arbeit, Leary & Levin, 1982; Epstein, Cline & Merigan, 1971a). It seems however that in some situations, macrophages may not be necessary (McCool, Catalona, Langford & Ratcliff, 1981; Sugiyama & Epstein, 1978) and in other cases, may inhibit IFN production (Arbeit *et al.*, 1982).

It has not been definitively shown which of these two cell types is the producer of IFN. The assumption is that the macrophage acts as an accessory, 'helper' cell for IFN production by the T-lymphocyte, based on the known accessory role of macrophages in other immune responses. Certainly, some cloned T-lymphocytes produce IFN in the absence of macrophages (Klein *et al.*, 1982; Marcucci *et al.*, 1981), and Epstein *et al.* (1971a) showed that freeze/thawing the lymphocytes affected IFN production to a much greater extent than did freeze/thawing macrophage cultures prepared by mixing together purified macrophages and lymphocytes. One report, by Neumann & Sorg (1977) indicated that IFN-free supernatants from purified, Con A-stimulated mouse lymphocytes could stimulate IFN production in purified mouse macrophages, suggesting that lymphocytes 'helped' macrophages to produce IFN, but in similar experiments by Epstein *et al.* (1971a), supernatants from PHA-treated human lymphocytes did not induce IFN in human macrophages.

The analysis of the interaction of macrophages with lymphocytes is complicated by the existence of numerous subpopulations of lymphocytes. The precise subpopulation(s) involved in IFN production apparently depends on the mitogen used and the anatomical source of the lymphocytes, but all three major lymphocyte classes – T, B and null cells – may be important.

Thus, Wallen, Dean & Lucas (1973) showed that treatment of mouse spleen cells with antiThy 1 antibody plus complement prevented IFN production in response to PHA and Con A, implying involvement of T-cells, but that the response to pokeweed mitogen (PWM) was much less affected, implying a role for non-T cells. Stobo, Green, Jackson & Baron (1974) using similar techniques also showed that T-cells were involved in PHA-stimulated IFN production.

Experiments with *nude* mice have suggested non-T lymphocytes also produce IFN. Homozygous *nu/nu* mice lack a normal thymus, and consequently lack *mature* T-cells and show few of the cellular immune responses – such as graft rejection – associated with mature T-cells. Spleen cells from *nu/nu* mice do, nevertheless, efficiently produce IFN in response to PHA (Wietzerbin *et al.*, 1978; Kirchner *et al.*, 1980) but not in response to Con A (Kirchner *et al.*, 1980). However, *nu/nu* mice do have a small number of lymphocytes which express the Thy 1 antigen (associated with T-cells) (Raff, 1973) and which may represent an immature 'T-cell' population able to support IFN production in response to PHA but not Con A. Human thymocytes, which are also immature 'T-cells' have been shown to produce IFN in response to mitogens (Reem, Cook, Henriksen & Vilček 1982). Epstein, Kreth & Herzenberg (1974), using fluorescence activated cell sorting to fractionate human lymphocytes, found that both T- and also B-cells produce IFN in response to PHA and PWM; Klimpel *et al.* (1977) using agglutination to sheep red blood cells to fractionate human lymphocytes into T-cells (E^+) and non-T-cells (E^-) also found both classes of cell produced IFN in response to PHA. On the other hand, Wiranowska-Stewart & Stewart (1981) found that human E^- lymphocytes produced little IFN in response to a number of mitogens (namely PHA, SEA, PWM). Other evidence for production of IFN-gamma by non-T-cells includes the finding of Epstein & Salmon (1974) that peripheral lymphocytes from patients with myeloma (a B-cell neoplasm) produce IFN. Kirchner *et al.* (1979a) demonstrated production of IFN-gamma by non-T, non-B cells upon stimulation with PHA and PWM. In this laboratory we have found that low-density lymphocytes from Percoll fractionation of human peripheral mononuclear cells lacking high-affinity sheep red blood cells, which are the cell type responsible for mediation of natural killer (NK) cell activity, produce large amounts of IFN in response to SEA. Ratliff, McCool & Catalona (1981) reported that a cell type which also has the surface characteristics of NK cells produced IFN in response to *Staphylococcus aureus* A protein.

Lymphoid cells from a variety of anatomical sites have all been shown to produce IFN upon mitogen stimulation, but the characteristics of IFN production vary with the source of cells. In addition to peripheral blood leucocytes both human tonsillar and thoracic duct lymphocytes produce IFN in response to PHA or SEA (Klimpel, Day & Lucas, 1975; Langford *et al.*, 1981b; Sugiyama, Yamamoto, Kinoshita & Kimura, 1977). Murine peripheral blood leucocytes produce IFN-gamma in similar amounts to spleen cells (C. Sutton and A. G. Morris, unpublished observations) and murine thymocytes can under certain circumstances produce IFN-gamma (Reem *et al.*, 1982). However, in one study reported by Neumann & Sorg (1978) murine lymph node cells did not produce IFN in response to Con A.

Since T-lymphocytes themselves are an heterogeneous population of cells, there have been attempts to determine which T-cell subsets are responsible for IFN-gamma production. Again, both fractionation techniques and T-cell cloning have been employed.

Different classes of T-cell may be distinguished by a number of criteria. Brodeur, Weinstein, Melmon & Merigan (1977) fractionated murine T-cells using adherence to a histamine column; adherent cells produced more IFN in response to PHA than did non-adherent. This IFN-producing T-cell population appeared to have suppressive activity. Archer, Smith, Ulrich & Johnson (1979) used the classification of Raff & Cantor (1971) who divide murine T-cells into T1 cells, which are short-lived, non-recirculating, and T2 cells which are longer-lived and which recirculate. Both T1 and T2 cells produce IFN in response to mitogens, although there are differences in the pattern of response to different mitogens (PHA, Con A and SEA). Epstein & Gupta (1981) fractionated human T-cells into those bearing receptors for the Fc portion of IgG (T gamma), IgM (T mu) or neither (T phi). All three classes of T-cell produced approximately equal amounts of IFN in response to PHA in the presence but not in the absence of macrophages.

The availability of antibodies (monoclonal or polyclonal) reactive with specific subsets of leucocyte has provided a powerful tool for fractionating T-cells and other leucocytes. A report by Chang *et al.* (1982) who used monoclonal antibodies to fractionate human peripheral blood mononuclear leucocytes indicated that $OKT4^+$ cells in combination with $OKM1^+$ cells produced IFN-gamma in response to mitogenic stimulation with the pan T monoclonal antibody OKT3. These cells correspond to the classic T-cell functional subset 'helper/inducer' ($OKT4^+$) and monocytes-macrophages ($OKM1^+$). $OKT8^+$ cells ('suppressor/cytotoxic') did not produce IFN-gamma, alone or in combination with $OKM1^+$, despite responding to $OKT3^+$ mitogenesis as strongly as $OKT4^+$ cells.

Similar experiments have been carried out by O'Malley *et al.* (1982). These authors fractionated human peripheral lymphocytes into sheep red blood cell rosetting T-cells after depleting phagocytic cells (monocytes and macrophages). The T-cells were then fractionated into T gamma, T mu and T phi (cf Epstein & Gupta, above) and $OKT4^+$, $OKT8^+$, $OKM1^+$, $OKT11a^+$ (cf. Chang *et al.*, above). Cell fractions were induced with PHA. It was found that IFN gamma was produced by T-cells characterized as E^+, Fc $gamma^+$, Fc mu^-, $OKM1^+$, $OKT4^-$, $OKT8^-$, $OKT11a^+$. It is not clear what may be the function of such a cell type, which appears to share markers of both T-cells (E^+, $OKT11a^+$) and monocytes ($OKM1^+$), yet lacks the functional T-cell subset markers, $OKT4^+/8^+$. The authors carried out no functional tests with these cells; it would be interesting for example, to determine whether they have natural killer cell activity, since it appears that at least some NK cells bear the E, Fc gamma and OKM1 markers. O'Malley *et al.* did not study the effect of adding back monocytes to their T-cell fractions and so their data do not exclude the likely possibility that other T-cell subsets may produce IFN-gamma in the presence of monocytes (cf. Chang *et al.*). The data are also in apparent conflict with the results of Epstein & Gupta, who found that monocytes augmented IFN-gamma production by T gamma cells; however, in their experiments, there was residual IFN production by T gamma cells in the absence of monocytes. This may perhaps be ascribed to monocyte-independent IFN production by a subset of T gamma cells as described by O'Malley *et al.* The difference in the results of O'Malley *et al.* and Chang *et al.* may also be ascribable to the different IFN inducers used.

In the mouse system, analogous studies have been carried out by Torres, Farrar & Johnson (1982). These authors treated mouse spleen cells with complement and monoclonal antibodies to Lyt 1 or Lyt 2 antigens. These treatments destroy classical 'helper' and 'suppressor/cytotoxic' T-cells respectively, with Lyt 1 and Lyt 2 being at least roughly equivalent to human OKT4 and OKT8. In both cases, the residual cells were unable to produce IFN-gamma. Recombination of the two cell fractions restored IFN-gamma production implying a role for *both* T-cell classes. This finding is in conflict with the results of Chang *et al.* (1982, discussed above) who found that *human* 'helper' T-cells were not necessary for IFN production. The requirement for Lyt 1 cells could be replaced by the addition of interleukin 2 (IL2) preparations. The authors conclude that the function of the Lyt 1^+ cells is to produce IL2 which is necessary for IFN production by Lyt 2^+ cells.

The experiments of Marcucci *et al.* (1982), referred to earlier, indicated that T-cell clones produce IFN-gamma in response to mitogens. However, the T-cell clones these authors employed had no known functional properties and lacked Lyt antigens so it was impossible to assign them to any T-cell subset. Klein *et al.* (1982) have however demonstrated IFN-gamma production by a clone of alloreactive cytotoxic T-cells. Clones of T-cells cytotoxic for influenza virus-infected cells also produce IFN-gamma upon mitogen stimulation (A. G. Morris & B. Askonas, unpublished). Krammer *et al.* (1982) report that most clones of Con A induced cytotoxic T-cells examined by limiting dilution analysis produced IFN, but that most non-cytotoxic clones did not. In these experiments, the functional phenotype of the non-cytotoxic cells was not defined.

Matsuyama *et al.* (1982) have studied Con A induced IFN-gamma production in a series of 39 human T-cell clones. These clones were divided into two broad groups: those with cytotoxic activity and those without. The cytotoxic T-cells were further distinguished on the basis of their target specificity: some killed homologous Epstein Barr Virus infected lymphoblastoid cells (used to stimulate growth of these T-cells), others killed the NK target K562 and a third group killed both. Some, but not all, clones from each group, produced IFN-gamma upon Con A stimulation. It is interesting to note that one clone with the target specificity of NK cells produced IFN: this has also been reported by Kedar *et al.* (1982) who found that murine NK-like cell lines produce IFN upon lectin stimulation.

Certain human and murine T-lymphoblastoid (tumour) cell lines produce IFN-gamma, either spontaneously (Morris, Morser & Meager, 1982; Ponzio, Fitzgerald, Vilček & Thorbecke, 1980) or upon mitogen stimulation (Landolfo, Arnold & Suzan, 1982a; Nathan *et al.*, 1981). Landolfo, Kirchner & Simon (1982b) report that 2 murine T-lymphoma lines bearing surface markers Thy 1^+, Lyt 1^+ and lacking Lyt 2, 3, or Ig produce IFN upon treatment with Con A and PHA. These cells may belong to the helper T-cell subset which is known to produce many other lymphokines such as IL-2, IL-3 and MIF.

To summarize, these data indicate that IFN is produced by various T-cell subsets and interaction between T-cell subsets may be important. It appears that cultured T-cells make larger amounts of IFN than do fresh leucocytes. This finding is not understood; is it due to changes in the differentiation state of the cells, loss of suppressor cells, or simply to enrichment for the producer cell type? Bearing on this last point, Wiranowska-Stewart & Stewart (1981) using a single cell assay technique involving overlaying IFN-producing cells onto a monolayer of indicator cells subsequently challenged with a virus, found that less than 1 in 10^4 *fresh* murine spleen cells produced IFN, whilst Krammer *et al.* (1982) using a limiting dilution technique found that roughly one in two one-week-old T-cell *clones* produced IFN. Yet the technique of Wiranowska-Stewart & Stewart may have been relatively insensitive and underestimated the proportion of IFN-producing cells.

Another point of difference between fresh and growing T-cells is that the latter very clearly do not require macrophages for IFN production. This may also reflect differences in differentiation state: and it may also be that factors in the conditioned medium used for T-cell growth can substitute for some macrophage functions.

A further (and final) complication of the cell biology of mitogen induction of IFN-gamma is that certain T-cells may suppress IFN-gamma production. This was reported by Johnson (1981) who found that murine spleen cells treated *in vivo* or *in vitro* with mitogen 24 h previously will inhibit IFN production by other spleen cells. The finding that suppression is reversed by IL2 suggests that suppressor cells may simply act by soaking up IL2.

To summarize these rather disparate data, the consensus is that the T-lymphocyte is an important cell type involved in mitogen-stimulated IFN-gamma production and is probably the major producing cell. This may in part be due to the preponderant use of mitogens specific for T-cells and the fact that T-cells are a large proportion of the total leucocytes. However, as will be discussed below, T-cells are also important in antigenic induction of IFN-gamma; and since T-cells play a major role in other immune responses, it is not unreasonable that the T-cell should be considered as central to IFN-gamma production. However, other cell types clearly are involved.

Induction of IFN by non-specific agents other than mitogens

Corynebacterium parvum and LPS (endotoxin) are two IFN inducers which seem to be somewhat ambiguous in their action. Both agents may act as lymphocyte mitogens but are generally considered also to stimulate the immune system *via* direct effects on macrophages. Both appear to induce mostly IFN alpha/beta, produced most probably by macrophages with no lymphocyte involvement, and may therefore act in a manner quite distinct from lymphocyte mitogens.

Neumann, Macher & Sorg (1980) have obtained good evidence that mouse macrophages produce IFN alpha/beta upon treatment with *Corynebacterium parvum.* Prior sensitization of the donor was not necessary for IFN production.

Kirchner and co-workers (Kirchner *et al.*, 1979a; Peter *et al.*, 1980) have shown that in human leucocyte cultures, the null lymphocyte or a monocyte precursor is probably the major producer of IFN (type-alpha). On the other hand, Sugiyama & Epstein (1978) find that in human leucocytes, the T-cell with no macrophage requirement is the main producer of IFN (type-gamma) but these authors caution that since human donors are often sensitized to *C. parvum* this induction may be immune specific. Vilček, Sulea, Volvovitz & Yip (1980) very clearly demonstrated that the product of human lymphocytes stimulated by *C. parvum* was IFN-alpha.

Similarly, there is a good evidence that macrophages produce IFN alpha/beta in response to stimulation with LPS (Fleit & Rabinowitch, 1980; Maehara & Ho, 1977; Neumann & Sorg, 1981).

INDUCTION OF IFN BY ANTIGEN-SPECIFIC STIMULATION OF THE IMMUNE SYSTEMS

As previously mentioned, primary immune responses to most conventional antigens generally involve only very small proportions of the lymphocyte pool, and at least partly for this reason, are undetectable *in vitro*, and thus IFN production in primary immune responses cannot easily be analysed. However, due to the 'memory' of the immune system, secondary responses are more vigorous, and the amplification means that such secondary responses can be detected *in vitro*. Generally IFN is produced when lymphoid cells from a donor sensitized to a particular antigen are re-exposed to that antigen in culture.

However, the exception to the above is that primary responses to allogeneic or some hapten-modified syngeneic major histocompatibility complex (MHC) antigens are easily detected *in vitro*. The immune system 'sees' antigen together with products of genes encoded within the MHC. In 'natural' responses, virus or other intracellular infectious agents are detected by presentation of their proteins with MHC products on the infected cell surface; whereas soluble antigens are 'processed' by macrophages, and similarly 'presented' associated with MHC products on the macrophage surface. Thus, in this type of response, the immune system recognizes 'self + foreign' or 'altered self'. Responses to allogeneic MHC products are extremely strong; this may possibly be due to cross-reactions of allo-MHC with a wide range of 'altered-self' specificities already in immune memory. IFN is produced in both primary and secondary responses to allogeneic cells in a mixed lymphocyte culture, and to trinitrophenyl (TNP)-modified syngeneic cells (M. Cooley, in preparation).

Because of the differences between responses to foreign MHC

antigens and other, 'conventional' antigens, we will consider them separately.

Production of IFN in response to conventional antigens

Early observations which strongly hinted at immune-specific induction of IFN were published by Glasgow (1966). He demonstrated that the yield of IFN from murine leucocytes challenged *in vitro* with Chikungunya virus depended on the immune status of the donors; if the mice were immune to the virus, very much more IFN was produced than if the mice were not immune. This response was specific in that the yield of IFN from the leucocytes upon challenge with other viruses was at the same low level irrespective of whether or not the donor was immune to Chikungunya. A large number of experiments have been carried out confirming induction of IFN by antigens of bacterial or viral origins in immune animals or humans. Generally, IFN-gamma is produced (Langford *et al.*, 1981a; Osborne *et al.*, 1980; Valle, Jordan, Haahr & Merigan, 1975b), although some early reports indicated that the IFN produced was sometimes acid stable, i.e. IFN-alpha/beta. In the case of induction with virus 'antigens', the antigen preparations used have often simply been killed virus preparations; since 'killed' viruses may induce IFN-alpha/beta in non-lymphoid cells by classical viral induction mechanisms, care must be exercised in the interpretation of such experiments (see for example Starr *et al.*, 1980).

Thus, Green, Cooperband & Kibrick (1969) and Green, Cooperband, Kleinman & Kibrick (1970) demonstrated that leucocytes from humans immune to tetanus or diphtheria toxoid responded *in vitro* in a specific manner to these antigens by producing IFN. This IFN was apparently relatively acid stable. Leucocytes from non-immune donors produced no IFN. Similar results were obtained by Epstein, Cline & Merigan (1971b) using purified protein derivative (PPD) from tuberculin as the antigen: leucocytes from immune donors produce IFN when treated with PPD. In this case, the IFN produced was acid labile (i.e. IFN-gamma). Similar results have been obtained in studies of IFN production by PPD-immune mice (Kato *et al.*, 1980; Milstone & Waksman, 1970; Salvin, Youngner & Leder, 1973; Salvin, Ribi, Granger & Youngner, 1975; Sonnenfeld, Mandel & Merigan, 1979; Stinebring & Absher, 1970; Youngner & Salvin, 1973).

In addition to the studies mentioned above with Chikungunya

virus, antigens of herpes simples, cytomegalo, influenza and vaccinia viruses all induce IFN in leucocytes from immune donors (Babiuk & Rowse, 1976; Ennis & Meager, 1981; Fujibayashi, Hooks & Notkins, 1975; Haahr, Rasmussen & Merigan, 1976; Kirchner, Zawatzky & Schirrmacher, 1978; Lodmell & Notkins, 1974; Rasmussen & Merigan, 1978; Rasmussen, Jordan, Stevens & Merigan, 1974; Starr *et al.*, 1980; Valle, Bobrove, Strober & Merigan, 1975a).

Cellular source of IFN produced in response to conventional antigens

As with mitogen induction, it appears that T-lymphocytes and macrophages are both involved in IFN-gamma production in response to antigens in fresh leucocyte cultures from immune individuals (Babiuk & Rowse, 1976; Epstein *et al.*, 1971b, 1972; Rasmussen *et al.*, 1974; Sonnenfeld *et al.*, 1979; Valle *et al.*, 1975a). There have, however, been reports that B-cells may also be involved (Rasmussen & Merigan, 1978; Sonnenfeld *et al.*, 1979). Once again, it is not possible to determine which of these interacting cell types actually produces the IFN-gamma; the T-cell is assumed to be the producer cell, but it is formally possible that macrophages produce IFN-gamma with 'help' from T-cells. As will be discussed below, growing T-cells will produce IFN when antigenically stimulated in the absence of macrophages, which suggests that in fresh cultures the T-cell also produces IFN.

Studies have been carried out in the murine (but not as yet human) system to determine which particular class of T-cell produce the IFN-gamma in response to stimulation with antigens.

Depletion techniques employing antibodies specific for T-cell subsets were used by Sonnenfeld *et al.* (1979) and the findings were that the main IFN producer cell in response to PPD was the Lyt 2^+ (suppressor/cytotoxic) T-cell, in agreement with the studies with mitogens described earlier (though Lyt 1^+ helper cells were apparently not required). The study of antigen induction of IFN in T-cell clones has confirmed that Lyt 2^+ T-cells with cytotoxic properties produce IFN-gamma upon antigenic stimulation. Morris *et al.* (1982) showed that a cloned T-cell line of Balb/c origin Thy 1^+, Lyt 1^- and Lyt 2^+ phenotype produced IFN-gamma when cultured with its specific target, influenza-infected P815 cells. The production of IFN-gamma showed specificity for the influenza antigenic type and H-2 restriction in the same way as the killing did. In other words the production of IFN was immune specific. As with mitogen induction of T-cell clones, macrophages were not necessary for IFN production. However, free virus did *not* induce IFN, presumably indicating that the antigen has to be 'seen' by the T-cells in the context of self-histocompatibility antigens. These may presumably be provided either by the target cell in the case of cytotoxic T-cell clones or by macrophages in the case of induction of fresh lymphocytes. In similar studies by McKimm-Breschkin, Mottram, Thomas & Miller (1982), two antigen specific T-cell lines produced IFN in response to treatment with the immunizing antigen. The antigen used was the hapten oxazalone conjugated to the surface of lymphocytes. IFN induction was antigen-specific and H2 restricted. Clones from these lines also produced IFN-gamma. No data were provided concerning surface markers of these lines, but both lines and IFN-producing clones from them mediated delayed type hypersensitivity *in vivo* and it was suggested that they belonged to the Lyt 1^+ helper T-cell subset.

There is therefore evidence for IFN-gamma production by both classical T-cell subsets (suppressor/cytotoxic + helper) in response to antigenic stimulation.

Production of IFN in response to alloantigenic stimulation

IFN is produced in mixed lymphocyte cultures (MLC) between donors differing in their histocompatibility antigens (Gifford, Tibor & Peavy, 1981; Manger, Kalden, Zawatzky & Kirchner, 1981; Perussia, Mangoni, Engers & Trinchieri, 1980; Virelizier, Allison & De Maeyer, 1977). In the murine MLC, highest yields are obtained if there are differences in class II cell surface antigens, coded for by the I-S-G region of the mouse MHC, although IFN is also produced if there are differences in class I cell surface antigens, coded for by the K and D regions of the MHC, or differences in minor histocompatibility antigens (Kirchner *et al.*, 1979c; Landolfo, Giovarelli, Capusso & Forni, 1979a; Landolfo, Marcucci, Schirrmacher & Kirchner, 1981a; Landolfo *et al.*, 1981b). Similarly in the human system, IFN is produced if there are differences in any of the major histocompatibility antigens, but again most if there are differences in class II antigens coded for by the HLA-DR region of the human MHC (analogous to I-S-G region of mouse) (Andreotti & Cresswell, 1981). It is not clear whether these differences in responses to class I or class II molecules have any great significance: class II antigens are present mainly on lymphoid cells and appear to be important in interactions between lymphoid cells, whilst class I antigens present on all cells may be more important in recognition of non-lymphoid cells by lymphoid cells.

IFN may also be produced in response to alloantigens on non-lymphoid cells; however, in this case prior sensitization is necessary (Ito *et al.*, 1980a).

Although IFN-gamma is produced in primary MLC, it is produced more rapidly and in larger amounts when the donor of the responder lymphocytes is sensitized to the stimulator lymphocytes i.e. in secondary MLC (Landolfo *et al.*, 1982b; Perussia *et al.*, 1980).

As with conventional antigenic induction, the main IFN type produced in response to alloantigenic stimulation is IFN-gamma (Osborne *et al.*, 1980; Perussia *et al.*, 1980). The cell types involved in IFN-gamma production are again T-cells (Ito *et al.*, 1979, 1980b; Kirchner *et al.*, 1979b; Landolfo *et al.*, 1979a, 1981a; Perussia *et al.*, 1980) and probably macrophages (Farrar *et al.*, 1981; Ito *et al.*, 1979; Landolfo *et al.*, 1979a) although macrophages are not always required (Ito *et al.*, 1981). Studies by Landolfo *et al.* (1982b) indicated that in the murine primary MLC 2 T-cell subsets characterized as Lyt 1^+, Qat 5^+ and Lyt 123^+, Qat 5^+, produced IFN. These subsets belong to the classical 'helper' subset of T-cells, and T-cells of the 'suppressor/cytotoxic' subset did not produce IFN. However, in *secondary* MLC, an additional subset, characterized as Lyt 23^+, Qat 5^-, *with* cytotoxic capacity, did produce IFN-gamma.

Alloreactive cytotoxic T-cell clones also produce IFN when co-cultured with their appropriate target cell (Klein *et al.*, 1982).

Thus, T-cell subsets involved in IFN-gamma production in secondary MLC to some extent resemble those involved in induction by conventional antigens.

INTERFERON INDUCTION BY TUMOUR CELLS

When fresh leucocytes from non-sensitized donors are mixed *in vitro* with certain tumour cell lines (mixed lymphocyte-tumour cell culture, MLTC) IFN is produced. This response is at least superficially similar to IFN production in MLC. However, it differs in several important ways: (1) IFN is produced independent of histocompatibility differences between responder and stimulator cells – syngeneic tumour cells (Djeu, Huang & Herberman, 1980; Olstad,

Degre & Seljelid, 1981; Trinchieri, Santoli, Dee & Knowles, 1978) or cells lacking alloantigens (e.g. K562: Klein *et al.*, 1976) both induce IFN; (2) in the human system but probably not the mouse system the main product is IFN-alpha rather than IFN-gamma; (3) there is clear evidence that the producer cell is the non-T, non-B ('null') cell; (4) accessory cells (macrophages) are not required; and (5) at least in the human system, production of IFN in the MLTC is much faster than in the MLC.

IFN production in MLTC was first reported by Trinchieri, Santoli & Knowles (1977) and Trincheri *et al.* (1978). These authors demonstrated that IFN was produced by human or murine leucocytes in response to stimulation by a variety of tumour or virus-transformed cells of syngeneic, allogeneic or xenogeneic origin. In some MLTCs very large amounts of IFN were produced, comparable to amounts produced by virus infection and far more than is usually produced in MLC. Most of the IFN produced was probably IFN-alpha but there may have been also a small amount of IFN-gamma produced. The cell type producing IFN was shown to be a non-B, non-T lymphocyte. A number of other reports have appeared confirming these data (Birke *et al.*, 1981; Blalock, Langford, Georgiades & Stanton, 1979; Djeu *et al.*, 1980; Ito *et al.*, 1981; Olstad *et al.*, 1981; Peter *et al.*, 1980; Potter, Moore & Morris, 1982). Timonen, Saksela, Virtanen & Cantell (1980) very elegantly demonstrated that IFN-alpha was produced by non-B, non-T lymphocytes morphologically characterized as large granular lymphocytes (LGL). They used a fluorescein-labelled monoclonal antibody to IFN-alpha (Secher & Burke, 1980) to label human lymphocytes interacting with tumour cells; they found that only LGLs fluoresced. This finding is particularly important since LGLs are thought to mediate cytotoxicity of natural killer (NK) cells (Timonen, Ortaldo & Herberman, 1981) which is stimulated by IFN; this suggests a closed feedback loop for NK cells regulation by IFN (Reid *et al.*, 1981; Saksela, 1981).

However, other authors find evidence for the involvement of other cell types in IFN production in MLTC: Olstad *et al.* (1981) found that macrophages produced IFN-gamma in response to co-culture with syngeneic tumour cells, and Weigent *et al.* found human B lymphocytes to be the major IFN-producing cell type in MLTC.

Although these experimental results seem clear cut, there are two areas of contention which bear on the estimation of their significance. One is the suggestion that mycoplasma contamination of the IFN-inducing target cells is responsible for IFN induction (Beck, Engler, Brunner & Kirchner, 1980; Birke *et al.*, 1981). It is clear that mycoplasma do induce IFN (Cole, Overall, Lombardi & Glasgow, 1976) and that many tumour cell lines used in MLTC are contaminated with mycoplasma. It is however quite certain that at least some mycoplasma-free tumour cell lines do induce IFN (e.g. Birke *et al.*, 1981; Potter *et al.*, 1982). A possibly more serious concern is that there have been no published reports that cells from fresh explants of naturally-occurring primary tumours induce interferon in autologous MLTC; however, unpublished data of Saksela indicate that in at least some cases autologous combinations of tumour cells and LGL fractions from the patient's blood produced interferon (E. Saksela, 1982, personal communication). These experiments are clearly important in order to be sure that tumour cell induction of IFN is not an artefact of long term *in vitro* culture of tumour cell lines. On the other hand, Hahn & Levin (1982) have reported that more than half of a group of 38 patients with various malignant diseases had significant amounts of IFN in their sera, compared with about 2% of normal donors, suggesting that IFN indeed is produced *in vivo* in response to tumours.

MECHANISMS OF IMMUNE INDUCTION OF INTERFERON

Mechanisms may be considered at two levels: the cellular, and the molecular. For the former, we need to understand the cell types involved in induction, and the nature of their interactions. For the latter, we need to know the molecular events which occur between stimulation of the IFN-producing cell and the secretion of IFN.

The data we have discussed above indicate that there is a bewildering chaos of information about cellular mechanisms involved in IFN production. However some generalizations about mechanisms at the cellular level, which seem reasonable to us, are: (1) Upon stimulation of fresh leucocytes with mitogens, conventional antigens, or alloantigens, T-cells interacting with macrophages produce IFN-gamma; however, mature, growing T-cells produce IFN-gamma without the help of macrophages (or any other accessory cell). (2) Stimulation of macrophages with agents such as *C. parvum* or lipopolysaccharide (LPS) results in their producing IFN alpha and/or beta, without help from other cells. (3) Tumour cells stimulate null cells to produce IFN-alpha.

To date, most available information relates to (1) and this situation we will discuss in some detail. Aspects which seem particularly interesting to us are the relevance of the differentiated state of the T-cell, the role of different T-cell subsets, and the role of macrophages.

It seems that the differentiated state of the T-cell is very important in amounts of IFN-gamma produced. Immature T-cells, such as thymocytes, produce IFN-gamma with difficulty. Yields of IFN-gamma from thymocytes, however, may be considerably boosted by a number of treatments which may be considered to influence their differentiation state, including treatment with thymic hormones, tumour promoters, or conditioned medium from cultures of stimulated T-cells. Similarly, growing T-cells (either cloned or not) generally produce larger amounts of IFN-gamma than do fresh, non-growing T-cells which are in a quite different differentiation state. One thing which is clear is that growth *per se* is not the important factor, since growing T-cells do not constitutively produce IFN-gamma and in fresh leucocyte cultures it has been repeatedly shown using different inducers and cell types that IFN-gamma production is often independent of cellular proliferation. If one can think, somewhat naively, of a differentiation sequence:

$$\text{Thymocyte} \xrightarrow[\text{hormone}]{\text{thymic (etc)}} \text{mature peripheral T-lymphocyte}$$

$$\xrightarrow[\text{mitogenic stimulation}]{\text{Antigenic or}} \text{T-lymphoblast}$$

then it appears that at each step the capacity to produce IFN-gamma is increased. Why this should be poses a fascinating problem for molecular biologists.

That a variety of T-cell subtypes may produce interferon is abundantly evident. There are, as yet, insufficient data to allow one to generalize which T-cells are quantitatively most important in IFN-gamma production in, say, animals undergoing a virus infection. Perhaps all T-cells have the potential for producing IFN-gamma, depending on the nature of the stimulating signal. Nor do we have sufficient data to allow us to draw conclusions about interactive networks between different T cells and other accessory cells involved in IFN-gamma induction. Bearing on this point are the interesting observations of Torres *et al.* (1982), and Farrar *et al.* (1981), discussed above. These authors suggest that in fresh leucocyte cultures helper T-cells interacting with macrophages produce soluble factors especially IL2 which allow mitogen or antigen stimulated cytotoxic/suppressor T-cells to produce IFN-gamma. This model requires further testing, but is attractive in view of immunologists' current interest in interacting cellular networks. However in view of the fact that helper T-cells also produce IFN-gamma, this model is not complete.

Macrophages may play several roles in IFN-gamma induction. Two likely roles are as 'antigen- (or mitogen-) presenting cell' (by analogy with its proposed role in generation of antibody) and the production of soluble factors. That an antigen-presenting cell is necessary for IFN-gamma production is suggested by the finding that free antigen (influenza virus) will not induce IFN-gamma production in a T-cell clone (Morris *et al.*, 1982) – in this case, macrophages are not required; the antigen presenting cell is instead the influenza-infected target cell. There are, however, no clear-cut experiments showing that macrophages act thus in fresh leucocyte cultures, although there are data indicating that the physical presence of macrophages is necessary for IFN-gamma induction (Epstein *et al.*, 1971a). Macrophages are known to produce soluble factors – especially important is interleukin 1 (IL1) – which are important in T-cell functioning. Farrar *et al.* (1981) have suggested that IL1 produced by macrophages stimulates IL2 production which is required for IFN-gamma production.

Turning now to molecular mechanisms of IFN-gamma production, we find that we are very far from a complete description of how IFN-gamma is produced. It can be assumed that receptors on the cell surface are important in turning on IFN-gamma production, but the nature of these receptors is unknown. The finding that galactose oxidase will induce IFN (Dianzani, Monahon, Scupham & Zucca, 1979) and that enzymatic cleavage of *N*-acetylneuraminic acid residues abolishes IFN induction (Dianzani, Monahon & Santiano, 1982) suggests the (unsurprising) conclusion that glycoproteins are involved. Calcium fluxes may be an important second step, since a calcium ionophore induces IFN-gamma and calcium depletion inhibits induction (Dianzani, Monahon, Georgiades & Alperin, 1980; Dianzani *et al.*, 1982). Recent reports have indicated that mitogen induction of fresh leucocytes results in increases (from undetectable levels) of the mRNA for IFN-gamma. However, these studies have barely begun; it is not known, for example, whether IFN-gamma synthesis is actively 'switched off' as IFN-alpha or IFN-beta synthesis appears to be, at the level of mRNA concentration. Since the gene for Hu-IFN-gamma has recently been cloned, we may expect in the near future a flood of information about its control during mitogen induction.

RELEVANCE OF IMMUNE INDUCTION OF IFN

The potential relevance of antigenic induction of IFN is obvious in the case of virus disease: many viruses, especially DNA viruses, are not particularly good inducers of interferon by classical mechanisms and a second route for IFN induction is clearly of potential importance. However, there are no reports of IFN-gamma production *in vivo* during a virus infection and no evidence whatever that IFN-gamma plays a role in resistance to virus infection. In the case of non-viral antigens, such as bacterial antigens, the relevance of IFN induction is far less obvious. There is clear evidence for production of IFN-gamma *in vivo* in response to bacterial antigens. On the other hand, mechanisms for an antibacterial effect of IFN are not immediately obvious: perhaps IFN acts in an immune-regulatory action, activating general antibacterial mechanisms such as phagocytic cell systems. Similar difficulties arise when considering a physiological role for mitogen induction of IFN-gamma – products of many pathogenic micro-organisms are mitogenic (e.g. staphylococcal enterotoxin A (SEA)) and so mitogenic induction of IFN-gamma may occur quite frequently *in vivo* – but again, what IFN-gamma would actually do in such circumstances is not clear.

Paradoxically, the one situation where there is some direct evidence for a role for immune induction of IFN is in that most recently discovered – tumour cell induction of IFN-alpha production by null cells. Since IFN serves to activate the natural cytotoxicity of such cells, there is a potential feedback control system mediated by IFN (see p. 166). Evidence that this operates *in vivo* comes from the studies of Reid *et al.* (1981) who demonstrated enhancement of tumour growth when antiserum to IFN-alpha/beta was administered to tumour-bearing mice.

The chapter in this volume by M. Moore documents in detail the effects exogenous IFN is known to exert on immune responses studied *in vivo* and *in vitro*. Our chapter discusses the mass of data concerning production of IFN during immune responses. It is, we feel, very unlikely that IFN induction and action are of no relevance to immune responses, being simply an accidental byproduct of such responses. However, it does seem somewhat unsatisfactory that there is no wealth of experimentation demonstrating that IFN does actually play a significant role in resistance to disease: perhaps the next phase of interferon research should aim at understanding what is the relevance of IFN in resistance to disease.

The authors wish to acknowledge Dr Margaret Cooley's careful appraisal of this article, and are very grateful for her useful suggestions and constructive criticism.

REFERENCES

Andreotti, P. E. & Cresswell, P. (1981). HLA control of interferon production in the human mixed lymphocyte culture. *Human Immunology*, **3**, 109–20.

Arbeit, R., Leary, P. L. & Levin, M. J. (1982). Gamma interferon production by combinations of human peripheral blood lymphocytes, monocytes and cultured macrophages. *Infection and Immunity*, **35**, 383–90.

Archer, D. L., Smith, B. G., Ulrich, J. T. & Johnson, H. M. (1979). Immune interferon induction by mitogens involves different T-cell populations. *Cellular Immunology*, **48**, 420–6.

Ascher, O., Apte, R. & Pluznik, D. (1981). Generation of LPS-induced interferon in spleen cell cultures. I. Genetic analysis and cellular requirements. *Immunogenetics*, **12**, 117–27.

Babiuk, L. & Rowse, B. T. (1976). Immune interferon production by lymphoid cells: role in the inhibition of herpes viruses. *Infection and Immunity*, **13**, 1567–78.

Beck, J., Engler, H., Brunner, H. & Kirchner, H. (1980). Interferon production in cocultures between mouse spleen cells and tumour cells. Possible role of mycoplasmas in interferon induction. *Journal of Immunological Methods*, **38**, 63–73.

Birke, C., Peter, H. H., Langenberg, H. A., Mueler-Hermes, J. P., Hinrich-Peters, J., Heitman, J., Leibold, W., Dallugge, H., Krapf, E. & Kirchner, H. (1981). Mycoplasma contamination in human tumour cell lines: effect on interferon induction and susceptibility to natural killing. *Journal of Immunology*, **127**, 94–8.

Blalock, J. E., Langford, M. P., Georgiades, J. & Stanton, G. J. (1979). Non-sensitized lymphocytes produce leukocyte interferon when cultured with foreign cells. *Cellular Immunology*, **43**, 197–201.

Brodeur, B. R., Weinstein, Y., Melmon, K. L. & Merigan, T. C. (1977). Reciprocal changes in interferon production and immune responses of mouse spleen cells fractionated over columns of insolubilized conjugates of histamine. *Cellular Immunology*, **29,** 363–72.

Chang, T. W., Testa, D., Kung, P. C., Perry, L., Dreskin, H. J. & Goldstein, G. (1982). Cellular origin and interactions involved in alpha-interferon production induced by OKT-3 monoclonal antibody. *Journal of Immunology*, **128,** 585–9.

Cole, B. C., Overall, J. C., Lombardi, P. S. & Glasgow, L. H. (1976). Induction of interferon in bovine and human lymphocyte cultures by mycoplasmas. *Infection and Immunity*, **14,** 88–94.

Committee on Nomenclature (1980). Interferon Nomenclature. *Nature*, **286,** 110.

De Ley, M., Van Darime, J., Claeys, H., Weening, H., Heine, J., Billiau, A., Vermylen, C. & De Somer, P. (1980). Interferon induced in human leukocytes by mitogens: production, partial purification and characterization. *European Journal of Immunology*, **10,** 877–82.

Dianzani, F., Monahon, T., Georgiades, A. & Alperin, J. (1980). Human immune interferon: induction in lymphoid cells by a calcium ionophore. *Infection and Immunity*, **29,** 561–3.

Dianzani, F., Monahon, T. M. & Santiano, M. (1982). Membrane alterations responsible for the induction of gamma interferon. *Infection and Immunity*, **36,** 915–17.

Dianzani, F., Monahon, T., Scupham, A. & Zucca, M. (1979). Enzymatic induction of interferon production by galactose oxidase treatment of human lymphoid cells. *Infection and Immunity*, **26,** 879–82.

Djeu, J. Y., Huang, K.-Y. & Herberman, R. (1980). Augmentation of mouse natural killer activity and induction of interferon by tumour cells *in vivo*. *Journal of Experimental Medicine*, **151,** 781–9.

Ennis, F. A. & Meager, A. (1981). Immune interferon produced to high levels by antigenic stimulation of human lymphocytes with influenza virus. *Journal of Experimental Medicine*, **154,** 1279–89.

Epstein, L. B., Cline, M. J. & Merigan, T. C. (1971a). The interaction of human macrophages and lymphocytes in the phytohaemagglutinin-stimulated production of interferon. *Journal of Clinical Investigation*, **50,** 744–53.

Epstein, L. B., Cline, M. J. & Merigan, T. C. (1971b), PPD stimulated interferon: *in vitro* macrophage–lymphocyte interaction in the production of a mediator of cellular immunity. *Cellular Immunology*, **2,** 602–13.

Epstein, L. B., Cline, M., J. & Merigan, T. C. (1972). The *in vitro* interaction of immune macrophages and lymphocytes in the antigen (PPD)-induced production of interferon. In *Proceedings of the 6th Annual Leukocyte Culture Conference*, ed. M. R. Schwartz, pp. 265–7. London, New York: Academic Press.

Epstein, L. B., & Gupta, S. (1981). Human T-lymphocyte subset production of immune (gamma) interferon. *Journal of Clinical Immunology*, **1,** 186–94.

Epstein, L. B., Kreth, H. W. & Herzenberg, L. A. (1974). Fluorescence-activated sorting of human T and B lymphocytes. II. Identification of the cell type responsible for interferon production and cell proliferation in response to mitogens. *Cellular Immunology*, **12,** 407–21.

Epstein, L. B. & Salmon, S. E. (1974). The production of interferon by malignant plasma cells from patients with multiple myeloma. *Journal of Immunology*, **112,** 1131–8.

Falcoff, R., Oriol, R. & Iskaki, S. (1972). Lymphocyte stimulation and interferon induction by 7S anti human lymphocyte globulins and their uni- and divalent fragments. *European Journal of Immunology*, **2,** 478–82.

Farrar, W. L., Johnson, H. M. & Farrar, J. J. (1981). Regulation of the production of immune interferon and cytotoxic T-lymphocytes by interleukin 2. *Journal of Immunology*, **126,** 1120–5.

FLEIT, H. & RABINOVITCH, M. (1980). Production of interferon by *in vitro* derived bone marrow macrophages. *Cellular Immunology*, **57,** 495–504.

FRIEDMAN, R. & COOPER, H. L. (1967). Stimulation of interferon production in human lymphocytes by mitogens. *Proceedings of the Society for Experimental Biology and Medicine*, **125,** 901–5.

FUJIBAYASHI, T., HOOKS, J. J. & NOTKINS, A. L. (1975). Production of interferon by immune lymphocytes exposed to Herpes simplex virus antibody complexes. *Journal of Immunology*, **115,** 1191–3.

FUSE, A., HEREMANS, H., OPDENAKKER, G. & BILLIAU, A. (1982). Messenger RNA of mouse immune interferon (Mu IFN gamma). *Biochemical and Biophysical Research Communication*, **105,** 1309–14.

GIFFORD, G. E., TIBOR, A. & PEAVY, D. L. (1981). Interferon production on mixed lymphocyte cultures. *Infection and Immunity*, **3,** 164–71.

GLASGOW, L. A. (1966). Leukocytes and interferon in the host response to viral infections. II. Enhanced interferon response of leukocytes from immune animals. *Journal of Bacteriology*, **91,** 2185–91.

GRAY, P. W., LEUNG, D. W., PENNICA, D., YELVERTON, E., NAJAVIAN, R., SIMONSEN, C., DERYNCK, R., SHERWOOD, P., WALLACE, D., BERGER, S., LEVENSON, A. D. & GOEDELL, D. (1982). Expression of human immune interferon cDNA in *E.coli* and monkey cells. *Nature*, **295,** 503–8.

GREEN, J. A., COOPERBAND, S. R. & KIBRICK, S. (1969). Immune-specific induction of interferon production in cultures of human blood lymphocytes. *Science*, **164,** 1415–17.

GREEN, J. A., COOPERBAND, S. R., KLEINMAN, L. F. & KIBRICK, S. (1970). Immune stimulation of interferon production in human leukocyte cultures by non-viral antigens. *Annals of the New York Academy of Sciences*, **173,** 736–40.

GRESSER, I., MAURY, C., KRESS, C., BLANGY, D. & MANOURY, M. (1979). Role of interferon in the pathogenesis of virus diseases in mice as demonstrated by the use of anti-interferon serum. VI. Polyoma virus infection. *International Journal of Cancer*, **24,** 178–83.

HAAHR, S., RASMUSSEN, L. & MERIGAN, T. C. (1976). Lymphocyte transformation and interferon production in human mononuclear cell microculture for assay of cellular immunity to Herpes simplex. *Infection and Immunity*, **14,** 47–51.

HAHN, T. & LEVIN, S. (1982). The interferon system in patients with malignant disease. *Journal of Interferon Research*, **2,** 97–104.

HO, M. (1964). Interferon-like viral inhibitor in rabbits after intravenous administration of endotoxin. *Science*, **146,** 1472–4.

HUANG, K.-Y., KIND, P. D., JAGODA, E. & GOLDSTEIN, A. L. (1981). Thymosin treatment modulating production of interferon. *Journal of Interferon Research*, **1,** 411–20.

ISAACS, D., CLARKE, J., TYRRELL, D., WEBSTER, A. & VALMAN, H. (1981). Deficient production of leucocyte interferon (interferon-alpha) *in vitro* and *in vivo* in children with recurrent respiratory tract infections. *Lancet*, **ii,** 950–2.

ISAACS, A. & LINDENMANN, J. (1957). Virus Interference. I. The interferon. *Proceedings of the Royal Society* Series B **147,** 258–67.

ITO, Y., AOKI, H., KIMURA, Y., TAKANO, M., MAENO, K. & SHIMOKATA, K. (1980a). Generation and maintenance of immune interferon-producing cells induced by allogeneic stimulation in mice. *Infection and Immunity*, **29,** 383–9.

ITO, Y., AOKI, H., KIMURA, Y., TAKANO, M., MAENO, K. & SHIMOKATA, K. (1980b). Enumeration of immune interferon-producing cells induced by allogeneic stimulation. *Infection and Immunity*, **28,** 542–5.

ITO, Y., AOKI, H., KIMURA, Y., TAKANO, M. & SHIMOKATA, K. (1981). Natural interferon-producing cells in mice. *Infection and Immunity*, **31,** 519–23.

ITO, Y., NISHIYAMA, Y., SHIMOKATA, K., TAKEGAMA, H. & KUMI, A. (1979). Immune interferon produced *in vitro* as a quantitative indicator of cell mediated immunity. *Microbiological Immunology*, **23,** 1169–77.

JOHNSON, H. M. (1981). Cellular regulation of immune interferon production. *Antiviral Research*, **1,** 37–46.

JOHNSON, H., STANTON, G. & BARON, S. (1977). Relative ability of mitogens to stimulate production of interferon by lymphoid cells and to induce suppression of the *in vitro* immune response. *Proceedings of the Society for Experimental Biology and Medicine*, **154,** 138–41.

KATO, N., NAKASHIMA, I., OHTA, M., NAGASE, F., YOKOCHI, T. & NAITO, S. (1980). Interferon and cytotoxic factor (cytoxin) released in the blood of mice infected with *Mycobacterium bovis* BCG. III. Interferon and cytoxin induced by the specific antigen as compared with those induced by bacterial lipopolysaccharide. *Microbiological Immunology*, **24,** 1043–51.

KEDAR, E., IKEJIRI, B., SREDNI, B., BONAVIDA, B. & HERBERMAN, R. (1982). Propagation of mouse cytotoxic clones with characteristics of natural killer (NK) cells. *Cellular Immunology*, **69,** 305–29.

KIRCHNER, H., FENKL, H., ZAWATZKY, R., ENGLER, H. & BECKER, H. (1980). Dissociation between interferon production induced by phytohaemaglutinin and concanavalin A in spleen cell cultures of nude mice. *European Journal of Immunology*, **10,** 224–5.

KIRCHNER, H., PETER, H. H., HIRT, H. M., ZAWATZKY, R., DALUGGE, H. & BRADSTREET, P. (1979a). Studies of the producer cell of interferon in human lymphocyte cultures. *Immunobiology*, **156,** 65–75.

KIRCHNER, H., ZAWATZKY, R., ENGLER, H., SHIRRNACHER, V., BECKER, H. & VON WUSSOW, P. (1979b). Production of interferon in the murine mixed lymphocyte culture. II. Interferon production is a T-cell-dependent function, independent of proliferation. *European Journal of Immunology*, **9,** 824–6.

KIRCHNER, H., ZAWATZKY, R. & HIRT, H. M. (1978). *In vitro* production of immune interferon by spleen cells of mice immunized with Herpes simplex virus. *Cellular Immunology*, **40,** 204–10.

KIRCHNER, H., ZAWATZKY, R. & SCHIRRMACHER, V. (1979c). Interferon production in the murine mixed lymphocyte culture. I. Interferon production caused by differences in the H-2K and H-2D region but not by differences in the I region or the M locus. *European Journal of Immunology*, **9,** 97–103.

KLEIN, J. R., RAULET, D. H., PASTERNAK, M. & BEVAN, M. J. (1982). Cytotoxic T-lymphocytes produce immune interferon in response to antigen or mitogen. *Journal of Experimental Medicine*, **155,** 1198–1203.

KLEIN, E., BEN-BASSAT, H., NEUMANN, H., RALPH, P., ZEUTHEN, J., POLLIACK, A. & VANKY, F. (1976). Properties of the K562 cell line, derived from a patient with chronic myeloid leukaemia. *International Journal of Cancer*, **18,** 421–31.

KLIMPEL, G. R., DAY, K. D. & LUCAS, D. O. (1975). Differential production of interferon and lymphotoxin by human tonsil lymphocytes. *Cellular Immunology*, **20,** 187–96.

KLIMPEL, G., DEAN, J., DAY K., CHEN, P. & LUCAS, D. O. (1977). Lymphotoxin and interferon production by rosette separated human peripheral blood leukocytes. *Cellular Immunology*, **32,** 293–301.

KRAMMER, P., MARCUCCI, F., WALLER, M. & KIRCHNER, H. (1982). Heterogeneity of soluble T-cell products. I. Precursor frequency and correlation analysis of cytotoxic and immune interferon (IFN-gamma) producing spleen cells in the mouse. *European Journal of Immunology*, **12,** 200–4.

LANDOLFO, S., ARNOLD, B. & SUZAN, M. (1982a). Immune (gamma) interferon production by murine T-cell lymphomas. *Journal of Immunology*, **128,** 2807–9.

LANDOLFO, S., KIRCHNER, H. & SIMON, M. (1982b). Production of immune interferon is regulated by more than one T-cell subset. Lyt-1, 2, 3 and Qat-5 phenotypes of murine T-lymphocytes involved in IFN-gamma production in primary and secondary mixed lymphocyte reaction. *European Journal of Immunology*, **12,** 295–9.

LANDOLFO, S., GIOVARELLI, M., CAPUSSO, A. & FORNI, G. (1979a). Lymphokine production in primary mixed lymphocyte culture (MLC). III. Alloantigen signals and cell-cell interactions involved in migration inhibitions factor and immune interferon release. *Bollettino del'Istituto Sierotera Pico Milanese*, **58,** 2.

LANDOLFO, S., MARENCCI, F., GIOVARELLI, M., VIANO, I. & FORNI, G. (1979b). Lymphokine production in the mouse mixed lymphocyte reaction (MLR). II. Tentative mapping of murine alloantigens activating migration inhibition factor and interferon release and their relationship with those activating proliferative response. *Immunogenetics*, **9,** 245–53.

LANDOLFO, S., MARCUCCI, F., SCHIRRMACHER, V. & KIRCHNER, H. (1981a). Characteristics of alloantigens and cellular mechanisms responsible for gamma-interferon production in primary murine MLC. *Journal of Interferon Research*, **1,** 339–45.

LANDOLFO, S., MARCUCCI, F., SCHIRRMACHER, V., SIMON, M. M. & KIRCHNER, H. (1981b). Characteristics of alloantigens and cellular mechanisms responsible for IFN-gamma production in primary murine MLC. In *The Biology of the Interferon System*, ed. E. de Maeyer, G. Galasso, H. Schellekens, pp. 261–4. London, New York, Amsterdam: Elsevier/North-Holland.

LANGFORD, M. P., GEORGIADES, J. A., STANTON, G. J., DIANZANI, F. & JOHNSON, H. M. (1979). Large-scale production and physicochemical characterization of human immune interferon. *Infection and Immunity*, **26,** 36–41.

LANGFORD, M. P., WEIGENT, D. A., GEORGIADES, J., JOHNSON, H. & STANTON, G. J. (1981a). Antibody to staphylococcal enterotoxin-A induced human immune interferon (IFN-gamma). *Journal of Immunology*, **126,** 1620–3.

LANGFORD, M., WEIGENT, D. A., POLLARD, R., FISH, J., FLYE, W. & STANTON, G. (1981b). Rapid production of high levels of IFN-gamma by human thoracic duct lymphocytes stimulated with SEA. *Journal of Clinical Hematology and Oncology*, **11,** 108.

LODMELL, D. L. & NOTKINS, A. L. (1974). Cellular immunity to Herpes simplex virus mediated by interferon. *Journal of Experimental Medicine*, **140,** 764–78.

MAEHARA, N. & HO, M. (1977). Cellular origin of interferon induced by bacterial lipopolysaccharide. *Infection and Immunity*, **15,** 78–83.

MAEHARA, N., HO, M. & ARMSTRONG, J. A. (1977). Differences in mouse interferons according to cell source and mode of induction. *Infection and Immunity*, **17,** 572–9.

MANGER, B., KALDEN, J., ZAWATZKY, R. & KIRCHNER, H. (1981). Interferon production in the human mixed lymphocyte culture. *Transplantation*, **32,** 149–52.

MARCUCCI, F., NOWAK, M., KRAMMER, P. & KIRCHNER, H. (1982). Production of high titres of interferon-gamma by cells derived from short-term cultures of murine spleen leukocytes in T-cell growth factor conditioned medium. *Journal of General Virology*, **60,** 195–8.

MARCUCCI, F., WALLER, M., KIRCHNER, H. & KRAMMER, P. (1981). Production of immune interferon by murine T-cell clones from long-term cultures. *Nature*, **291,** 79–81.

MATSUYAMA, M., SUGAMURA, K., KAWADE, Y. & HINUMA, Y. (1982). Production of immune interferon by human cytotoxic T-cell clones. *Journal of Immunology*, **129,** 450–1.

MCCOOL, R. E., CATALONA, W. J., LANGFORD, M. P. & RATCLIFF, T. L. (1981).

Induction of human gamma interferon by protein A from *Staphylococcus aerueus. Journal of Interferon Research*, **1,** 473–81.

McKimm-Breschkin, J. L., Mottram, P. L. Thomas, W. R. & Miller, J. F. A. P. (1982). Antigen-specific production of immune interferon by T-cell lines. *Journal of Experimental Medicine*, **155,** 1204–9.

Milstone, L. M. & Waksman, B. H. (1970). Release of virus inhibitor from tuberculin sensitised peritoneal cells stimulated by antigen. *Journal of Immunology*, **105,** 1068–72.

Morris, A. G., Lin, Y.-L. & Askonas, B. A. (1982). Immune interferon release when a cloned cytotoxic T-cell line meets its correct influenza-infected target cell. *Nature*, **295,** 150–2.

Morris, A. G., Morser, M. J. & Meager, A. (1982). Spontaneous production of gamma interferon and induced production of beta interferon by human T-lymphoblastoid cell lines. *Infection and Immunity*, **35,** 533–6.

Nathan, I., Groopman, J., Quan, S., Berson, N. & Golde, D. (1981). Immune (gamma) interferon produced by a human T-lymphoblast line. *Nature*, **292,** 842–4.

Neta, R. (1981). Mechanisms in the *in vivo* release of lymphokines. II. Regulation in *in vivo* release of type II interferon IFN-gamma. *Cellular Immunity*, **60,** 100–8.

Neumann, C., Macher, E. & Sorg, C. (1980). Interferon production by *Corynebacterium parvum* and BCG-activated murine spleen macrophages. *Immunobiology*, **157,** 12–23.

Neumann, C. & Sorg, C. (1977). Immune interferon. I. Production by lymphokine activated murine macrophages. *European Journal of Immunology*, **7,** 719–25.

Neumann, C. & Sorg, C. (1978). Immune interferon. II. Different cellular site for the production of murine macrophage migration inhibitory factor and interferon. *European Journal of Immunology*, **8,** 582–9.

Neumann, C. & Sorg, C. (1981). Heterogeneity of murine macrophages in response to interferon inducers. *Immunobiology*, **158,** 320–9.

Olstad, R., Degre M. & Seljelid, R. (1981). Production of immune interferon (type II) in co-cultures of mouse peritoneal macrophages and syngeneic tumour cells. *Scandinavian Journal of Immunology*, **13,** 605–8.

O'Malley, J., Nussbaum-Blumenson, A., Sheedy, D., Grossmayer, B. J. & Ozer, H. (1982). Indentification of the T-cell subset that produces human gamma interferon. *Journal of Immunology*, **128,** 1522–6.

Osborne, L., Georgiades, J. & Johnson, H. (1980). Classification of interferons with antibody to immune interferon. *Cellular Immunology*, **53,** 65–70.

Pang, R. H. L., Yip, Y. K. & Vilček, J. (1981). Immune interferon induction by a monoclonal antibody specific for human T-cells. *Cellular Immunology*, **64,** 304–11.

Perussia, B., Mangoni, L., Engers, H. D. & Trinchieri, G. (1980). Interferon production by human and murine lymphocytes in response to alloantigens. *Journal of Immunology*, **125,** 1589–95.

Peter, H., Dallugge, H., Zawatzky, R., Euler, S., Leibold, W. & Kirchner, H. (1980). Human peripheral null lymphocytes. II. Producers of type I interferon upon stimulation with tumor cells, Herpes simplex virus and *Corynebacterium parvum. European Journal of Immunology*, **10,** 547–55.

Ponzio, N. M., Fitzgerald, K. L., Vilček, J. & Thorbecke, G. J. (1980). Spontaneous production of T (type II) interferon by a murine reticulum cell sarcoma. *Annals of the New York Academy of Sciences*, **350,** 157–67.

Potter, M. R., Moore, M. & Morris, A. G. (1982). Stimulation of natural cytotoxic activity in mixed cell cultures and by culture supernatants containing interferon. *Immunology*, **46,** 406–9.

RAFF, M. (1973). Theta-bearing lymphocyte in nude mice. *Nature*, **246,** 350–1.
RAFF, M. C. & CANTOR, H. (1971). *Progress in Immunology*, ed. B. Amas, p. 83. New York: Academic Press.
RASMUSSEN, L., JORDAN, G. W., STEVENS, D. & MERIGAN, T. C. (1974). Lymphocyte interferon production and transformation after Herpes simplex infections in humans. *Journal of Immunology*, **112,** 728–36.
RASMUSSEN, L. & MERIGAN, T. C. (1978). Role of T-lymphocytes in cellular immune responses during Herpes simplex infection in humans. *Proceedings of the National Academy of Sciences of the USA*, **75,** 3957–61.
RATLIFF, T., MCCOOL, R. & CATALONA, W. (1981). Interferon induction and augmentation of natural-killer activity by Staphylococcus protein A. *Cellular Immunology*, **57,** 1–12.
REEM, G. H., COOK, L. A., HENRIKSEN, D. M. & VILČEK, J. (1982). Gamma interferon induction in human thymocytes activated by lectins and B-cell lines. *Infection and Immunity*, **37,** 216–21.
REID, L. M., MINATO, N., GRESSER, I., HOLLAND, J., KADISH, A. & BLOOM, B. (1981). Influence of antimouse interferon serum on the growth and metastasis of tumour cells persistently infected with virus and of human prostatic tumours in athymic *nude* mice. *Proceedings of the National Academy of Sciences of the USA*, **78,** 1171–5.
RYTEL, M. W. & HOOKS, J. J. (1977). Induction of immune interferon by murine cytomegalovirus. *Proceedings of the Society for Experimental Biology and Medicine*, **155,** 611–18.
SAKSELA, E. (1981). Interferon and natural killer cells. *Interferon*, **3,** 44–63.
SALVIN, S. B., YOUNGNER, J. S. & LEDER, W. J. (1973). MIGRATION INHIBITORY FACTOR and interferon in the circulation of mice with delayed hypersensitivity. *Infection and Immunity*, **7,** 68–75.
SALVIN, S. B., RIBI, E., GRANGER, D. & YOUNGNER, J. S. (1975). Migration inhibitory factor and type II interferon in the circulation of mice sensitized with mycobacterial components. *Journal of Immunology*, **114,** 354–61.
SECHER, D. S. & BURKE, D. C. (1980). A monoclonal antibody for large-scale purification of human leukocyte interferon. *Nature*, **285,** 446–50.
SHOHAM, J., ESHEL, I., ABOUD, M. & SALZBERG, S. (1980). Thymic hormonal activity on human peripheral blood lymphocytes *in vitro*. II. Enhancement of the production of immune interferon by activated cells. *Journal of Immunology*, **125,** 54–8.
SONNENFELD, G., MANDEL, A. D. & MERIGAN, T. C. (1979). *In vitro* production and cellular origin of murine type II interferon. *Immunology*, **36,** 883–90.
STARR, S. E., DALTON, B., GARRABRANT, T., PANCKER, K. & PLOTBIN, S. A. (1980). Lymphocyte blastogenesis and interferon production in adult human leukocyte cultures stimulated with cytomegalovirus antigen. *Infection and Immunity*, **30,** 17–22.
STEWART, W. E. II (1979). In *The Interferon System*, pp. 27–57. Wien and New York: Springer-Verlag.
STINEBRING, W. R. & ABSHER, P. M. (1970). Production of interferon following an immune response. *Annals of the New York Academy of Sciences*, **173,** 714–25.
STOBO, J., GREEN, I., JACKSON, L. & BARON, S. (1974). Identification of a subpopulation of mouse lymphoid cells required for interferon production after stimulation with mitogens. *Journal of Immunology*, **112,** 1589–93.
SUGIYAMA, M. & EPSTEIN, L. B. (1978). Effect of *Corynebacterium parvum* on human T-lymphocyte interferon production and T-lymphocyte proliferation *in vitro*. *Cancer Research*, **38,** 4467–73.

SUGIYAMA, M., YAMAMOTO, K., KINOSHITA, Y. & KIMURA, S. (1977). Studies on the capacity of human tonsillar lymphocytes to produce IFN. *Acta oto-laryngologica*, **84,** 296–305.

TIMONEN, T., ORTALDO, J. R. & HERBERMAN, R. B. (1981). Characteristics of human large granular lymphocytes and relationship to natural killer and K cells. *Journal of Experimental Medicine*, **153,** 569–82.

TIMONEN, T., SAKSELA, E., VIRTANEN, I. & CANTELL, K. (1980). Natural killer cells are responsible for the interferon production induced in human lymphocytes by tumor cell contact. *European Journal of Immunology*, **10,** 422–7.

TORRES, B. A., FARRAR, W. L. & JOHNSON, H. M. (1982). Interleukin 2 regulates immune interferon (IFN-gamma) production by normal and suppressor cell cultures. *Journal of Immunology*, **128,** 2217–19.

TRINCHIERI, G., SANTOLI, D. & KNOWLES, B. B. (1977). Tumour cell lines induce interferon in human lymphocytes. *Nature*, **270,** 611–13.

TRINCHIERI, G., SANTOLI, D., DEE, R. R. & KNOWLES, B. B. (1978). Antiviral activity induced by culturing lymphocyte with tumour-derived or virus-transformed cells. Identification of the antiviral activity as interferon and characterization of the human effector lymphocyte subpopulation. *Journal of Experimental Medicine*, **147,** 1299–1313.

TRINCHIERI, G. & SANTOLI, D. (1978). Anti-viral activity induced by culturing lymphocytes with tumour-derived or virus-transformed cells. Enhancement of human natural killer cell activity by interferon. *Journal of Experimental Medicine*, **147,** 1314–21.

TYRING, S. K. & LEFKOWITZ, S. S. (1980). Induction of interferon by levamisole and concanavalin A in HA/ICR and NZB/W mouse spleen cells. *Experientia*, **36,** 1323–4.

VALLE, M. J., BOBROVE, A. M., STROBER, S. & MERIGAN, T. C. (1975a). Immune specific production of interferon by human T-cells in combined macrophage-lymphocyte cultures in response to Herpes simplex antigen. *Journal of Immunology*, **114,** 435–46.

VALLE, M. J., JORDAN, G. W., HAAHR, S. & MERIGAN, T. (1975b). Characteristics of immune interferon produced by human lymphocyte cultures compared to other human interferons. *Journal of Immunology*, **115,** 130–3.

VAN DAMME, J., DE LEY, M., CLAEYS, H., BILLIAU, A., VERMYLEN, C. & DE SOMER, P. (1982). Production, partial purification and characterization of interferon components induced in human leukocytes by concanavalin A. *Developments in Biological Standardization*, **50,** 369–74.

VILČEK, J., SULEA, I., VOLVOVITZ, F. & YIP, Y. K. (1980). Characteristics of interferons produced in cultures of human lymphocytes by stimulation with *Corynebacterium parvum* and phytohaemaglutinin. In *Biochemical Characterization of Lymphokines*, pp. 323–9. New York: Academic Press.

VIRELIZIER, J. L., ALLISON, A. C. & DE MAEYER, E. (1977). Production by mixed lymphocyte cuture of a Type II interferon able to protect macrophages against virus infections. *Infection and Immunity*, **17,** 282–5.

WALLEN, W. C., DEAN , J. M. & LUCAS, D. O. (1973). Interferon and the cellular immune response: separation of interferon-producing cells from DNA synthetic cells. *Cellular Immunology*, **6,** 110–22.

WEIGENT, D., LANGFORD, M., SMITH, E. M., BLALOCK, J. E. & STANTON, G. J. (1981). Human B-lymphocytes produce leukocyte interferon after interaction with foreign cells. *Infection and Immunity*, **32,** 508–12.

WHEELOCK, E. F. (1965). Interferon-like virus-inhibitor induced in human leukocytes by phytohaemagglutinin. *Science*, **149,** 310–11.

WIETZERBIN, J., STEFANAS, S., FALCOFF, R., LUCERO, M., CATINOT, L. & FALCOFF,

E. (1978). Immune interferon induced by phytohaemagglutinin in nude mouse spleen cells. *Infection and Immunity*, **21,** 966–72.

WILKINSON, M. & MORRIS, A. (1983). Interferon with novel characteristics produced by human mononuclear leukocytes. *Biochemical and Biophysical Research Communication*, **111,** 498–503.

WIRANOWSKA-STEWART, M. (1981). Heterogeneity of human gamma interferon preparations: evidence for presence of alpha interferon. *Journal of Interferon Research*, **1,** 315–21.

WIRANOWSKA-STEWART, M. & STEWART, W. E. II (1981). Determination of human leukocyte populations involved in production of interferon alpha and gamma. *Journal of Interferon Research*, **1,** 233–44.

YIP, Y. K., PANG, R., URBAN, C. & VILČEK, J. (1981a). Partial purification and characterization of human gamma (immune) interferon. *Proceedings of the National Academy of Sciences of the USA*, **78,** 1601–5.

YIP, Y. K., PANG, R., OPPENHEIM, J., NADIBAR, M. S., HENRIKSEN, D., ZEREBECKYJ-ECKHARDT, I. & VILČEK, J. (1981b). Stimulation of human gamma interferon production by diterpene esters. *Infection and Immunity*, **34,** 131–9.

YOUNGNER, J. S. & SALVIN, S. B. (1973). Production and properties of migration inhibitory factor and interferon in the circulation of mice with delayed hypersensitivity. *Journal of Immunology*, **11,** 1914–22.

INTERFERON AND THE IMMUNE SYSTEM 2: EFFECT OF IFN ON THE IMMUNE SYSTEM

M. MOORE

Paterson Laboratories, Christie Hospital and Holt Radium Institute, Manchester M20 9BX, UK

(1) INTRODUCTION

Interferons (IFNs), first described as antiviral proteins, express a multiplicity of other biological properties, foremost among which are their effects on the immune response (Stewart, 1979). Three approaches have been adopted to elucidate the role of IFNs in this context. (i) The effects of *exogenous* material on components of the immune system *in vitro*, and *in vivo* responses following administration to normal and immunodeficient hosts (including man), have been studied. (ii) *Endogenous* IFN has been generated *in vivo* or *in vitro* by a variety of inducers (viruses, poly I : C, antigens, mitogens) and the effects on several immune parameters analysed. (iii) The effects of antisera to IFN have been studied both as an adjunct to (i) and (ii) and in the attempt to elucidate the biological role of IFN in natural viral infection and host resistance to tumours.

The majority of these studies have been conducted with virus-induced IFNs (Type I; IFN-α, IFN-β), the *primary* function of which is probably direct inhibition of virus replication. Fewer studies are yet available based on immune IFN (Type II; IFN-γ), which is by definition a lymphokine, produced as part of an immune response to both viral and non-viral antigens and thus a more 'natural' candidate for an immunoregulatory molecule. (For recent comment on this see Basham & Merigan, 1982.)

Although the antiviral and immunomodulatory properties of the two major classes of IFNs largely overlap, there is firm evidence for quantitative differences and in some systems even of qualitative differences in these properties. There are few data yet available on biological activity as a function of individual IFN subtypes.

Although electrophoretically pure (Gresser, De Maeyer-Guignard, Tovey & De Maeyer, 1979) and gene-cloned products (Weissman, 1981) are now readily available many of the studies using exogenous IFN have been conducted with partially purified, or even crude material, which in the light of contemporary knowledge almost certainly contained more than one type and presumably, at least in the case of IFN-α, several subtype IFN proteins. The evidence that IFN was the active principle in the phenomena observed is reasonably sound, if not entirely conclusive as far as IFN-α/β are concerned, but much less persuasive where IFN-γ is implicated. Most investigators adopted the criteria that (i) the factor inducing the effect(s) responded to chemical or physical treatment in a manner parallel to the antiviral activity in the IFN-containing preparations (i.e. response to trypsin, low pH, heat, ultracentrifugation); (ii) the factor, in common with the antiviral activity, exhibited the property of defined host range; (iii) heterologous IFNs or material prepared without inducers that serve as controls, were ineffective; (iv) the magnitude of the effect closely paralleled that of the antiviral activity (irrespective of the purity of the preparations); and (v) the effect was abrogated by antisera prepared against the type of IFN utilized or implicated.

Where IFN is generated endogenously in the response to antigenic or mitogenic stimulation which involves the elaboration not only of more than one type of IFN but of other pharmacologically active non-antibody macromolecules (lymphokines) as well, the interplay between the various mediators and the cells upon which they act is particularly complex (Cohen & Bigazzi, 1980), and the possibility of synergistic interactions of IFN with other molecules may also need to be taken into account.

The properties of IFN relevant to its capacity to modulate the immune response include antiproliferative activity, the ability to influence differentiation and to effect a variety of cell surface changes (Taylor-Papadimitriou, 1980). As already implied their classical antiviral activity *in vivo* is believed to comprise in many instances, an IFN-induced immunomodulatory component. In several cases, injection of anti-IFN globulin into animals infected with a wide variety of viruses results in a severely altered course of infection, indicative of the involvement of IFN acting directly upon viral replication, as well as the immune system (Sonnenfeld & Merigan, 1979). The fact that IFNs can influence so many cellular properties and functions implies the existence of IFN receptors, the expression of which is pervasive among lymphoid cells (Mogensen *et al.*, 1981). While the mere existence of receptor sites at this level has no predictive value for response to IFN, it may be that cells responding to IFN will have distinct binding characteristics with respect to concentration and type (and subtype) of IFN bound. IFN has the potential to affect several components of the immune response simultaneously and theoretically, at least, by more than one mechanism. The effects of IFN *in vivo* are likely to be very complex in this respect.

The current wave of interest in the IFNs as biological response modifiers coincides with the delineation of the control of the immune response at the molecular level. Many experiments involving exogenous material have recently been repeated with the purified products, now more readily available (Knight, 1980). The realization that many IFN genes code the subtypes of Hu-IFN-α (presently, 15 proteins) means that much work will need to be done to relate biological activity with IFN subspecies at the molecular level.

Space forbids an exhaustive treatment of this topic. At most this chapter attempts to describe select immunological phenomena upon which IFN has an unequivocal effect and to advance some mechanisms of action, insofar as these are known. Also it concentrates for the most part, on the effects of exogenous IFNs on isolated (or semi-isolated) components of the immune response. The role of IFN in the pathogenesis of various clinical disorders is likewise beyond its scope.

Classically, IFNs have been thought of as immunosuppressive agents accounting for the marked immunodepression which is a concomitant of viraemia. However, analysis of their effect on the various components of the immune response discloses a remarkable biological versatility which embraces an immunoenhancing capability as well.

The following brief review is divided into five main sections (numbered 2–6) in which the evidence for an effect of IFN on a given cellular component of the immune response is discussed. Such a subdivision, although convenient is of course, arbitrary, since the immune response, particularly in its early inductive phases, involves the interactions of more than one cell type.

(2) REGULATION OF B CELL RESPONSES

The effect of IFN on antibody formation *in vivo* (Braun & Levy, 1972) and *in vitro* (Gisler, Lindahl & Gresser, 1974) has been the subject of several recent reviews (Epstein, 1979; Johnson, 1977a) and only the major conclusions will be recapitulated here.

In vivo treatment of mice with IFN-α/β causes a suppression of the primary and secondary antibody response to the thymus-dependent antigen, sheep red blood cells (SRBC) and of the primary response to a thymus-independent antigen, lipopolysaccharide (LPS). The dose of IFN, time of administration and dose of the challenge antigen are critical for the level of inhibition. Maximum suppressive effects on the primary response occur when IFN is given 4–48 h prior to antigen challenge; low doses of IFN given after antigen have a slight enhancing effect. Both IgM and IgG responses to SRBC are affected. In the suppression of the primary response to LPS, IFN appears to act directly on B cells. IFN can also affect memory cells, as treatment with IFN shortly before the time of primary immunization with SRBC has a suppressive effect on a subsequent secondary response. IFN-α/β may also inhibit the IgE-mediated heterologous adoptive cutaneous anaphylaxis reaction by mouse cells in the rat and enhance the release of histamine from basophils in the presence of ragweed antigen or anti IgE (Ngan, Lee & Kind, 1976).

IFNs also affect both primary and secondary antibody responses generated *in vitro*. The effects are apparently exerted on some early event in the non-cycling, late-responding, antibody-forming cell precursors, as in some experiments exposure of the cultures to IFN for four hours is sufficient for an effect on the PFC response five days later. By deploying separated cell populations to study the murine response to SRBC, IFN has been claimed to affect B-lymphocytes primarily, and not macrophages (though these are difficult to exclude) or helper T lymphocytes (T_H). The dose of IFN and the time of administration relative to antigen challenge is critical in determining the nature of the effect; suppression is maximal when IFN is added concomitantly with the antigen. IFN-α/β can have a slight immunoenhancing effect on the primary antibody response *in vitro* if used at low doses or if given at various intervals late after antigen challenge or just before enumeration of the PFC response.

The general picture which emerges from these studies is of a population of precursor B lymphocytes that are capable of responding to a given antigen (SRBC). The cells are at various stages of differentiation, which in turn determine their relative abilities to differentiate into antibody-producing plasma cells in the presence of various concentrations of IFN-α/β and perhaps IFN-γ. The lesser-differentiated precursor cells are more susceptible to inhibition by IFN than are the highly-differentiated precursor cells. Precursor B cells that are sufficiently differentiated undergo normal clonal expansion for a limited time in the presence of IFN, but any interaction of IFN with T_H cells in this system is considered to be minimal (Johnson, 1977a).

Studies of the effect of IFNs on human B cell differentiation have utilized the pokeweed mitogen system (Choi, Lim & Saunders, 1981; Härfast *et al.*, 1981). This mitogen mimics T-dependent antigens like SRBC in that it requires T_H cells for the generation of optimal responses. According to Choi *et al.* (1981) the effect of Hu-IFN-α is highly dose-dependent, low doses (10^2 U/ml) enhancing and high doses (10^4 U/ml) suppressing B cell differentiation as assessed by polyclonal antibody production in the haemolytic plaque assay. In accordance with the observations made on murine lymphocytes, the action of IFN-α is upon an early stage of B-cell differentiation. However, the PWM-triggered differentiation of human B cells requires T-cells and monocytes in the culture. Pretreatment of the separated cellular components with IFN-α prior to the assay revealed that enhancement is mediated by activation of monocytes and suppression by inhibition of T_H proliferation. At high dose, B cell proliferation is also directly affected. In their experiments, Härfast *et al.* (1981) also found that, at different times of exposure Hu-IFN-α enhanced or suppressed PWM-induced IgG synthesis in peripheral blood lymphocytes. However, they obtained no evidence that IFN acted upon cells other than B lymphocytes.

As stated above (section 1), the experiments upon which many of the above conclusions have been based were all conducted with impure IFN preparations though there is probably little reason to doubt that IFN-α/β was the active principle. Moreover, inhibition of murine antibody formation *in vitro* has been confirmed in experiments utilizing pure IFN-α/β (Gresser *et al.*, 1979). By contrast the role of IFN-γ in the regulation of antibody responses is decidedly less clear. Sonnenfeld, Mandel & Merigan (1977) reported that antigen-specific murine IFN-γ has about 250 times the immunosuppressive potency of IFN-α/β when added to spleen cell cultures 24 h before antigen and that this immunoregulatory action, like that of the latter IFNs, is time and dosage dependent. Thus, if added to spleen cell cultures hours after addition of antigen, IFN-γ will cause enhancement of the PFC response (Sonnenfeld, Mandel & Merigan, 1978).*

* In these experiments it is important to realize that the source of Mu-IFN-γ was, in fact, serum from a BCG-infected mouse and that consequently, other factors (lymphokines) may have contributed to the effects observed.

A second approach has been the attempt to correlate the capacity of various mitogens to inhibit the *in vitro* antibody response with their ability to generate the production of endogenous IFN-γ in murine spleen cell cultures (Johnson, 1977a). It was found that the ability to inhibit the PFC response to SRBC was proportional to the capacity of these mitogens to induce IFN in the cultures. Staphylococcal enterotoxin A (SEA) was the most effective inhibitor of the PFC response, and the best inducer of IFN, followed by Con A, with PHA-P being the least effective. Adenosine 3′5′-cyclic monophosphate (cAMP) has an inhibitory effect on the immunological and inflammatory functions of lymphocytes (Johnson, 1977b; Bourne *et al.*, 1974). From this and other studies evidence has been obtained which suggests that there is a close relationship between cAMP, IFN-γ activity, suppressor T-cell (Ts) activity and regulation of the immune response.

Dibutyryl cAMP effectively protects PFC responses from mitogen-induced suppression and it has been shown that the effect operates at the level of IFN-γ induction, rather than upon established antiviral activity. The concentrations of cyclic nucleotide that blocked the development of suppressor activity correlated with those which blocked the production of IFN in spleen cell cultures stimulated by the T-cell mitogens.

Cholera toxin raises the endogenous level of cAMP by stimulating adenylate cyclase activity. The methyl xanthine, 3-isobutyl-1-methyl-xanthine raises the endogenous cAMP level by inhibiting phosphodiesterase activity; an effect also achieved by theophylline. Both agents block SEA suppression of the *in vitro* anti-SRBC PFC response in an amount similar to that observed for dibutyryl cAMP.

Further, they block SEA stimulation of IFN-γ production in mouse spleen cell cultures at concentrations that block SEA suppressor activity. Cumulatively, the data suggest that mitogen-induced IFN is associated with mitogen-induced suppressor cell activity, and that IFN may possibly be a mediator of such activity (Johnson, 1977b).

Comparative studies on the antiviral and *in vitro* immunosuppressive properties of IFN-α/β and IFN-γ suggest that different mechanisms may be involved. For example, 2-mercaptoethanol (2-ME) blocks the suppressive effects of IFN-α/β in spleen cell cultures, but enhances the suppressive activity of mitogen-induced IFN-γ (Johnson, 1977a). The suggestion that the suppressive effects of IFN-γ cannot be inhibited by 2-ME (unlike IFN-β) suggest that different mechanisms and mediators are involved. However, this conclusion should probably be regarded as provisional.

The mechanisms(s) by which IFN-β suppresses antibody responses by murine spleen cells to SRBC *in vitro* has been recently pursued to another level (Aune & Pierce, 1982). IFN-β mediated suppression is partially or completely prevented by catalase, 2-mercaptoethanol and certain peroxidase substrates (ascorbic acid, potassium iodide and tyrosine). These reagents also inhibit suppression by mediators from Con A-activated murine suppressor T cells, soluble immune response suppressor (SIRS), and macrophage (Mφ)-derived suppressor factor (Mφ-SF) and act by inactivating Mφ-SF or preventing the generation of Mφ-SF from SIRS. These experiments suggest that IFN-β may act by inducing the production of a molecule that has the properties of SIRS. Treatment of spleen cells with IFN-β leads to the generation of a population of Lyt 2^+ suppressor T cells that act by elaborating a soluble factor. This IFN-β induced suppressor T-cell factor (IFN-TsF) has properties in common with SIRS: (i) both SIRS and IFN-TsF suppress antibody responses with the same characteristic kinetic pattern; responses initiate normally but terminate prematurely after day four of culture. (ii) IFN-TsF and SIRS are of comparable size (45 K–55 K) and are converted to Mφ-SF by low (1 μM) concentrations of H_2O_2 or by Mφ. (iii) Mφ-SF obtained from IFN-TsF or SIRS is inactivated by similar concentrations of reagents such as ascorbic acid, potassium iodide and 2-ME. These data show that the immunosuppressive properties of IFN-β are due, at least in part, to its ability to activate suppressor T cells that produce mediators that appear to be analogous to those in the SIRS/Mφ-SF pathway of immunosuppression.

(3) REGULATION OF T-CELL FUNCTIONS

IFN has been shown to affect the proliferative phase of the immune response in a number of *in vitro* systems, including suppression of murine and human lymphocytes in response to mitogens and antigens (e.g. PPD, mumps, tetanus), and in the one-way mixed lymphocyte reaction (MLR). Maximum effects are observed when IFN is added at the initiation of the cultures, from which it would appear that IFN acts upon lymphocytes during the early stages of the cell cycle. *In vivo* administration of IFN also depresses the blastogenic response of murine spleen cells *in vivo* and this effect is maximal if IFN is given 24 h prior to sacrifice.

A differential sensitivity to IFN on the part of different subclasses of T and B lymphocytes would appear to be indicated by the observation that the doses required to suppress PHA-stimulated cells are apparently much lower ($\simeq 10$ U/ml) than those required to suppress the PWM (2×10^4) or MLR (10^3–10^4) responses. Differential effects on the part of different types of IFN (IFN-α and IFN-β) on lymphocyte blastogenesis, attributed to the mode of action or the receptor-binding properties of the two IFNs, have also been reported (Miorner *et al.*, 1978).

In addition to modulating T–B cell interactions discussed in the previous section, IFN has an important role in T–T cell interaction. For example, there is evidence that mitogen-induced suppression of lymphocyte activation *in vitro* is mediated by IFN: (i) there is a correlation between the suppressive activities of Con A-treated modulator lymphocytes and their ability to produce IFN *in vitro*. (ii) There is a parallel between the antiviral activity of supernatants of Con A-stimulated lymphocytes and their suppressive activities. (iii) Culture supernatants containing IFN induced by virus, tumour cell coculture (see section 5), or antigen (PPD) mimic the effect of mitogen-induced suppressor cells and supernatants. (iv) The suppressive activities of Con A-stimulated lymphocytes or their supernatant are inhibited by two different antisera against Hu-IFN-α (Kadish *et al.*, 1980).

Delayed hypersensitivity (DH) is one manifestation of cell-mediated immunity and its essential effectors are T-cells (De Maeyer & De Maeyer-Guignard, 1980). The deployment of inbred mice with diminished capacity for IFN production has permitted use of inducers to examine the action of IFN upon both sensitization and elicitation of DH. IFN-α can affect both afferent and efferent pathways of DH to antigens of viral and non-viral origin, although this observation by itself does not signify that IFN is a normal component of the DH reaction. Indeed, IFN-α is mainly though not exclusively, induced by viruses and most antigens do not trigger its production. When administered to mice 24 h before immunization, IFN can inhibit sensitization, either to SRBC or NDV. This timing is critical since when given a few hours after sensitization, IFN enhances sensitization. The question arises as to whether IFN depresses or stimulates T-cell activity, depending on whether it acts before or after sensitization. The antiproliferative effect of IFN, as witnessed, for example, by inhibition of blast formation *in vivo* could explain the inhibition of sensitization that is observed when IFN is given 24 h before antigen since upon antigen stimulation, T-cell progenitors within lymphoid tissue transform into blast cells, divide and give rise to progeny with effector function. More difficult to explain is the stimulation of sensitization which occurs when IFN is administered *after* the antigen. If the antiblastogenic effect of IFN is the sole mechanism for this phenomenon, it must presumably be directed against suppressor cell proliferation.

IFN-α also inhibits the proliferation of allogeneic mouse spleen or syngeneic bone marrow cells in irradiated mice (Cerottini *et al.*, 1973), the proliferative response associated with graft-versus-host (GVH) disease, as well as the activation of leukaemia virus which is often a concomitant of the GVH reaction (Hirsch, Ellis, Proffitt & Black, 1973). IFN prolongs allograft survival (across both major and minor histocompatibility barriers) in mice by several days (Mobraaten, De Maeyer & De Maeyer-Guignard, 1973; Hirsch *et al.*, 1974). Congeneic mice which are high IFN producers (in response to appropriate inducers) tolerate allografts more readily than those which are low producers, underlining the role of endogenous IFN in allograft prolongation.

IFN-α/β enhances the specific cytotoxic capacity of previously sensitized murine lymphocytes for certain tumour targets (e.g. L1210 cells) but whether their effect in some systems is to accelerate the response, recruit more cells or enhance the cytotoxicity of individual effectors is not known (Lindahl, Leary & Gresser, 1972). In this model at least, any effect on target cells could be eliminated by pretreatment of the isolated effectors with IFN prior to test.

The effect of IFNs on MLRs in which allospecific cytotoxic T-cells and NK-like effectors are generated is relatively well-documented. The MLR consists essentially of two phases: a proliferative phase in which the lymphocytes of one donor divide in response to lymphocyte-determined (LD) antigens of the other allogeneic donor; and a cytotoxic phase in which the lymphocytes from one donor actually destroy the lymphocytes of the other. Along with the other lymphokines, IFN is itself a product of the murine and human MLR (Kirchner *et al.*, 1979) and peak production occurs when the cytotoxic phase is just beginning. Exogenous IFN-α added at the initiation of culture inhibits the proliferative phase of the response but the cytotoxic capacity of the killer cells is markedly enhanced in a dose-dependent fashion (up to 500 U/ml). Heron, Berg & Cantell

(1976) and Zarling *et al.* (1978), in their respective systems described the killer cells as T-cells, the latter partly on the grounds that they were not cytotoxic for the HLA-negative erythroleukaemic K562 targets which are so exquisitely sensitive to peripheral blood NK cells. The study of Zarling *et al.* (1978) is important not least because it was conducted with purified IFN-β induced by poly I : C. Several possible mechanisms could account for their observations. In MLC, helper cells, cytotoxic T-cells and suppressor cells develop in the course of allogeneic stimulation (for review see Bach *et al.*, 1977) as well as a diversity of lymphokines, including IFN. (i) IFN may decrease the generation of suppressor cells that may normally regulate the level of the cytotoxic response generated. (ii) IFN may exert its effect on a human analogue of the murine Ly $1,2,3^+$ regulatory cell (Tada, 1977; Eardley, Shen, Cantor & Geshon, 1977), which in turn may govern the magnitude of the resulting cytotoxic response. (iii) IFN may directly increase the responsiveness of cytotoxic T precursors, perhaps by increasing the receptor-stimulating cell interaction. (iv) IFN may augment cytotoxic T-cell response by increasing the expression, on the stimulating lymphocytes and/or macrophages, of cell-surface antigens that elicit cytotoxic responses (see section 4).

Autologous lymphocyte-mediated cytotoxicity (ALC) against tumour biopsy cells is demonstrable in about 30% of all cancer patients coming to surgery. The characteristics of the cytolysis differ in several important respects from natural killing; the effector cells are T-cells, and the interaction generally accepted to reflect specific immunological recognition of tumour cells. However, this autoreactivity is apparently not enhanced by pretreatment of the effector cells with IFN-α (Vanky, Argov, Einhorn & Klein, 1980). The reason why ALC is uninfluenced by IFN is unclear. It could be the result of the low number of antigen-sensitized lymphocytes and/or the weak expression of the relevant antigens on the human targets. IFN-α does not act on FcγR-negative T-cells (Klein, Masucci, Masucci & Vanky, 1981) wherein the lymphocytes triggered for proliferation by confrontation with antigens (Masucci, Masucci, Klein & Berthold, 1980) are to be found (see section 5).

In MLC, concomitant cytotoxicity is also induced against a spectrum of tumour cell lines, including even autologous lymphoblastoid cell lines (LCL) (Poros & Klein, 1978; Seeley & Golub, 1978). This latter phenomenon has been described as 'anomalous killing' and lymphocytes cultured with allogeneic normal lymphocytes or lymphoid cell lines exert comparable, seemingly indiscriminate killing of appreciable magnitude (Jondal & Targan, 1978; Potter & Moore, 1981). These effectors lyse K562 and are thus distinguishable from allospecific cytotoxic T-cells and have features in common with the natural killer (NK) cells of peripheral blood (Poros & Klein, 1979).

The generation of anomalous killers in mixed cell systems appears to be associated in some way with T-cell activation. However, its expression precedes that of allospecific cytotoxicity and the proliferation of T-cells on which the latter is dependent. The generation of anomalous cytotoxicity would thus appear to occur directly either by polyclonal activation or by more restricted action upon an NK precursor of T-cell lineage; or indirectly via the release of soluble mediators. Several investigators have shown that factors capable of enhancing NK-like activity are produced and released by T-cells in response to stimulation by allogeneic lymphocytes of LGL (Koide & Takasugi, 1978; Potter, Moore & Morris, 1982). The similarity of these effects to that of exogenous IFN on effector cells of apparently similar type has suggested a role for endogenous IFN in the induction of augmentation of NK-like activity in mixed cell systems. These supernatants contain IFNs of more than one type – IFN-α, the major producers of which are B-lymphocytes, null lymphocytes (including NK cells) and macrophages, and IFN-γ the major producers of which are T lymphocytes. Notwithstanding the presence of different types of IFN in MLC/LCL supernatants capable of enhancing native cytotoxicity, a cause-and-effect relationship for the appearance of anomalous cytotoxicity *in vitro* has yet to be established.

As already stated, the activation of murine alloantigen-specific cytotoxic T lymphocyte precursors is dependent upon the presence of both macrophages and T_H cells or regulatory molecules derived from these facilitative cells. Three biochemically distinct helper factors have been identified: Interleukin-1 (Mφ-derived); Interleukin-2 (T-cell derived) and immune IFN (IFN-γ). All three factors are found in the supernatants of mixed lymphocyte cultures (MLC). The removal of Mφ from these cultures affects their production of these factors as well as the induction of cytotoxic T lymphocytes (CTL) (Farrar, Johnson & Farrar, 1981). The addition of IL-2 to these Mφ-depleted cultures restores the ability of responder T-cells to: (i) bypass the requirement for Mφ soluble function; (ii) produce IFN-γ; and (iii) generate CTL. The kinetics and dose response of

IFN-γ production in response to IL-2 correlates with the generation of CTL. The production of IFN-γ as well as the generation of CTL requires T-cells, alloantigen and IL-2. Furthermore, the induction of CTL by IL-2 may be neutralized by the addition of anti-IFN-γ serum. These data thus suggest that (i) the regulation of IFN-γ production is based on a T-T cell interaction mediated by IL-2; and (ii) IFN-γ production may be required for IL-2 induction of CTL. These findings are consistent with the hypothesis that the induction of CTL involves a linear cell-factor interaction in which IL-1 (Mφ-derived) stimulates T-cells to produce IL-2, which in turn stimulates other T-cells to produce IFN-γ and become cytotoxic. The IFN-γ generated in these cultures is apparently involved in the specific T lymphocyte response to the primary alloantigen and does not augment NK-like killing. These data are consistent with the observations of Kirchner, Zawatzky & Schirrmacher (1979) who showed that MLC-derived IFN-γ was always associated with the development of CTL and may be used as an early (48 h) means of detecting H-2 differences.

(4) EFFECTS ON CELL SURFACE ANTIGENS

Many of the effects of IFN on the cellular components of the immune system may occur through changes induced in the membranes, not only of effector cells, but also of target cells (Friedman, 1979). For instance, IFN is known to increase FcγR expression in murine and human lymphocytes, lymphocyte-derived cell lines (Itoh, Inoue, Kataoka & Kumagi, 1980; Fridman *et al.*, 1980) and in murine macrophages (Hamburg, Manejias & Rabinovitch, 1978). In addition IFN affects the expression of a number of target cell surface structures, though not as was first thought in a non-specific way. Thus IFN enhances the expression of histocompatibility antigens (Lindahl, Gresser, Leary & Torey, 1976; Vigneaux & Gresser, 1977; Heron, Hockland & Berg, 1978), β_2 microglobulin (β_2m) (Fellous, Kamoun, Gresser & Bono, 1981), carcinoembryonic antigen (Attallah, Needy, Noguchi & Elisberg, 1979), and tumour-associated antigens (Liao, Kwong, Khosravi & Dent, 1982) but not that of Thy-1, Ia or HLA-DR antigens, at least on lymphocytes (cf. Mφ section 6).

It is not at present known whether this apparent differential effect of IFN on the expression of various antigens has biological significance or whether it reflects the inherent dynamics of antigen synthesis and shedding. Whatever the mechanism, these effects of IFN are probably important to the host since IFN does increase the expression of H-2 antigens *in vivo* (Lindahl *et al.*, 1976; Vigneaux & Gresser, 1977) and an increase in serum levels of β_2m is observed in patients treated with Hu-IFN-α (Fridman *et al.*, 1980). While the effects of the major IFNs are broadly similar, some quantitative and qualitative differences have also emerged. For instance, the increased expression of H-2D and H-2K antigens on mouse thymocytes appears to require only three units of IFN-γ compared with 600 of IFN-α/β to achieve the same effect. In addition IFN-γ has been shown to affect a wider spectrum of antigens than IFN-α/β; increasing production not only of H-2D and H-2K antigens, but also of Thy-1, Lyt-1 and Lyt 2,3 antigens on the surface of young murine thymocytes. Recently it was shown that IFN-γ from stimulated human lymphocytes and from monkey cells transfected by cloned human IFN-γ cDNA induced HeLa Class I antigen and β_2m mRNAs or their products at concentrations 100 times lower than those needed to induce the (2′–5′) oligo(A)synthetase and the antiviral state. This difference was not found with IFN-α/β (Wallach, Fellous & Revel, 1982).

Another property of IFNs involving an apparently direct effect on cell surface structures is their capacity to confer protection on certain targets against lysis by natural killers (NK) (Trinchieri & Santoli, 1978). The structures expressed on cell targets which are recognized by NK cells have yet to be defined, though they appear not to identify with any of the known cell surface antigens. The receptor for transferrin is currently a possible candidate. The basis of the protective role of IFN in these circumstances is difficult to define; modulation (decreased expression) of the hypothetical target structure may be envisaged, or an influence on the cellular capability to effect membrane repair. Interestingly, the IFN-induced protective effect is not confined to non-transformed cells (Moore, White & Potter, 1980). The fact that neoplastic cells are also affected means that under *in vivo* conditions the effects of IFN could be mutually antagonistic – enhancing certain effector mechanisms such as NK on the one hand, and protecting tumour targets from lysis on the other.

It should be noted that IFN may also exert other effects on cell surfaces pertinent to effector-target cell interactions (Friedman, 1979) including, for example, physically detectable changes such as alterations in electrophoretic behaviour and interference with toxin-receptor interactions, increases in Con A receptors and receptors for synthetic polypeptide antigens on T lymphocytes. IFNs also influence the transport of thymidine and uridine across the cell membrane and induce a very rapid increase (in one minute) in cGMP.

In summary it is apparent that IFN exerts a plethora of effects on the plasma membrane of a diversity of cell types. Certain of these are associated with the enhancement of the effector status of cells such as cytolytic T-cells, macrophages and NK cells, while others affect target cells and may either facilitate or obstruct effector target cell interactions depending on the nature of the cell surface structures and the corresponding effector mechanisms involved.

(5) REGULATION OF NATURAL KILLER ACTIVITY

The other major components of the cellular immune response upon which IFNs may exert considerable (even an immunoregulatory) influence are natural killer (NK) cells and cells of the mononuclear phagocytic system (MPS).

Unlike cytotoxic T lymphocytes, NK cells appear to lack immunological memory and histocompatibility restriction and are characterized operationally by an ability to lyse a wide variety of target cells, including those which are syngeneic, allogeneic and xenogeneic to the NK cell donor. Certain tumour cells and cell lines persistently infected with different viruses are particularly sensitive to NK attack. There is considerable circumstantial evidence that NK cells interact with several different cell types *in vivo* and that this property extends to the surveillance of tumour and virus-infected cells (for a comprehensive treatment of this topic see Herberman, 1980, 1982).

The cellular lineage of NK cells remains a matter of considerable debate. In the mouse, a number of serological markers are reported to be expressed on NK cells including NK_1 (Glimcher, Shen & Cantor, 1977), Ly-5 (Cantor *et al.*, 1979), Qa-2, Qa-4 and Qa-5 (Koo & Hatzel, 1980; Chun *et al.*, 1979), asialo GM_1 (Kasai *et al.*, 1979) and Thy-1 (Herberman, Nunn & Holden, 1978), many of which would be consistent with a T-cell lineage. However, NK cells have also been ascribed to the monocyte-macrophage series (Lohmann-Matthes, Domzig & Roder, 1979) by virtue of their expressing the

Mφ-1 antigen. The conflicting evidence on the phenotypes and characteristics of NK cells are best reconciled if they are regarded not as a single cell type derived from a common lineage, but as a heterogeneous population sharing the capacity for selective cytotoxicity.

Recently, Minato, Reid & Bloom (1981) delineated at least four phenotypically distinct classes of murine NK effector cells. One subset (NK_1) possesses the phenotype Thy-1^-, Ly-2^-, Qa-5^+, and lyses targets persistently infected with measles virus, and corresponds to the NK cells capable of lysing lymphoma targets. A second (NK_T) of comparable target cell specificity is a thymus-independent cell (Thy-1^+, Lyt-2^-, Qa-5^+, Ly-5^+). A third (TK) designated killer T-cells (Thy-1^+, Lyt-2^+, Qa-5^-, Ly-5^+), originating in conventional but not *nude* mouse spleens is lytic for P815 mastocytoma cells, but not virus-infected targets and is phenotypically indistinguishable from conventional, antigen-specific cytotoxic T lymphocytes. The fourth (NK_m), derived primarily from bone marrow cultures, is cytotoxic for measles virus-infected targets, but not P815 and expresses only Ly-5^+ of the markets tested. The naturally cytotoxic (NC) cell described by Lattime, Pecoraro & Stutman (1981) which lyses solid tumours but not lymphomas is possibly related to the NK_m subset.

The phenotypic heterogeneity of the murine NK population is also reflected in the differential response to IFN. Exogenous IFN-α/β selectively induced NK, and NK_T activity without augmenting NK_m or TK activity. It appears that IFN stimulates non-cytotoxic Qa-5^+ Ly-5^- precursor cells to differentiate into Qa-5^+ Ly-5^+ NK_1 effectors; converts a portion of NK_1 to NK_T cells, and increases the efficacy of some NK cells to lyse certain targets. Interestingly, IL-2 may also augment NK activity but acts on different subsets. Thus, IFN-free IL-2 preparations enhance the activities of the TK and NK_T subsets, but not those of the NK_1 or the NK_m subset. IL-2 augments NK activity of spleen cells from both conventional and *nu/nu* mice, but is without effect on spleens of *nu/nu* mice depleted of Thy-1^+ cells, which suggests among other observations, that IL-2 acts primarily if not exclusively on Thy-1^+ cells. It has thus been postulated that IFN induces the differentiation of precursor cells into NK_1 effector NK cells and subsequently converts a portion of them into NK_T effector cells, which are Thy-1^+, which might then be regulated and expanded by IL-2.

The interrelationship between these various subsets and their responses to the two biological regulators have important far-reaching implications for questions relating to the differentiation of functional cytotoxic T-cells via thymic and extrathymic pathways. However, in the present context, a more immediate question is whether the NK_1 and NK_T subsets operate *in vivo* since systems have been described in which IFN induces resistance to tumour growth or virus infection *in vivo*, under conditions where it is ineffective directly *in vitro*, thereby implicating some host mechanism(s) (Gresser, Maury & Brouty-Boye, 1972). In fact, anti-IFN serum suppresses the resistance of *nu*/*nu* mice to virus-infected tumour cells and these locally invasive tumours sometimes metastasize as well (Reid *et al.*, 1981). However, the relative contributions of the IFN-responsive NK subsets is far from clear.

In man, the principal cell type mediating natural cytotoxicity is the large granular lymphocyte (LGL) of human peripheral blood, so-called on account of the cytoplasmic azurophilic granules in these low density cells of high nuclear : cytoplasmic ratio (Timonen, Ortaldo & Herberman, 1981). The same dichotomy over lineage pervades studies of human NK cells which also express both T lymphocyte and myelomonocytic serological markers (Ortaldo, Sharrow, Timonen & Herberman, 1981). This implies that (i) LGL derive from a precursor cell common to both lymphocytes and monocytes; (ii) LGL constitute a distinct and separate lineage expressing determinants present on both cell types; or (iii) LGL are heterogenous comprising cells of both lymphocyte and monocyte provenance.

It is well established that LGL which functionally are non-phagocytic, non-adherent cells bearing receptors for the Fc portion of IgG, constitute the major IFN-responsive population in human peripheral blood and spleen (Timonen *et al.*, 1981). All types of IFN are active in this respect (Claeys *et al.*, 1982) and similar enhancement is demonstrable after *in vivo* administration (Pape, Hadam, Gisenburg & Riethmuller, 1981). Moreover, in the interests of elucidating their lineage and the clonal distribution, or otherwise, of their receptors for target structures, LGL have now been cloned in lymphocyte-conditioned medium. The cells retain their morphology, exhibit moderate ADCC (see below) and respond to IFN (Hercend *et al.*, 1982). In *in vivo* effector-target cell interactions, LGL themselves are the producers of IFN, since IFN can be visualized in their cytoplasm (by indirect immunofluorescence using monoclonal anti-α antibody) shortly after target cell conjugation

(Saksela, 1981). Two possible mechanisms for the enhancement of NK activity by IFN can thus be envisaged. Either IFN increases the cytolytic activity of each individual effector or increases the number of active cells. Experiments at the single-cell level indicate that augmentation of NK activity occurs predominantly via recruitment of a non-cytolytic 'pre-NK' pool into full cytotoxic activity. However, there is evidence for activation of target bound 'pre-NK' as well. The fact that mature NK cells bind avidly to targets and release IFN on contact, which in turn activates 'pre-NK' to become fully cytotoxic, suggests the existence of a positive feedback type of amplification loop operating within the NK system itself. Although K562 cells are capable of inducing IFN production in LGL, kinetic studies suggest that *basal* NK activity at least may actually be independent of the endogenous production of IFN (Copeland, Koren & Jensen, 1981).

Peripheral blood lymphocytes pretreated with IFN-α lyse a wider spectrum of targets than their untreated counterparts (Moore & Potter, 1980) and this extends not only to relatively refractory B cell lines, but also to fresh solid tumour cells (Vose & Moore, 1980), leukaemias (Zarling *et al.*, 1979; Moore, Taylor & White, 1982) and lymphomas (Pattengale *et al.*, 1981). Such findings may have important implications for the *in vivo* activation of NK cell activity by other lymphokines which in addition to IFN are generated as part of the adaptive immune response to tumour-associated antigens.

Recent studies in the mouse have indicated that the effects of IFN-β and the lymphokine IL-2 on spleen cell cytotoxicity *versus* lymphoma targets are at least additive (Kuribayashi, Gillis, Kern & Henney, 1981). An increased ability on the part of spleen cells stimulated with poly I : C to absorb IL-2 implies that IFN increases the level of IL-2 receptor expression on NK cells. Similar conclusions have been reached for human LGL and coincidentally small T-cells where limiting dilution analyses have shown that the proliferation frequency of LGL in response to lectin-free IL-2 is increased five-fold by pretreatment with IFN-α/β (500 I.U./ml); whereas the presence of IFN throughout the assay period (seven days) has a contrary effect. This latter is mediated by irradiated T-cells in the feeder populations and indicates that IFN may exert an immunomodulatory effect on both NK and T-cell proliferation and that it may function both by induction of IL-2 receptors and activation of suppressor T-cells (B. M. Vose, personal communication).

It is probable that LGL also comprise the principal mediators of antibody-dependent cellular cytotoxicity (ADCC) otherwise known as killer (K) cells (Timonen *et al.*, 1981). Unlike NK activity the question whether this function is potentiated by IFN is controversial. Many targets are sensitive to both effector mechanisms and under these conditions the relative contributions of the two lytic effects are difficult to analyse. Where NK-resistant targets (e.g. human red blood cells) are deployed with antibody of the appropriate specificity (anti D), no enhancement of K cell function is observed, so that it would appear that NK and K cell functions, though probably mediated by the same cells, may be operationally distinguished by this agent (Kimber & Moore, 1981).

In common with experience in the mouse, IFN has proved a useful reagent with which to probe human NK cell heterogeneity. Palatine tonsils for example, possess a low intrinsic NK activity, compared with peripheral blood, spleen and even lymph nodes, but they are apparently devoid of LGL and totally unresponsive to IFN-α (Kimber & Moore, 1982). However, they do respond to lectin-pulsed lymphocyte supernatants and while IL-2 has not been formally implicated, the phenomenon bears a superficial resemblance to data indicating a differential capacity of murine NK subsets to respond to the two biological response modifiers.

In summary, it is evident that the role of IFN in the regulation of NK activity is central, but that other lymphokines too are involved. Further delineation of the intricate immunological circuitry involving IFN is an important prerequisite for a clearer understanding of the biological role of this substance in the *in vivo* regulation of the immune response.

(6) EFFECTS ON MONONUCLEAR PHAGOCYTES

Macrophages form an important part of the immune defence principally by removing micro-organisms from blood and tissue. Their ability to phagocytose and pinocytose many different kinds of antigen underlines their participation in early immune inductive events. They also respond to external stimuli which derive from activated lymphocytes and/or microbial agents and perform in cell-mediated types of immune reactions.

Since the demonstration by Gresser *et al.* (1970) of enhanced phagocytosis of RC14 tumour cells by intraperitoneal macrophages in mice treated daily with partially-purified IFN, several functions of cells of the mononuclear phagocyte system (MPS) are now known to be affected by IFN. In common with T and B cells and NK cells, under appropriate conditions Mφ may produce IFN as well as respond to it.

Many of the properties of Mφ treated with IFN are superficially similar to those following exposure to various inflammatory stimuli, infection with intracellular pathogens, or incubation with killed micro-organisms (or fractions thereof) and include an increased ability to spread over inert surfaces, enhanced immunological and non-immunological phagocytosis, and augmented microbicidal, tumouricidal and secretory activities. The effects are usually demonstrable following recovery of Mφ from hosts treated with IFN (or IFN inducers) *in vivo* as well as *in vitro* after addition of IFN-enriched exogenous material. Many of the induced changes are common to those associated with macrophage 'activation'. Early experiments by Huang demonstrated enhanced pinocytic uptake of colloidal carbon by Mφ stimulated *in vivo* with Newcastle Disease virus or *in vitro* with IFN-rich preparations. Mouse Mφ elicited fresh or cultivated *in vitro* with IFN rapidly exhibit enhanced phagocytosis of IgG-coated erythrocytes.

Fc-mediated phagocytosis appears to occur in two main stages: (i) attachment to the Mφ surface, mediated by the recognition of the Fc portion of IgG bound to the particle; and (ii) ingestion, triggered by sequential recognition of the ligands on the surface of the particle, resulting in a localized cytoskeletal reorganization and formation of an endocytic vacuole. Both steps are FcR-dependent. The enhanced phagocytic activity of Mφ exposed to IFN could be the consequence of IFN effects on any of the steps involved in the endocytic process. Thus, binding would be enhanced if more functional FcR were available or if the activity of the FcR were increased. Increased ingestion could result from metabolic changes, from changes in cytoskeletal elements such as the state of active polymerization, or from increased mobility of FcR in the plane of the membrane. There is evidence for at least two FcR on Mφ. FcR I is sensitive to proteolytic cleavage by trypsin and binds to the Fc region of IgG_{2a}, whereas FcR II is trypsin-resistant and binds to either IgG_1 or IgG_{2b}. Until recently the FcR could be quantified only by indirect methods. However, the availability of a monoclonal antibody to FcR II has facilitated a more direct enquiry into the mechanism of IFN-mediated enhancement of phagocytosis. In summary, the data indicate that the process does not primarily involve the synthesis of additional FcR, though the possibility that new more avid FcR are synthesized after addition of IFN and prior to assay of the phagocytic response, cannot be formally excluded (Hamburg, Fleit, Unkeless & Rabinovitch, 1980).

Macrophage activation for cytotoxicity involves a sequence of events. The highly phagocytic inflammatory Mφ is not cytotoxic, but it can acquire this ability when exposed to small concentrations of endotoxins, serum fractions or lymphocyte products at concentrations that do not activate resting, non-inflammatory cells. Interestingly, inflammatory Mφ induced by thioglycollate or protease peptone can be further stimulated (as estimated by further enhancement of phagocytosis) by IFN *in vitro*, as could Mφ stimulated with LPS. However, Mφ from hosts treated with IFN inducers cannot be further stimulated by IFN. LPS-prestimulated and normal Mφ show similar time-course and dose-response curves to IFN, indicating that the mechanism of stimulation is probably similar in both cell types (Hamburg *et al.*, 1980).

A related FcR-dependent property exhibited by Mφ is the capacity to mediate antibody-dependent cellular cytotoxicity (ADCC) against erythrocyte targets (Kimber & Moore, 1981). Both intra- and extra-cellular killing appear to be involved and the phenomenon is enhanced by effector cell pretreatment with IFN. Whether the important underlying change in the Mφ membrane, which is responsible for this enhancement involves the FcR has not been settled (Fridman *et al.*, 1980). It should be noted that purified IFN-α can also activate human monocytes to greater cytolytic capacity against non-erythrocyte targets in the absence of antibody (Herberman *et al.*, 1980).

Mφ-mediated tumour destruction observed for some experimental neoplasms results predominantly from a non-phagocytic, contact-mediated event. The process of Mφ activation for cytotoxicity probably involves a sequence of events. Several agents, particularly synthetic polyanionic IFN inducers, render Mφ tumouricidal through direct exposure. It has been postulated that this type of direct activation may be mediated by IFN produced by the Mφ themselves. In support of this, anti-IFN globulin neutralizes the ability of various inducers of IFB-α/β to render Mφ cytotoxic. The same antibody, however, has no apparent effect on the ability of Con A-induced supernatant (containing IFN-γ) to elicit Mφ cytotoxicity (Chirigos, Schultz & Stylos, 1980).

It has been postulated that Mφ activation is a self-limiting process since the cells release high levels of PGE_2, which along with PGE_1 abrogate the ability of IFN to activate Mφ to the tumouricidal state. On this model, prostaglandins act in a negative feedback inhibition to limit cellular activity.

In addition to participating in cytotoxic events, Mφ exhibit regulatory functions in several lymphocyte activities, some of which are proliferation-dependent such as responses to antigens and mitogens, and others which are proliferation-independent, such as production of lymphokines (for review, see Nelson, 1976). Activated Mφ are particularly well-endowed with suppressive activity, in comparison with unstimulated Mφ, and the same *in vivo* activating stimulae enhance both suppressive and cytolytic functions of Mφ (Russell, Gillespie & McIntosh, 1977). *In vivo* Mφ activation is a complex phenomenon involving a multiplicity of factors, but a major component is the lymphokine, Mφ-activating factor (MAF) (Evans & Alexander, 1971), which enhances Mφ cytolytic activity through a mechanism clearly distinguishable from that of IFN (Boraschi & Tagliabue, 1981). Recently it was shown in the mouse that the enhancement of Mφ cytotoxicity and the augmentation of Mφ suppressive activity could also be clearly distinguished (Boraschi, Soldateschi & Tagliabue, 1982). MAF, which enhance tumouricidal activity of inflammatory but not of resident Mφ *in vitro*, was without effect on the suppressive activity of both resident and inflammatory peritoneal Mφ. By contrast, IFN-β abolished or greatly reduced suppression of both lymphoproliferation and lymphokine production by either population. Like MAF, IFN-β increased Mφ cytotoxicity against tumour cells, an effect achieved for both resident and inflammatory Mφ, thus confirming different activation mechanisms for MAF and IFN-β. The influence of IFN on Mφ function could thus be envisioned along the following lines: activated lymphocytes develop the ability to proliferate and produce MAF and IFN. The action of IFN on Mφ might relieve their suppressive activity on lymphocytes, thus enhancing lymphoproliferation and lymphokine production. The arrest of this reaction could then take place through the action on IFN of an IFN-inhibitor produced by lymphocytes (Fleischmann *et al.*, 1979) acting as a feedback regulator. The recovery of macrophage-suppressive capacity due to IFN blocking would normalize the enhanced lymphocyte activities.

Interferons induce other changes in macrophages which are important in the immune response. There is a marked increase in expression of human monocyte HLA Class II antigens in response to HU-IFN-β (Rhodes & Stokes, 1982). T lymphocytes recognize foreign determinants in association with HLA-Class II antigens (DR products) or rodent Ia regions of the macrophage membrane. The polymorphism of these products accounts for the genetic differences in immune responsiveness. Scher, Beller & Unanue (1980) have described a T-cell factor which induces Ia-positive macrophage populations *in vivo*. IFN-γ, as a product of a subset of T lymphocytes, could conceivably have a role in this type of regulation (section 6). It is of interest that IFN-α/β does not apparently increase HLA Class II antigen expression by lymphoid cells, although Class I antigens are enhanced (Fellous *et al.*, 1979).

The mechanism by which IFN induces changes in the monocytic membrane is unknown but altered ratios of intracellular cyclic nucleotides may be instrumental, since IFN increases GMP levels and the opposite effect, in the form of increased cAMP levels, is known to inhibit macrophage FcR expression.

(7) CONCLUSIONS

The control of the immune response is highly complex, involving interactions of various lymphocyte subpopulations and collaboration of lymphocytes with macrophages, either via the secretion of soluble mediators such as lymphokines or monokines, or antigen-specific suppressor factors. It is important to realize that not every substance capable of influencing immune responses necessarily participates in their regulation since many substances are known to have immunomodulatory activity, which are not normal components of immune reactions. The multiplicity of interactions of IFNs with the various cellular components of the immune system is indicative of a *potential* for immunoregulation which is uncommon if not unique, among biological macromolecules. Such observations are important and interesting, but do not necessarily mean that these macromolecules participate in the regulation of immunity. Indeed, IFN-α/β, upon which the vast majority of work in this context has been done to date, are induced predominantly by viruses and are generated only in minute quantities, if at all, during immune stimulation by non-viral antigens. This is not so for IFN-γ, however, which is produced in response to a wide variety of substances including mitogens, bacterial and viral antigens, and tumour cells. Not surprisingly therefore, there is currently much interest in the candidacy of IFN-γ as an immunoregulatory molecule.

As in the delineation of the immunomodulatory capability of IFN-α/β an essential prerequisite for the elucidation of the role of IFN-γ in the immune response has been the availability of the material in appropriate purity. Considerable progress toward this goal has been facilitated by the recent advent of recombinant IFN-γ. Experiments with this product have upheld the provisional conclusion (section 4) based on tests with impure IFN-γ contaminated with other lymphokines, that IFN-γ is approximately 250 times more potent in immune cell regulation than the other IFNs. Not only has it been confirmed that IFN-γ has greater potency for the induction of HLA Class I antigen on lymphoid cell surfaces, but there is also a differential response between the dose of recombinant IFN-γ required for HLA induction and the enzyme (2′–5′) oligoadenylate synthetase, which inhibits protein synthesis and is related to the antiviral activity of IFN. By contrast IFN-α/β induced both of these cellular proteins at the same concentration. This observation has given rise to the hypothesis that the induction of HLA antigens is not related to the antiviral activity of IFN-γ and that this molecule is the most important type of IFN in immunoregulation.

The function of IFN-induced enhancement of HLA Class I antigen expression is at present speculative, but could conceivably be related to cell–cell recognition. The most obvious role is in potentiating the cytotoxic T-cell response of a virus-infected host for the elimination of HLA-compatible infected cells. Another effect apparently related to increased expression of HLA is the protection of uninfected host cells from lysis by NK cells. Again, however, the demonstration that pure IFN-γ can affect components of the immune response at lower unitage than IFN-α/β is not conclusive evidence *per se* for an immunoregulatory role. Such a claim might be more compelling if IFN-γ were shown to have a differential effect on the expression of immune response antigens that are critical in antigen presentation by macrophages to lymphocyte subpopulations. There is already evidence for this, at least in the mouse, where the expression of the human HLA-DR analogue, Ia, is increased, but much work in this important area remains to be done (see section 6). Likewise, the capacity of pure IFN-γ to influence other immune functions (sections 2–6) at the inductive and/or effector phases requires examination.

With the advent of genetic engineering, pure preparations of biosynthetic IFNs should enable investigators to advance beyond the stage of 'phenomenology' to determine the mechanism whereby the different IFNs exert their various activities both within and without the immune system and to elucidate their biological and clinical roles.

REFERENCES

ATALLAH, A. M., NEEDY, C. F., NOGUCHI, P. D. & ELISBERG, B. L. (1979). Enhancement of carcinoembryonic antigen expression by interferon. *International Journal of Cancer*, **24,** 49–52.

AUNE, T. M. & PIERCE, C. W. (1982). Activation of a suppressor T cell pathway by interferon. *Proceedings of the National Academy of Sciences of the USA*, **79,** 3808–12.

BACH, F. H., GRILLOT-COURVALIN, C., KUPERMAN, O. J., SOLLINGER, H. W., HAYES, C., SONDEL, P. M., ALTER, B. J. & BACH, M. L. (1977). Antigenic requirements for triggering of cytotoxic T lymphocytes. *Immunological Reviews*, **35,** 76–96.

BASHAM, T. & MERIGAN, T. C. (1982). Immunoregulation by gamma-interferon? *Nature*, **299,** 778.

BORASCHI, D., SOLDATESCHI, D. & TAGLIABUE, A. (1982). Macrophage activation by interferon: dissociation between tumouricidal capacity and suppressive activity. *European Journal of Immunology*, **12,** 320–6.

BORASCHI, D. & TAGLIABUE, A. (1981). Interferon-induced enhancement of macrophage-mediated tumour cytolysis and its difference from activation by lymphokines. *European Journal of Immunology*, **11,** 110–14.

BOURNE, H. R., LICHTENSTEIN, L. M., MELMON, K. L., HENNEY, C. S., WEINSTEIN, Y. & SHEARER, G. M. (1974). Modulation of inflammation and immunity by cyclic AMP. *Science*, **184,** 19–28.

BRAUN, W. & LEVY, H. B. (1972). Interferon preparations as modifiers of immune responses. *Proceedings of the Society for Experimental Biology and Medicine*, **141,** 769–73.

CANTOR, H., KASAI, F., SHEU, F. W., LECLERC, J. C. & GLIMCHER, L. (1979). Immunogenetic analysis of natural killer activity in the mouse. *Immunological Reviews*, **44,** 1–12.

CEROTTINI, J.-C., BRUNNER, K. T., LINDAHL, P. & GRESSER, I. (1973). Inhibitory effect of interferon preparations and inducers on the multiplication of transplanted allogeneic spleen cells and syngeneic bone marrow cells. *Nature New Biology*, **242,** 152–3.

CHIRIGOS, M. A., SCHULTZ, R. M. & STYLOS, W. A. (1980). Interaction of interferon, macrophage and lymphocyte tumoricidal activity with prostaglandin effect. In *Regulatory Functions of Interferons*, eds. J. Vilcek, I. Gresser and T. C. Merigan. *Annals of the New York Academy of Sciences*, **350,** 91–101.

CHOI, Y. S., LIM, K. H. & SAUNDERS, F. K. (1981). Effect of interferon α on pokeweed mitogen-induced differentiation of human peripheral blood B lymphocytes. *Cellular Immunology*, **64,** 20–8.

CHUN, M., PASAREN, V., HAMMERLING, U., HAMMERLING, G. AND HOFFMAN, M. K. (1979). Tumour necrosis serum induces a serologically distinct population of NK cells. *Journal of Experimental Medicine*, **150,** 426–31.

Claeys, H., van Danone, J., DeLey, M., Vermylen, C. & Billiau, A. (1982). Activation of natural cytotoxicity of human peripheral blood mononuclear cells by interferon: a kinetic study and comparison of different interferon types. *British Journal of Haematology*, **50,** 85–94.

Cohen, S. & Bigazzi, P. E. (1980). Lymphokines, cytokines and interferon(s). In *Interferon II*, ed. I. Gresser, pp. 81–95. London: Academic Press.

Copeland, C. S., Koren, H. S. & Jensen, P. J. (1981). Natural killing can be independent of interferon generated *in vitro. Cellular Immunology*, **62,** 220–5.

De Maeyer, E. & De Maeyer-Guignard, J. (1980). Immunoregulatory action of Type I interferon in the mouse. In *Regulatory Functions of Interferons*, ed. J. Vilcek, I. Gresser & T. C. Merigan, *Annals of the New York Academy of Sciences*, **350,** 1–11.

Eardley, D. D., Shen, F. W., Cantor, H. & Geshon, R. K. (1977). Feedback induction of suppression by in vitro educated Ly 1 T helper cells. In *Immune System: Genetics and Regulation*, ed. E. E. Sercerz, L. A. Herzenberg & C. F. Fox, pp. 525–31. New York: Academic Press.

Epstein, L. B. (1979). The comparative biology of classical (Type I) and immune (Type II) interferon. In *Biology of the Lymphokines*, ed. S. Cohen, E. Pick & J. J. Oppenheim, pp. 443–14. New York: Academic Press.

Evans, R. & Alexander, P. (1971). Rendering macrophages specifically cytotoxic by a factor released from immune lymphoid cells. *Transplantation*, **12,** 227–9.

Farrar, W. L., Johnson, H. M. & Farrar, J. J. (1981). Regulation of the production of immune interferon and cytotoxic T lymphocytes by interleukin 2. *Journal of Immunology*, **126,** 1120–5.

Fellous, M., Kamoun, M., Gresser, I. & Bono, R. (1979). Enhanced expression of HLA antigens and beta 2 microglobulin on interferon-treated human lymphoid cells. *European Journal of Immunology*, **9,** 446–9.

Fleischmann, Jr., W. R. Georgiades, J. A., Osborne, L. C., Dianzani, F. & Johnson, H. M. (1979). Induction of an inhibitor of interferon action in a mouse lymphokine preparation. *Infection and Immunity*, **26,** 949–55.

Fridman, W. H., Gresser, I., Bandu, M. T., Agnet, M. & Neauport-Santes, C. (1980). Interferon enhances the expression of Fc γ receptors. *Journal of Immunology*, **124,** 2436–41.

Friedman, R. M. (1979). Interferons: interactions with cell surfaces. In *Interferon I*, ed. I. Gresser, pp. 53–74. London: Academic Press.

Gisler, N. H., Lindahl, P. & Gresser, I. (1974). Effects of interferon on antibody synthesis *in vitro*. *Journal of Immunology*, **113,** 438–44.

Glimcher, L., Shen, F. W. & Cantor, H. (1977). Identification of a cell surface antigen selectively expressed on the natural killer cells. *Journal of Experimental Medicine*, **145,** 1–9.

Gresser, I., Bourali, C., Chouroulinkov, D., Fontaine-Brouty-Boyé, D. & Thomas, M. (1970). Treatment of neoplasia in mice with interferon preparations. *Annals of the New York Academy of Sciences*, **173,** 694–705.

Gresser, I., De Maeyer-Guignard, J., Tovey, M. and De Maeyer, E. (1979). Electrophoretically pure mouse interferon exerts multiple biologic effects. *Proceedings of the National Academy of Sciences of the USA*, **76,** 5308–12.

Gresser, I., Maury, C. & Brouty-Boye, D. (1972). On the mechanism of the antitumous effects of interferon in mice. *Nature*, **239,** 167–8.

Hamburg, S. I., Fleit, H. B., Unkeless, J. C. M. & Rabinovitch, M. (1980). Mononuclear phagocytes: Responders to and producers of interferon. In *Regulatory Functions of Interferons*, ed. J. Vilcek, I. Gresser & T. C. Merigan, *Annals of the New York Academy of Sciences*, **350,** 72–90.

HAMBURG, S. I., MANEJIAS, R. E. & RABINOVITCH, M. (1978). Macrophage activation: increased ingestion of IgG coated erythrocytes after administration of interferon inducers to mice. *Journal of Experimental Medicine*, **147,** 593–8.

HÄRFAST, B., HUDDLESTONE, J. R., CASALI, P., MERIGAN, T. C. & OLDSTONE, M. B. A. (1981). Interferon acts directly on human B lymphocytes to modulate immunoglobulin synthesis. *Journal of Immunology*, **127,** 2146–50.

HERBERMAN, R. (ed.). (1980). *Natural Cell-Mediated Immunity Against Tumours.* New York & London: Academic Press.

HERBERMAN, R. (ed.). (1982). *NK Cells and Other Natural Effector Cells.* New York & London: Academic Press.

HERBERMAN, R., NUNN, M. E. & HOLDEN, H. T. (1978). Low density of Thyl antigen of mouse effector cells mediating natural cytotoxicity against tumour cells. *Journal of Immunology,* **121,** 304–9.

HERBERMAN, R. B., ORTALDO, J. R., DJEU, J. Y., HOLDEN, H. T., JETT, J., LANG, N. P., RUBINSTEIN, M. & PESTKA, S. (1980). Role of interferon in regulation of cytotoxicity by natural killer cells and macrophages. In *Regulatory Functions of Interferons*, ed. J. Vilcek, I. Gresser & T. C. Merigan, *Annals of the New York Academy of Sciences*, **350,** 63–71.

HERCEND, T., MEUER, S., REINHERZ, E. L. SCHLOSSMAN, S. F. & RITZ, J. (1982). Generation of a cloned NK cell line derived from the 'null cell' fraction of human peripheral blood. *Journal of Immunology*, **129,** 1299–1305.

HERON, I., BERG, K. & CANTELL, K. (1976). Regulatory effect of interferon on T cells *in vitro. Journal of Immunology*, **117,** 1370–3.

HERON, I., HOKLAND, M. & BERG, K. (1978). Enhanced expression of β_2-microglobulin and HLA antigens on human lymphoid cells by interferon. *Proceedings of the National Academy of Sciences of the USA*, **75,** 6215-19.

HIRSCH, M. S., ELLIS, D. A., BLACK, P. H., MONACO, A. P. & WOOD, M. L. (1974). Immunosuppressive effects of an interferon preparation *in vivo. Transplantation*, **17,** 234–6.

HIRSCH, M. S., ELLIS, D. A., PROFFITT, M. R. & BLACK, P. H. (1973). Effects of interferon on leukaemia virus activation in graft versus host disease. *Nature New Biology*, **244,** 102–3.

HUANG, K. (1977). Effect of interferon on phagocytosis. In *The Interferon System*, ed. S. Baron & F. Danzani, *Texas Reports on Biology and Medicine*, **35,** 350–6.

ITOH, K., INOUE, M., KATAOKA, S. & KUMAGAI, K. (1980). Differential effect of interferon expression of IgG and IgM Fc receptors on human lymphocytes. *Journal of Immunology*, **124,** 2589–95.

JOHNSON, H. M. (1977a). Effect of interferon on antibody formation. In *The Interferon System*, ed. S. Baron & F. Dianzani, *Texas Reports on Biology and Medicine*, **35,** 357–69.

JOHNSON, H. M. (1977b). Cyclic AMP regulation of mitogen-induced interferon production and mitogen suppression of immune response. *Nature*, **265,** 154–5.

JONDAL, M. & TARGAN, S. (1978). *In vitro* induction of cytotoxic effector cells with spontaneous killer cell specificity. *Journal of Experimental Medicine*, **147,** 1621–36.

KADISH, A. S., TANSEY, F. A., YU, G. S. M., DOYLE, A. T. & BLOOM, B. R. (1980). Interferon as a mediator of human lymphocyte suppression. *Journal of Experimental Medicine*, **151,** 637–50.

KASAI, M., IWAMORI, Y., NAGAI, K., OKAMURA, K. & TADA, T. (1979). A glycolipid on the surface of mouse natural killer cells. *European Journal of Immunology*, **10,** 175–80.

KIMBER, I. & MOORE, M. (1981). Selective enhancement of human mononuclear leucocyte cytotoxic function by interferon. *Scandinavian Journal of Immunology*, **13,** 375–81.

KIMBER, I. & MOORE, M. (1982). The natural cytotoxic capacity of human tonsillar lymphocytes. In *Current Concepts in Human Immunology and Cancer Immunomodulation*, ed. B. Serrou *et al.*, pp. 355–64. New York, Amsterdam: Elsevier Biomedical.

KIRCHNER, H., ZAWATZKY, R., ENGLER, H., SCHIRRMACHER, V., BECKER, H. & VON WUSSOW, P. (1979). Production and interferon in the mixed lymphocyte culture. II. Interferon production is a T cell-dependent function independent of proliferation. *European Journal of Immunology*, **9,** 824–6.

KIRCHNER, H., ZAWATZKY, R. & SCHIRRMACHER, V. (1979). Interferon production in the murine mixed lymphocyte culture. I. Interferon production caused by differences in the H-2K and H-2D region but not by differences in the I region or the M locus. *European Journal of Immunology*, **9,** 97–9.

KLEIN, E., MASUCCI, G., MASUCCI, M. G. & VANKY, F. (1981). Natural and activated killer lymphocytes. Interpretation of the results in short-term cell-mediated cytotoxic tests. In *NK Cells: Fundamental Aspects and Role in Cancer*, ed. R. B. Herberman, C. Rosenfeld & B. Serrou, *Human Cancer Immunology*, vol. 6. Amsterdam: Elsevier, North-Holland.

KNIGHT, JR., E. (1980). Purification and characterisation of interferons. In *Interferon II*, ed. I. Gresser, pp. 1–12. London: Academic Press.

KOIDE, Y. & TAKASUGI, M. (1978). Augmentation of human natural cell-mediated cytotoxicity by a soluble factor. I. Production of N-cell activating factor (NAF). *Journal of Immunology*, **121,** 872–9.

KOO, G. C. & HATZEL, A. (1980). Antigenic phenotype of mouse natural killer cells. In *Natural Cell-Mediated Immunity Against Tumours*, ed. R. B. Herberman, pp. 105–16. New York: Academic Press.

KURIBAYASHI, K., GILLIS, S., KERN, D. E. & HENNEY, C. S. (1981). Murine NK cultures: effects of interleukin-2 and interferon on cell growth and cytotoxicity. *Journal of Immunology*, **126,** 2321–7.

LATTIME, E. C., PECORARO, G. A. & STUTMAN, O. (1981). Natural cytotoxic cells against solid tumours in mice. III. Comparison of effector cell antigenic phenotype and target cell recognition structures with NK cells. *Journal of Immunology*, **126,** 2011–14.

LIAO, S-K., KWONG, P. C., KHOSRAVI, M. & DENT, P. B. (1982). Enhanced expression of melanoma-associated antigens and β_2-microglobulin on cultured human melanoma cells by interferon. *Journal of the National Cancer Institute*, **68,** 19–25.

LINDAHL, P., GRESSER, I., LEARY, P. & TOREY, M. (1976). Interferon treatment of mice: Enhanced expression of histocompatibility antigens on lymphoid cells. *Proceedings of the National Academy of Sciences of the USA*, **73,** 1284–7.

LINDAHL, P., LEARY, P. & GRESSER, I. (1972). Enhancement by Interferon of the specific cytotoxicity of sensitized lymphocytes. *Proceedings of the National Academy of Sciences of the USA*, **69,** 721–5.

LOHMANN-MATTHES, M. L., DOMZIG, W. & RODER, J. (1979). Promonocytes have the functional characteristics of natural killer cells. *Journal of Immunology*, **123,** 1883–6.

MASUCCI, M. G., MASUCCI, E., KLEIN, E. & BERTHOLD, W. (1980). Target selectivity of interferon induced human killer lymphocytes related to their Fc receptor expression. *Proceedings of the National Academy of Sciences of the USA*, **77,** 3620–4.

MINATO, N., REID, L. & BLOOM, B. R. (1981). On the heterogeneity of murine natural killer cells. *Journal of Experimental Medicine*, **154,** 750–62.

MIORNER, H., LANDSTROM, L. E., LARNER, E., LARSSON, I., LUNDGREN, E. & STRANNEGARD, O. (1978). Regulation of mitogen-induced lymphocyte DNA synthesis by human interferon of different origins. *Cellular Immunology*, **35,** 15–24.

MOBRAATEN, L. E., DE MAEYER, E. & DE MAEYER-GUIGNARD, J. (1973). Prolongation of allograft survival in mice by inducers of interferon. *Transplantation*, **16,** 415–20.

MOGENSEN, K. E. BANDU, M-T., VIGNAUX, F., AGUET, M. & GRESSER, I. (1981). Binding of ^{125}I-labelled human α interferon to human lymphoid cells. *International Journal of Cancer*, **28,** 575–82.

MOORE, M. & POTTER, M. R. (1980). Enhancement of human natural cell-mediated cytotoxicity by interferon. *British Journal of Cancer*, **41,** 378–87.

MOORE, M., TAYLOR, G. M. & WHITE, W. J. (1982). Susceptibility of human leukaemias to cell-mediated cytotoxicity by interferon-treated allogeneic lymphocytes. *Cancer Immunology and Immunotheräpy*, **13,** 56–61.

MOORE, M., WHITE, W. J. & POTTER, M. R. (1980). Modulation of target cell susceptibility to human natural killer cells by interferon. *International Journal of Cancer*, **25,** 565–72.

NELSON, D. S. (1976). In *Immunobiology of the Macrophage*, ed. D. S. Nelson, pp. 235–57. New York: Academic Press.

NGAN, J., LEE, S. H. S. & KIND, L. S. (1976). The suppressive effect of interferon on the ability of mouse spleen cells synthesising IgE to sensitized rat skin for heterologous adoptive cutaneous anaphylaxis. *Journal of Immunology*, **117,** 1063–6.

ORTALDO, J. R., SHARROW, S. O., TIMONEN, T. & HERBERMAN, R. B. (1981). Determination of surface antigens on highly purified human NK cells by flow cytometry with monoclonal antibodies. *Journal of Immunology*, **127,** 2401–9.

PAPE, G. R., HADAM, M. R., GISENBURG, J. & RIETHMÜLLER, G. (1981). Kinetics of natural cytotoxicity in patients treated with human fibroblast interferon. *Cancer Immunology and Immunotherapy*, **11,** 1–000.

PATTENGALE, P. K., GIDLUND, M., NILSSON, K., SANDSTROM, C., ORN, A. & WIGZELL, H. (1981). Lysis of human B-lymphocyte-derived lymphoma/leukaemia cells of established cell lines by interferon-activated natural killer (NK) cells. *International Journal of Cancer*, **28,** 459–68.

POROS, A. & KLEIN, E. (1978). Cultivation with K562 cells leads to blastogenesis and increased cytotoxicity with changed properties of the active cells when compared to fresh lymphocytes. *Cellular Immunology*, **41,** 240–55.

POROS, A. & KLEIN, E. (1979). Distinction of anti-K562 and anti-allocytotoxicity in *in vitro* stimulated populations of human lymphocytes. *Cellular Immunology*, **46,** 57–68.

POTTER, M. R. & MOORE, M. (1981). *In vitro* augmentation of human natural cytotoxic activity. *Clinical and Experimental Immunology*, **44,** 332–41.

POTTER, M. R., MOORE, M. & MORRIS, A. G. (1982). Stimulation of natural cytotoxic activity in mixed cell cultures and by culture supernatants containing interferon. *Immunology*, **46,** 401–9.

REID, L., MINATO, N., GRESSER, I., KADISH, A. & BLOOM, B. R. (1981). Influence of anti-mouse Interferon serum on the growth and metastasis of virus persistently-infected tumor cells and human prostatic tumors in athymic nude mice. *Proceedings of the National Academy of Sciences of the USA*, **78,** 1171–5.

RHODES, J. & STOKES, P. (1982). Interferon-induced changes in the monocyte membrane: inhibition by retinol and retinoic acid. *Immunology*, **45,** 531–6.

RUSSELL, S. W., GILLESPIE, G. Y. & MCINTOSH, A. T. (1977). Inflammatory cells in solid murine neoplasms. III. Cytotoxicity mediated *in vitro* by macrophages recovered from disaggregated regressing Moloney sarcomas. *Journal of Immunology*, **118,** 1574–9.

SAKSELA, E. (1981). Interferon and natural killer cells. In *Interferon III*, ed. I. Gresser, pp. 45–63. London: Academic Press.

SCHER, M. G., BELLER, D. I. & UNANUE, E. R. (1980). Demonstration of a soluble mediator that induces exudates rich in Ia-positive macrophages. *Journal of Experimental Medicine*, **152,** 1684–98.

SEELEY, J. & GOLUB, S. H. (1978). Studies on cytotoxicity generated in human mixed lymphocyte cultures. I. Time course and target spectrum of several distinct concomitant cytotoxic activities. *Journal of Immunology*, **120,** 1415–22.

SONNENFELD, G., MANDEL, A. D. & MERIGAN, T. C. (1977). The immunosuppressive effect of Type II mouse interferon preparations on antibody production. *Cellular Immunology*, **34,** 193–206.

SONNENFELD, G., MANDEL, A. D. & MERIGAN, T. C. (1978). Time and dosage dependence of immunoenhancement by murine Type II interferon preparations. *Cellular Immunology*, **40,** 285–93.

SONNENFELD, G. & MERIGAN, T. C. (1979). The role of interferon in viral infections. *Springer Seminar on Immunopathology*, **2,** 311–38.

STEWART II, W. E. (1979). Varied biologic effects of interferon. In *Interferon I*, ed. I. Gresser, pp. 29–51. London: Academic Press.

TADA, T. (1977). Regulation of the antibody response by T cell products determined by different I subregions. In *Immune System: Genetics and Regulation*, ed. E. E. Sarcarz, L. A. Herzenberg & C. F. Fox, pp. 345–61. New York: Academic Press.

TAYLOR-PAPADIMITRIOU, J. (1980). Effects of interferons on cell growth & function. In *Interferon II*, ed. I. Gresser, pp. 13–46. London: Academic Press.

TIMONEN, T., ORTALDO, J. R. & HERBERMAN, R. B. (1981). Characteristics of human large granular lymphocytes and relationship to natural killer and K cells. *Journal of Experimental Medicine*, **153,** 569–82.

TRINCHIERI, G. & SANTOLI, D. (1978). Anti-viral activity induced by culturing lymphocytes with tumour-derived or virus-transformed cells. Enhancement of human natural killer cell activity by interferon and antagonistic inhibition of susceptibility of target cells to lysis. *Journal of Experimental Medicine*, **147,** 1314–33.

VANKY, F., ARGOV, S. A., EINHORN, S. A. & KLEIN, E. (1980). Role of alloantigens in natural killing. Allogeneic but not autologous tumour biopsy cells are sensitive for interferon-induced cytotoxicity of human blood lymphocytes. *Journal of Experimental Medicine*, **151,** 1151–65.

VIGNEAUX, F. & GRESSER, I. (1977). Differential effects of interferon on the expression of H-2K, H-2D and Ia antigens on mouse lymphocytes. *Journal of Immunology*, **118,** 721–3.

VOSE, B. M. & MOORE, M. (1980). Natural cytotoxicity in humans: Susceptibility of freshly isolated tumour cells to lysis. *Journal of the National Cancer Institute*, **65,** 257–63.

WALLACH, D., FELLOUS, M. & REVEL, M. (1982). Preferential effect of γ interferon on the synthesis of HLA antigens and their mRNAs in human cells. *Nature*, **299,** 833–6.

WEISSMAN, C. (1981). The cloning of interferon and other mistakes. In *Interferon III*, ed. I. Gresser, pp. 101–34. London: Academic Press.

ZARLING, J. M., ESKRA, L., BORDEN, E. C. HOROSZEWICZ, J. & CARTER, W. A. (1979). Activation of human natural killer cells cytotoxic for human leukaemic cells by purified interferon. *Journal of Immunology*, **123,** 63–70.

ZARLING, J. M., SASMAN, J., ESKRA, L., BORDEN, E. C., HOROSZEWICZ, J. S. AND CARTER, W. A. (1978). Enhancement of T cell cytotoxic responses by purified human fibroblast interferon. *Journal of Immunology*, **121,** 2002–4.

THE PRODUCTION OF INTERFERON IN BACTERIA AND YEAST

S. M. KINGSMAN AND A. J. KINGSMAN

Department of Biochemistry, South Parks Road, Oxford OX1 3QU, UK

INTRODUCTION

Until recently the large-scale production of human interferons relied upon natural synthesis following virus stimulation of leucocytes, lymphoblastoid cells or fibroblasts (Stewart, 1979). Yields of naturally produced interferon are low. In the case of interferon-alpha (IFN-α), which is produced by buffy coat cell suspensions yields are limited by the availability of cells so that the total annual production from one major centre of supply is about 10^{11} units (Cantell & Hirvonen, 1977). Theoretical yields of interferon-beta (IFN-β) are higher because fibroblasts can be cultured and about 10^8 units can be produced from 10^{10} cells, that is, about 100 roller bottle cultures (Stewart, 1979). However, optimal production occurs only from limited passage cells and the management of large scale monolayer cultures presents problems. A continuous lymphoblastoid cell line, the Namalwa cell line, derived from a Burkitt's lymphoma patient can be grown in vat cultures to yield about 10^9 units per 800 litre culture (Bridgen *et al.*, 1977; Finter & Fantes, 1980). However, the clinical acceptability of interferon derived from a tumour cell line may be questionable. It is therefore difficult to obtain naturally produced IFN-α and IFN-β on a large scale and there is no system for the large-scale production of interferon-gamma (IFN-γ). A further problem with naturally derived interferon is its heterogeneity, for example, primary cultures of human buffy coat cells stimulated with Sendai virus produce predominantly IFN-α but a small amount of IFN-β is also produced (Havell *et al.*, 1975). Up to 20% of the interferon activity produced by Namalwa cells after induction with Newcastle disease virus (NDV) is IFN-β (Havell, Yip & Vilček, 1978), and in human diploid fibroblasts induced with NDV between 2 and 20% of the total interferon activity is IFN-α and the rest is IFN-β (Hayes, Yip & Vilček, 1979). In addition IFN-α comprises at least ten different but closely related gene products (Nagata, Mantei & Weissman, 1980a; Goedel *et al.*,

1981; Brack, Nagata, Mantei & Weissman, 1981); eight of these have been differentiated by nucleotide and/or amino acid sequence analysis and have been designated IFN-αA,B,C,D,E,F,G and H (Goeddel *et al.*, 1981). (Several of these polypeptides have an alternative nomenclature: IFN-αA is considered to be the same as IFN-α2 (Streuli, Nagata & Weissman, 1980) since they differ in only one amino acid at position 23; while IFN-αD is considered to be the same as IFN-α1 (Nagata *et al.*, 1980b) since they differ only at amino acid position 114; and IFN-αF is considered to be most like the predominant interferon in NDV-induced lymphoblastoid cells (Zoon *et al.*, 1980) and has been designated IFN-α3 (Streuli *et al.*, 1980). IFN-αB,C,G and H are most like IFN-αD and have also been classified as IFN-αl (Weck, Apperson, May & Stebbing, 1981a). IFN-αE is a pseudogene which is transcribed but not translated into a functional protein.) The relative amounts of the different IFN-α polypeptides vary in different cell types, for example IFN-αA and IFN-αD are the most abundant interferon proteins in the myeloblastoid KG-1 cell line (Goeddel *et al.*, 1980b) and in human leucocyte interferon (Nagata *et al.*, 1980b), whereas Namalwa cells produce predominantly IFN-αF when stimulated with NDV (Zoon *et al.*, 1980) and a mixture of at least five different polypeptides, but predominantly IFN-αA and IFN-αD, when stimulated with Sendai virus (Allen & Fantes, 1980). It is becoming apparent (see later) that the different interferon polypeptides singly or in various combinations have distinct biological properties. In order therefore to analyse the diverse biological properties of the interferons, for example as antiviral, anticellular and immunoregulatory agents (Stewart, 1979), to assess structure-function relationships of the different polypeptides and to define their full clinical potential it is important to improve the large scale synthesis of interferons and moreover to synthesize each polypeptide individually. One approach has been to manipulate microorganisms to synthesize the different interferons from cloned interferon genes. The interferon genes can be introduced into, and stably maintained in, dividing cultures of bacteria and yeast by incorporating them into plasmid vectors which replicate in these host organisms. However, the interferon genes are expressed only if they are linked to control sequences recognized by the host organisms as signals for transcription of the interferon gene and translation of its messenger RNA (mRNA). In this review we will describe some basic plasmid vectors and gene expression signals which have been used to introduce different interferon coding sequences into *Escherichia coli* and *Saccharomyces cerevisiae* and to direct the synthesis of 10^8 to 10^{10} units of an interferon from a single litre of simple batch culture. We will then describe what is known about the biological properties of the different interferon polypeptides produced in microorganisms and show how interferon genes can be manipulated to allow the synthesis in microorganisms of novel interferons which could not be produced naturally.

GENERAL PROPERTIES OF INTERFERON POLYPEPTIDES

All three types of interferon, IFN-α, IFN-β and IFN-γ are synthesized as preinterferons (preIFN) with a hydrophobic signal sequence of 23, 21 and 20 amino acids respectively, which is removed by cleavage to produce the mature polypeptide which has 166 (165 for IFN-αA), 166 and 146 amino acids respectively (e.g. Goeddel *et al.*, 1981; Taniguchi, Ohno, Fujii-Kuriyama & Muramatsu, 1980c; Gray *et al.*, 1982). IFN-β (Tan *et al.*, 1979) and probably IFN-γ (Yip, Pang, Urban & Vilcek, 1981) are glycoproteins and potential sites for glycosylation are at amino acid residue 80 in IFN-β (Derynck *et al.*, 1980a) and at amino acid position 29 and 101 in IFN-γ (Gray *et al.*, 1982). It is unlikely that IFN-α polypeptides are glycosylated (Allen & Fantes, 1980) and it seems that glycosylation is unnecessary for most if not all biological properties of the interferons (Bose *et al.*, 1976). The formation of intramolecular disulphide bridges is however important for the biological activity of IFN-α and IFN-β; disulphide bridges are formed between the cysteine residues at positions 1 and 98 and between 29 and 138 in IFN-αA (Wetzel, 1981) and a single disulphide bridge is probably formed between residues 31 and 141 in IFN-β (Shepard, Leung, Stebbing & Goeddel, 1981). When the mature polypeptide amino acid sequences of the different IFN polypeptides are compared (Streuli *et al.*, 1980; Goeddel *et al.*, 1981) there is about 84% homology between the IFN-α polypeptides, and about 29% homology between the IFN-α and IFN-β polypeptides (Taniguchi *et al.*, 1980b). However, only 12 out of 146 amino acid positions are shared or show only conservative changes when IFN-γ is compared with IFN-α and IFN-β (Epstein, 1982). The homology between IFN-α polypeptides and IFN-β is localized in two conserved domains, I and II, in the mature sequence (Taniguchi *et al.*, 1980b). Domain I is located between amino acid residues 28 and 80 and about 41% of the amino acids are congruent; a similar degree of congruence is seen when all the different IFN-α polypeptides are compared (Goeddel *et al.*, 1981). Domain II is located between amino acid residues 115 and 151, there is 84% congruence between the different IFN-α polypeptides in this region (Goeddel *et al.*, 1981) and about 54% congruence between IFN-α and IFN-β. In addition amino acids 132 to 151 in the second domain show almost 100% congruence between the IFN-α polypeptides (Allen & Fantes, 1980; Goeddel *et al.*, 1981). These conserved domains may define functionally significant regions of the polypeptides and it is perhaps significant that the limited number of shared amino acids between IFN-γ and IFN-α occur predominantly in domain II (Gray & Goeddel, 1982). Homologies between IFN-α and IFN-β amino acid sequences are also reflected in the nucleotide sequences which show an average homology of 45% in the coding sequence and the conserved domains are still evident although not as marked as in the amino acid sequence (Taniguchi *et al.*, 1980b). In addition there are a number of features of the nucleotide sequence such as repeated sequences and palindromes which may be important in forming secondary structures in the DNA or mRNA which might affect the efficiency of gene expression (Edge *et al.*, 1981). (For a detailed discussion of the structure of different interferons and chromosomal genes and their flanking regions see the article by J. Collins, p. 35).

THE PRODUCTION OF INTERFERON IN BACTERIA

Plasmid vectors for the introduction of foreign genes into bacteria have been reviewed (Old & Primrose, 1981) and are generally based on the plasmid pBR322 (Bolivar *et al.*, 1977). They have DNA sequences which allow the plasmid to replicate autonomously in *E.coli* and various antibiotic resistance genes e.g. for ampicillin or tetracycline resistance to allow detection and selection of the plasmid in *E.coli.* Interferon genes inserted into these basic vectors will not be expressed at high levels in *E.coli* because precise bacterial sequences are required to direct accurate and efficient transcription of the gene and translation of the mRNA. In addition, because there is no splicing activity in *E.coli*, the coding sequence for the interferon polypeptide must not contain introns. Therefore to express IFN-γ, which has three introns (Gray & Goeddel, 1982), the complementary DNA (cDNA) sequence must be used. IFN-α and IFN-β have no introns (e.g. Brack *et al.*, 1981; Tavernier, Derynck & Fiers, 1981) and therefore either genomic or cDNA sequences can be used.

The way genes are expressed and regulated in *E.coli* is largely understood. Nucleotide sequences which control gene expression are found adjacent to the start of the mRNA and within the transcribed leader sequence (L) of the mRNA which precedes the amino acid coding sequence. The basic sequences controlling transcription are a promoter (P) at which RNA polymerase binds to initiate transcription and an operator (O) which interacts with regulatory proteins to either stimulate or inhibit transcription. Bacterial promoter regions contain two conserved sequences, the first, the Pribnow box has the consensus sequence TATPuATG and is found 5 to 10 bp upstream from the start of the mRNA (Pribnow, 1975). The second conserved sequence is located 27 to 37 bp upstream from the mRNA start and has the consensus sequence TTGACA (Siebenlist, 1979). Both these sequences define major sites of interaction between the DNA and the RNA polymerase (Siebenlist, Simpson & Gilbert, 1980). The operator region is usually between 20 and 40 bp long and centres around an axis of two-fold rotational symmetry. Regulatory proteins bind to the operator and either prevent transcription (negative control) or stimulate transcription (positive control). Transcription is initiated upstream from the coding sequence and the precise nucleotide sequence of this leader may be important in determining the efficiency of gene expression (Ganoza, Fraser & Neilson, 1978). This may relate to a requirement for the formation of secondary structure for initiation of translation (Gold *et al.*, 1981). The sequence largely controlling the efficiency of translation is the ribosome binding site (RB) which comprises sequences located in the leader region of the mRNA and in the coding sequence surrounding the AUG or GUG which defines the amino-terminal (N-terminal) methionine initiation codon of the polypeptide. In particular there is a sequence 3 to 9 bp long referred to as the Shine-Dalgarno (SD) sequence located 3 to 11 bp upstream from the ATG for the N-terminal methionine. This is complementary to the 3′ end of the 16S ribosomal RNA (rRNA) and probably promotes ribosome binding via pairing with this rRNA (Shine & Dalgarno, 1975). The entire sequence containing the RNA polymerase binding site, the operator and the SD sequence is often referred to as a promoter but will be referred to here as an expression element. There may also be sequences located adjacent to the end of the coding sequence in a transcribed trailer sequence which play a role in stabilizing the mRNA by the formation of secondary structures (e.g. Gottesman, Oppenheim & Court, 1982). Additional regulatory features may also be superimposed on these basic control sequences.

The spatial arrangement of these prokaryotic transcription and translation signals is shown in Fig. 1 for the tryptophan and lactose operons of *E.coli* which have both been used to direct interferon expression. The mRNA start for the *trp* operon lies 162 bp upstream from the N-terminal ATG of the *trp*E coding sequence within a 222 bp expression element. Immediately adjacent and upstream is a 60 bp promoter containing the two conserved sequences at 7 bp and 26 bp upstream from the mRNA start, although the Pribnow box sequence, TTAACT, is somewhat different from the consensus sequence. In the presence of high concentrations of tryptophan, expression of the operon is reduced 70-fold by a repressor which binds at the operator site located within this 60 bp promoter region. Transcription is also reduced about 10-fold by transcription termination at a site, termed attenuator (A), which is located within the transcribed leader approximately 140 bp downstream from the mRNA start site. There is a ribosome-binding site at the beginning of the leader sequence and an open reading frame for a 14 amino acid leader polypeptide. The location of ribosomes translating the leader determines the formation of secondary structure in the mRNA which results in transcription termination in the presence of high concentrations of tryptophan (for a detailed discussion of attenuation see Yanofsky, 1981). Maximum expression directed by the *trp* expression element is achieved by inactivating the repressor either by growing the culture in the absence of tryptophan or in the presence of competitive inhibitors of tryptophan such as indole acrylic acid (IAA) or indole-propionic acid (IPA) (for a detailed account of the tryptophan operon see Platt, 1980). The lactose operon contains a 128 bp expression element consisting of a 44 bp leader sequence containing a SD sequence and an operator sequence but no attenuator, and an 84 bp promoter sequence which contains the two conserved elements and an additional sequence (C). This sequence binds a protein (CAP, catabolite gene activator protein) which is necessary for the formation of a stable transcrip-

(*a*) Tryptophan

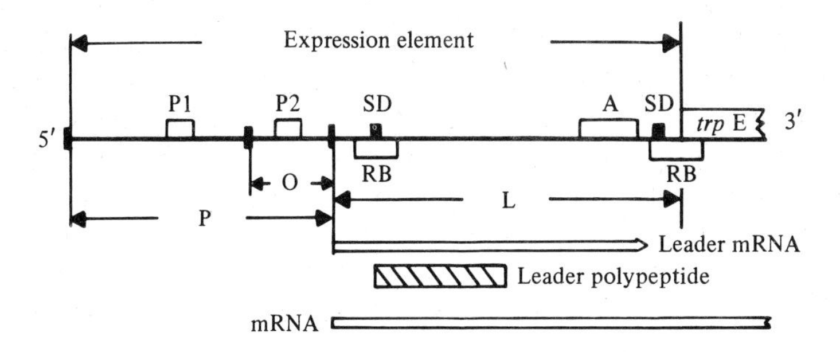

(*b*) Lactose

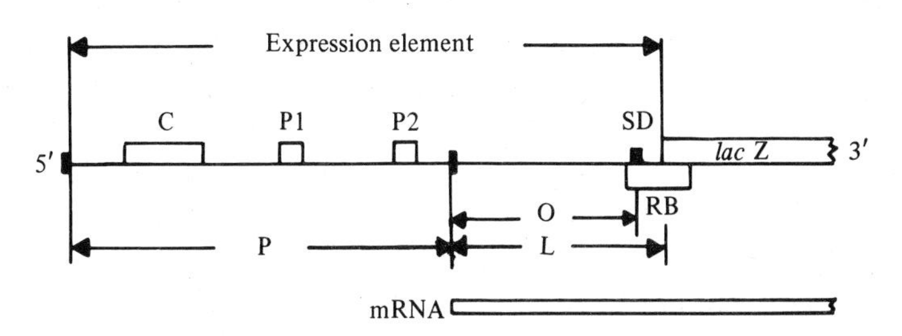

Fig. 1. Expression elements of the trytophan and lactose operons of *E.coli.* The spatial relationship of control features is given as a line drawing to represent the sense strand of the DNA. The drawings are not to scale; for details of nucleotide sequence refer to relevant articles in Miller & Reznikoff (1980). A, attenuator; C. catabolite gene activator protein-binding site; L, leader sequence; O, operator; P, promoter; P1, conserved sequence at −27 to −37; P2, Pribnow box; SD, Shine-Dalgarno sequence; RB, ribosome-binding site.

tion initiation complex. CAP will only bind in the presence of cyclic AMP (cAMP), and because cAMP levels are low during growth on glucose the operon is repressed under these conditions (catabolite repression). A mutation, UV5 (Miller & Reznikoff, 1980) which alters the Pribnow box sequence from TATGTTG to TATAAT allows a stable initiation complex to form in the absence of CAP binding and renders the operon insensitive to catabolite repression. The lactose operon is induced by growth on lactose or, more efficiently, with the non-metabolized analogue iso-propyl β-D thiogalactoside (IPTG) (for a detailed discussion of the lactose operon see Reznikoff & Abelson, 1980).

The basic regulatory features described above have been manipulated to maximize expression of interferon in *E.coli.* Efficient transcription of the interferon gene has been achieved by using efficient prokaryotic expression elements such as those from the tryptophan or lactose operons (Fig. 1), or the strong promoter P_L from bacteriophage lambda (e.g. Remaut, Stanssens & Fiers, 1981). In addition, transcription levels can be increased by inactivating repressor proteins, by increasing the number of expression elements preceding the interferon gene or by increasing the copy number of the expression plasmid. In order to ensure efficient translation of the interferon mRNA it is important to provide a ribosome-binding site. If the complete bacterial ribosome-binding site is used this must include at least the N-terminal methionine initiation codon of the bacterial polypeptide and probably a few additional bacterial codons. The bacterial codons are therefore linked to the interferon codons and are translated to form a hybrid polypeptide. An expression vector which uses all the prokaryotic transcription signals and a complete bacterial ribosome-binding site is referred to as a transcription/translation fusion vector as it produces both a hybrid mRNA and a hybrid polypeptide. Interferons produced by such vectors will be fusion proteins and are referred to as f-IFN. To produce an interferon polypeptide unlinked to any bacterial codons it is necessary to use the prokaryotic transcription signals but only a part of the bacterial ribosome-binding site up to but not including the ATG initiation codon for the bacterial polypeptide. The ATG for translation initiation must therefore derive from the interferon gene to create a hybrid ribosome-binding site composed of a bacterial SD sequence and a eukaryotic ATG and surrounding sequences. In this configuration a hybrid mRNA is produced but translation initiates on the AUG of the interferon coding sequence to generate an authentic interferon polypeptide. This type of expression vector is termed a transcription fusion vector. Both the distance between the SD and the ATG and the actual nucleotide sequence in a hybrid ribosome-binding site are critical for maximum expression of the eukaryotic gene (Roberts, Kacich & Ptashne, 1979). However, these parameters vary for different 'promoter'-heterologous gene combinations and must be determined empirically. The general configurations of bacterial expression vectors are outlined in Fig. 2 and the same principles are applicable to the yeast expression vectors discussed later.

Both types of expression configurations have been used to direct the expression of interferon in bacteria. If the entire interferon-coding sequence is used then a preIFN polypeptide comprising the signal and mature peptide is produced. In order to express interferon without the signal sequence the interferon gene must be (*a*) Homologous bacterial gene expression

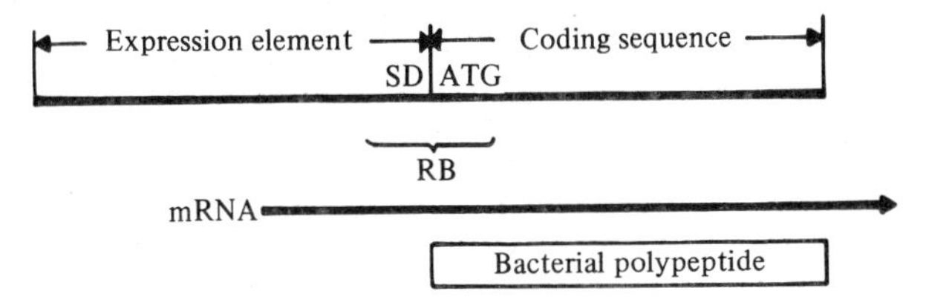

(*b*) Heterologous gene expression

(1) Transcription and translation fusion vector

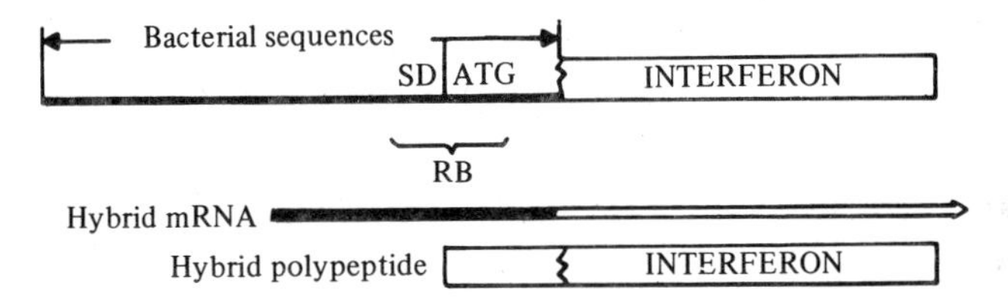

(2) Transcription fusion vector

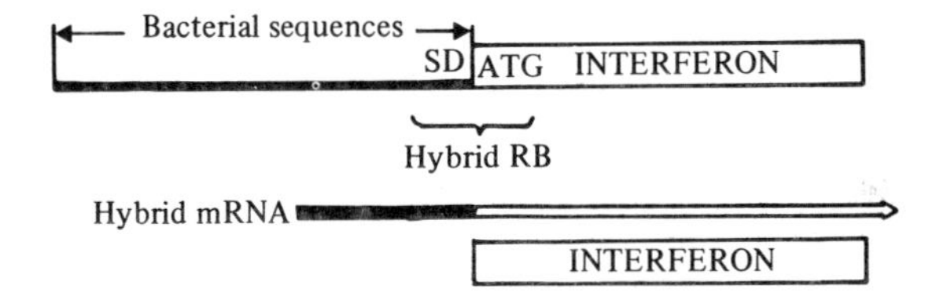

Fig. 2. Generalized expression configurations. The spatial arrangement of bacterial expression elements and homologous coding sequences (*a*) and bacterial expression elements and interferon coding sequences in a transcription and translation fusion vector (*b*1) and a transcription fusion vector (*b*2) configuration is shown. See text for explanation. SD, Shine-Dalgarno sequence; RB, ribosome-binding site; ATG, initiation codon.

manipulated *in vitro*, firstly to remove the nucleotides encoding the signal peptide and secondly to provide an ATG for translation initiation. In the case of IFN-β the mature polypeptide starts with a methionine residue and this can be used as the initiation codon for translation in transcription fusion vectors. However, the mature IFN-α and IFN-γ polypeptides both have cysteine as the first amino acid and therefore an initiator ATG codon has to be added to the sequence *in vitro* to ensure translation of the polypeptide in a transcription fusion vector (e.g. Goeddel *et al.*, 1980b). This modified interferon is referred to as met-interferon (met-IFN) and the product in bacteria often retains the methionine residue at its amino terminus (Wetzel *et al.*, 1981) so that the bacterial product may not always be identical to the natural interferon.

The first interferon to be expressed in bacteria was IFN-αD. An entire cDNA fragment containing the sequence coding for preIFN flanked by 5′ and 3′ non-coding sequences was inserted into the β-lactamase gene of deletion derivatives of pBR322 in all three reading frames to generate a potential fusion protein between β-lactamase and preIFN (Nagata *et al.*, 1980b). This transcription/translation fusion vector in fact produced preIFN rather than a f-IFN presumably as a result of translation initiation at the N-terminal AUG of the preIFN coding sequence. Yields were low, of the order of 20 000 u/litre. Goeddel *et al.* (1980b) inserted a cDNA sequence for preIFN-αA and non-coding flanking sequences into a transcription/translation fusion vector containing expression elements derived from the tryptophan operon. These were the promoter-operator region and 50 bp of the leader region to include the ribosome-binding site and eight amino acids of the leader polypeptide to generate a potential leader polypeptide-preIFN-αA fusion protein. Again the only product detected was preIFN-αA and yields were low, about 480 000 u/litre. The low yields in these two studies could be explained if only translation products that initiated at the *N*-terminal AUG of preinterferon were stable and/or had interferon activity and this initiation event was probably rare because no correct bacterial ribosome binding site flanked this AUG. Translation may also have initiated at the bacterial AUG for β-lactamase or for the *trp* leader polypeptide respectively but the resulting fusion polypeptide may have been rapidly degraded or inactive. Several transcription/translation fusion constructions have however yielded fusion proteins, with biological properties identical to naturally derived interferons and in reasonable yields. Expression signals from part of the *lac* operon derived from the bacteriophage lambda *plac*5 fragment in a derivative of the single stranded phage M13, M13mp7 (Messing, Crea & Seeburg, 1981) have been used to produce a *lac*Z-IFN-α2 fusion polypeptide (Slocombe, Easton, Boseley & Burke, 1982). The IFN-αA had part of the signal sequence removed so that the fusion protein had 10 N-terminal amino acids derived from β-galactosidase, the N terminal methionine being removed, eight amino acids derived from the signal sequence (S16–S23) and the 165 amino acids of the mature

IFN-αA polypeptide. The nature of these N-terminal amino acids was confirmed by amino acid sequence analysis (R. M. King, D. C. Burke, F. Northrop & D. S. Secher, personal communication). Expression from the *lac* promoter was induced with IPTG to obtain up to 2×10^9 u/litre of the fusion protein. The reason for these high yields is probably due in part to the high copy number of M13 phage, about 200–300 copies/cell as compared to pBR322 with 20 copies per cell. In this case the bacterial and signal amino acids did not markedly reduce the stability of the fusion protein. IFN-β has also been expressed as a fusion protein, Derynck *et al.* (1980b) have fused either 82 amino acids of the β-lactamase polypeptide or 98 amino acids of the replicase protein of phage MS2 to preIFN-β and part of its 5′ flanking sequence. The flanking sequence contributed one additional amino acid and 27 additional amino acids in the respective constructions and the expected fusion proteins were produced as indicated by molecular weight in SDS-polyacrylamide gel electrophoresis. Expression was controlled by the bacteriophage lambda promoter P_L and was regulated by introducing the vector into a defective lysogen which synthesizes a temperature-sensitive repressor protein (Remaut *et al.*, 1981). At low temperatures (28 °C) there is no expression of the IFN-β fusion polypeptide but raising the temperature to 42 °C inactivates the repressor and allows transcription. The yields of the f-IFN-β were relatively low, in part due to poor recovery because of non-specific binding to bacterial components and to sequestration of the protein in the periplasmic space. Lysis with SDS and 5M urea improved recoveries and yields were about 10 000 u/litre. Analysis of the molecular weight distribution of the products showed that some degradation had occurred but there was no strong evidence for any specific cleavage at the signal-mature junction. Some polypeptides the size of mature IFN-β were detected but this could have represented rare translation initiation at the AUG at the signal-mature junction. It is possible that low yields resulted from an increased turnover of f-IFN-β due to the presence of bacterial and signal sequences. However bacterial sequences *per se* do not seem to cause extreme lability of IFN-β fusion polypeptides. When 16 amino acids of the *trp*E protein are fused to mature IFN-β a reasonably stable, active fusion protein is produced (Houghton *et al.*, 1981). It is most likely that the signal amino acids in IFN-β make the product unstable and preIFN-β is in fact degraded more rapidly than mature IFN-β in *E.coli* when expressed in transcription fusion vectors (Taniguchi *et al.*, 1980a).

On the whole more reproducible results in terms of yield have been obtained with transcription fusion vectors. A 330 bp expression element containing the *trp* promoter-operator region and part of the leader sequence including the SD but stopping 4 bp short of the ATG for the leader polypeptide has been used to express various interferons in *E.coli* (Goeddel *et al.*, 1981; Shepard, Yelverton & Goeddel, 1982). This 'portable promoter' was positioned adjacent to and upstream of the coding sequence for met-IFN-αA to create a hybrid ribosome-binding site with an 11 bp gap between the bacterial SD and the interferon ATG. Expression was maximized by growing the culture in the absence of tryptophan and yields were about 2.5×10^8 u/litre. This *trp* 'portable promoter' has also been used to direct expression of met-IFN-αB to give 8×10^7 u/litre (Yelverton *et al.*, 1981), of met-IFN-γ to give approximately 2.5×10^3 u/litre (Gray *et al.*, 1982) and of mature IFN-β to give 2×10^7 units/litre (Goeddel *et al.*, 1980a; Shepard *et al.*, 1982). Expression of IFN-β was increased four fold by placing three copies of the 'promoter' fragment upstream from the IFN-β coding sequence (Goeddel *et al.*, 1980a). To obtain maximum yields of interferon in transcription vectors it has been essential to optimize the structure of the hybrid ribosome-binding site. A distance of nine nucleotides between the SD and the ATG was optimal for expression of both met-IFN-αA and IFN-β, and yields relative to the optimum were for example, 0.4% with a 2 bp gap, 50% with an 8 bp gap and 1.6% with a 15 bp gap (Shepard *et al.*, 1982). In general the configurations which allowed the formation of significant secondary structure involving the AUG were most efficient except where the intervening sequence was GC rich which tended to reduce translation. Taniguchi *et al.* (1980a) have shown that a gap of 7 bp between the SD of the *E.coli lac*Z gene and the interferon ATG gives expression of both preIFN-β and mature IFN-β although levels of mature IFN-β were relatively low with this system. Higher levels of mature IFN-β directed by a *lac* promoter have been obtained using a double *lac*UV5 fragment (Goeddel *et al.*, 1980a). The fragment contains two 95 bp *lac* UV5 'promoter fragments' in the same orientation separated by a 94 bp heterologous fragment. Each *lac* fragment contains a *lac*Z SD sequence but stops two nucleotides away from the ATG of the *lac*Z gene (Backman & Ptashne, 1978). This fragment was placed upstream and adjacent to a mature IFN-β coding sequence with one additional nucleotide preceding the ATG to give a 6bp space between the *lac* SD and the IFN-β ATG and yielded about 2250 molecules/cell. A comparison of the structure of various hybrid ribosome-binding sites and the relative interferon yields is given in Table 1.

Recently a complete met-IFN-αD gene has been chemically synthesized and where possible, nucleotides were chosen to effect a more favourable codon bias for expression in *E.coli*. In particular the number of CG residues was increased (Edge *et al.*, 1981). The synthetic gene has been expressed in *E.coli* under control of a 95 bp *lac* UV5 'promoter fragment' creating a hybrid ribosome-binding site with 10 bp between the *lacZ* SD and the ATG of the synthetic coding sequence (De Maeyer *et al.*, 1982). The synthetic gene produces biologically active interferon with the same characteristics as bacterially produced met-IFN-αD derived from the natural sequence. Maximum yields were about 1.3×10^6 units/1.5×10^{10} cells which is comparable to yields obtained with the natural sequence in other expression systems. It might be interesting to compare the interferon yield of the synthetic gene with the natural gene in the same expression system to show how the altered codon bias of the synthetic gene affects yields, if at all. The ability to synthesize an interferon gene which produces a fully biologically active polypeptide opens the way for synthesis of altered polypeptides which may have novel properties (see later).

We have attempted to summarize some of the information about yields of interferon produced in bacteria to indicate the parameters used to calculate these yields (Table 2). It is hard to assess the efficiencies critically because different interferon polypeptides have been expressed in different systems and titres and may vary considerably depending on the cell type and virus used in the assay. Standardization of the data to molecules per cell is also difficult because specific activity values for such calculations are derived for naturally produced and therefore heterogeneous interferon and it is becoming apparent that individual interferon polypeptides differ markedly in their specific activities depending upon the assay system used (see later). Overall, the *trp* promoter seems to give higher yields than the *lac* or P_L promoters except where the *lac* promoter has been used on a very high copy number vector (Slocombe *et al.*, 1982). The *trp* promoter may be most useful commercially because expression can be maximized simply by growing the culture in the absence of tryptophan, and it is unlikely that systems requiring high temperatures or chemical analogues would be favoured for economic production. Yields of the IFN-α

Table 1. *The structure of hybrid ribosome-binding sites in* E.coli *transcription expression vectors*

	SD/ATG Distance	Interferon type	Yield[a]	Reference
(a) *lac* expression signals				
AGGAAACAGCTATG	7			Normal *lac* RB
AGGAAACAGGATCCATG	10	Synthetic IFN-αD	1.3×10^6U/1.5×10^{10} Cells	De Maeyer *et al.*, 1982
AGGAAACAGCCATG	7	IFN-β	50 mol/cell	Taniguchi *et al.*, 1980*b*
AGGAAACAGACATG	7	preIFN-β	–	Taniguchi *et al.*, 1980*b*
AGGAAACAGAATTCATG	10	IFN-β	2250 mol/cell	Goeddel *et al.*, 1980*a*[b]
(b) *trp* expression signals				
AAGGGTATCGACAATG	7			Normal *trp* RB
AAGGGTATCGAATTCATG	9	metIFN-αA	10^5mol/cell	Shepard *et al.*, 1982
AAGGGTATCTAGCT AGAATTCATG	15	metIFN-αA	1.6×10^3 mol/cell	Shepard *et al.*, 1982
AAGGGTATAATTCATG	7	metIFN-αA	4.4×10^4 mol/cell	Shepard *et al.*, 1982
AAGGGTATCTTTCCATG	8	IFN-β	10^7 U/5×10^{11} cells	Houghton *et al.*, 1981
AAGGGTATCCATG	4	IFN-β	undetectable	Houghton *et al.*, 1981
AAGGGTATCTACTAGATG	9	IFN-β	2×10^4 mol/cell	Shepard *et al.*, 1982
AAGGGTATCTAGATG	6	IFN-β	7.4×10^3 mol/cell	Shepard *et al.*, 1982
AAGGGTATATG	2	IFN-β	80 mol/cell	Shepard *et al.*, 1982

[a] Interferon yields are given as reported in the cited reference and no attempt has been made to standardize data. U = Units of interferon as defined in the cited reference. mol/cell = molecules/cell

[b] A double *lac* promoter was used in this study.

The SD and ATG are underlined in each ribosome-binding sequence.

Table 2. *Levels of interferons produced in* E.coli

Expression system	Molecule expressed	Units/l[a]	Cells/l	Molecules/cell[b]	Reference
β-lactamase	pre IFN-αD	2×10^4	–	–	Nagata *et al.*, 1980*b*
trp	pre IFN-αA	4.8×10^5	3.5×10^{11}	120[d]	Goeddel *et al.*, 1980*b*
lac	lacZ-pre IFN-αA	2×10^9	–	–	Slocombe *et al.*, 1982
P_L	MS2-pre IFN-β	10^4	3×10^{10}	–	Derynck *et al.*, 1980*b*
trp (7)	met IFN-αA	–	3.5×10^{11}	10^5[c]	Shepard *et al.*, 1982
lac UV5	IFN-β	–	–	50[c]	Taniguchi *et al.*, 1980*b*
Double *lac* UV5	IFN-β	9×10^6	3.5×10^{11}	2.25×10^3[d]	Goeddel *et al.*, 1980*a*
trp (6)	IFN-β	1.8×10^7	3.5×10^{11}	4.5×10^3[d]	Goeddel *et al.*, 1980*a*
Triple *trp* (6)	IFN-β	8×10^7	3.5×10^{11}	2×10^4[d]	Goeddel *et al.*, 1980*a*
trp (9)	IFN-β	–	3.5×10^{11}	2×10^4[c]	Shepard *et al.*, 1982
lac UV5	Synthetic met IFN-αD	1.3×10^6	1.5×10^{10}	–	De Maeyer *et al.*, 1982
trp	met IFN-γ	2.5×10^4	3.5×10^{11}	–	Gray *et al.*, 1982

[a] Units of interferon are as defined in the reference cited.
[b] Values are taken directly from the reference cited.
[c] A specific activity of 2×10^8 units/mg assumed.
[d] A specific activity of 4×10^8 units/mg assumed.
Numerals in parentheses denote distance between the SD and ATG
A dash indicates data not available.

polypeptides are generally better than of IFN-β or IFN-γ and the maximum reported yield of IFN-α is 10^5 molecules per cell (Shepard *et al.*, 1982), i.e. about 1–10 mg of IFN-α from a litre of a simple batch culture. The maximum reported yield of IFN-β is about 2×10^4 molecules per cell (Shepard *et al.*, 1982). Possible reasons for the lower yields of IFN-β may be that the protein is degraded in *E. coli* and this seems to be most marked if the signal sequence is retained (Taniguchi *et al.*, 1980c; Derynck *et al.*, 1980b). However, significant degradation is also seen with mature IFN-β (Taniguchi *et al.*, 1980a). It is also possible that recovery of active IFN-β is more difficult because of binding to *E.coli* proteins and sequestration in the periplasmic space (Derynck *et al.*, 1980b) although this latter has also been observed with IFN-αD (Mantei *et al.*, 1980). Constructions which contain all or most of the IFN-α signal sequence (Nagata *et al.*, 1980b; Goeddel *et al.*, 1980b) also give lower yields than constructions with most or all of the signal sequence removed (e.g. Shepard *et al.*, 1982; Slocombe *et al.*, 1982). The reasons for the low yields of IFN-γ in *E.coli* are not yet clear but it is possible that the lack of post-translational modifications may reduce the detectable biological activity (Gray *et al.*, 1982).

Some preliminary studies have shown that interferon can be expressed in bacteria other than *E.coli*, e.g. *Methylophilus methylotrophus* (De Maeyer *et al.*, 1982) and *Bacillus subtilis* (cited in Hardy, Stahl & Kupper, 1981). These organisms might have advantages over *E.coli* in industrial processes as, for example, *M.methylotrophus* uses cheap substrates such as methanol and ammonia, and *B.subtilis* secretes many products.

EXPRESSION OF INTERFERON IN YEAST

Although reasonably high yields of interferon have been produced in bacteria there is considerable interest in developing alternative expression systems and one of the approaches has been to use *Saccharomyces cerevisiae* (yeast) (Hitzeman *et al.*, 1981; Tuite *et al.*, 1982; Dobson *et al.*, 1982a). It is possible that the expression of a eukaryotic gene in a eukaryotic host may result in higher yields of interferon and may provide a more economical means of production. A further reason for turning to yeast is derived from the desire to produce interferon for clinical use. Bacteria contain toxic and pyrogenic factors necessitating stringent purification of the interferon before use whereas yeast has no pathogenic relationship with man and may yield a more acceptable interferon for clinical use. It is also clear that although bacteria can process some eukaryotic products (Talmadge, Kaufman & Gilbert, 1980) they probably fail to process preIFN and in order to synthesize mature IFN-α and mature IFN-γ an N-terminal methionine codon has to be added to the mature sequence. This methionine residue is retained in at least 50% of the met-mature IFN-α produced in bacteria (Wetzel *et al.*, 1981) and it may subtly alter the biological properties of the interferon but perhaps more significantly it may render interferon antigenic in man. It is possible that yeast may process either preIFN correctly or may remove the N-terminal methionine from met-mature interferons because few yeast proteins retain their own N-terminal methionine residues. In addition, because the secretory pathways in yeast are generally similar to those in higher eukaryotes (Novick, Field & Schekman, 1980) it is possible that interferon will be secreted into the culture medium. This could be useful if the protein is liable to intracellular degradation and it may facilitate product purification.

Techniques for introducing DNA into yeast have been described (Hinnen, Hicks & Fink, 1978) and several plasmid vectors are available (Beggs, 1978; Kingsman, Clarke, Mortimer & Carbon, 1979; Struhl, Stinchcomb, Scherer & Davis, 1979). The essential features of these vectors are a system to allow the plasmid to replicate autonomously in yeast and a selective marker to allow detection and selection of the plasmid. Most yeast vectors also incorporate parts of an *E.coli* plasmid, usually pBR322, to allow plasmid replication and selection in *E.coli* which is the preferred host for manipulating genes and preparing plasmid DNA. Two yeast replication systems are available. The first uses a replication origin derived from an endogenous yeast plasmid, the 2 μm circle (Broach, 1982); either the whole plasmid (Beggs, 1978) or a part carrying the replication origin (e.g. Dobson *et al.*, 1982b) have been used. The 2 μm based plasmids are maintained at a high copy number, from 20 to 200 per cell, and can be stably maintained for at least forty generations in the absence of selection (M. J. Dobson, A. J. Kingsman and S. M. Kingsman, unpublished data). The second system uses replication origins derived from yeast chromosomal DNA known as autonomous replication sequences (*ARS*). A typical *ARS*-based vector is plasmid YRp7 (Struhl *et al.*, 1979). These *ARS*-based plasmids replicate autonomously but inefficiently, being lost from replicating yeast cells even in the presence of selection so that less than 80% of cells in a culture contain plasmid (Kingsman *et al.*, 1979). The replication of these plasmids can be stabilized by the addition of a second class of chromosomal DNA sequences (CEN). These are centromeric or centromere-related sequences which cause *ARS* plasmids to be maintained as low copy number minichromosomes (Clarke & Carbon, 1980). Various selective markers have been used; genes for nutritional selection e.g. *LEU2* (e.g. Beggs, 1978), and *TRP1* (e.g. Kingsman *et al.*, 1979) are used most widely but it is also possible to use dominant drug resistance markers such as the genes for resistance to the antibiotic G418 (Jiminez & Davies, 1980) or the Herpes virus thymidine kinase gene which confers resistance to folate antagonists (McNeil & Friesen, 1981). There are to date few studies of heterologous eukaryotic gene expression in yeast (Beggs, van den Berg, van Ooyen & Weissman, 1980; Hitzeman *et al.*, 1981; Dobson *et al.*, 1982a; Kiss *et al.*, 1982; McNeil & Friesen, 1981; Tuite *et al.*, 1982; Valenzuela *et al.*, 1982) but it appears that there are two major considerations. Firstly the gene must not contain introns because although some yeast genes contain introns (Gallwitz & Sures, 1980; Bollen *et al.*, 1982), heterologous introns on plasmid borne genes are apparently not processed correctly (Beggs *et al.*, 1980). Secondly yeast gene expression signals must be used to direct transcription initiation as heterologous signals are apparently not functional (Beggs *et al.*, 1980; Kiss *et al.*, 1982). Very little is known about the DNA sequence requirements for gene expression in eukaryotes but it has been assumed that as with prokaryotes important regulatory features will be located in the DNA immediately flanking the coding sequences. At least two sequences seem to be important for transcription initiation in higher eukaryotes, the first, the TATA box, has the canonical sequence TATAT/AAT/A and is usually located 30 bp upstream from the start of the mRNA (Gannon *et al.*, 1979). The second, the CAAT box with the canonical sequence GGC/TAATCT is located about 80 bp upstream from the start of the mRNA (Benoist, O'Hare, Breathnach & Chambon, 1980). Both sequences seem to be important for efficient transcription *in vivo* (e.g. Grosveld, de Boer, Shewmaker & Flavell, 1982). Certain features are also thought to be important for efficient translation; Kozak (1981) has suggested that the nucleotide environment around the ATG may be important and in some cases possible ribosome-binding sequences have been identified (Zalkin & Yanofsky, 1982).

The 5′ flanking regions of several yeast genes have been characterized and they seem to display different features depending on whether the gene encodes an abundant or rare gene product. The nucleotide sequence and general features of the 5′ expression elements of two yeast genes is shown in Fig. 3. The first is the *PGK* (phosphoglycerate kinase) gene (Dobson *et al.*, 1982b) whose product is present in the cell as 1–5% of total cell mRNA and protein and the second is the *TRP*1 (N-(5′-Phosphoribosyl)-anthranilate isomerase) gene (Dobson *et al.*, 1982a) which encodes a relatively rare gene product, about 0.01% of total cell protein and mRNA.

Most yeast genes contain a TATA box but the CAAT box is often absent (Dobson *et al.*, 1982a). The *PGK* gene has both TATA and CAAT elements but the distance of these elements from the mRNA start site is greater than seen with higher eukaryotic genes. A third element which may be important comprises a pyrimidine-rich block (CT block) on the coding strand followed about 10 bp downstream by the sequence CAAG which may define the mRNA start site. This CT-CAAG structure is found in all abundant yeast genes analysed to date but is either absent from poorly expressed genes or the spacing between the CT block and the CAAG is large, between 30 and 50 bp (Dobson *et al.*, 1982a). This structure may be important for determining high efficiency gene expression. The *TRP*1 5′ region contains three possible TATA boxes located 31, 224 and 262 bp upstream from the ATG. There is no CAAT box and although both a CT block and a CAAG structure are present they are 47 bp apart. However with this gene the CT block forms part of a region of dyad symmetry which generates a stem loop structure in the region of the mRNA start (Dobson *et al.*, 1982a). A stem loop has been found in a similar location relative to the mRNA start in the 5′ non-coding regions of other yeast genes for amino acid biosynthetic enzymes (Struhl, 1981; Donahue, Farabaugh & Fink, 1982; Zalkin & Yanofsky, 1982) and may therefore have implications for the control of transcription. In addition there is an open reading frame extending from position –429 to –258 (Dobson *et al.*, 1982a) and a short potential coding sequence is also found upstream of the *TRP*5 gene (Zalkin & Yanofsky, 1982).

Certain sequences have been implicated as important for efficient translation of yeast mRNAs. These are a sequence CACACA located 13 bp upstream from the ATG in the *TRP*1 gene and in a similar location in other yeast genes, and a related sequence

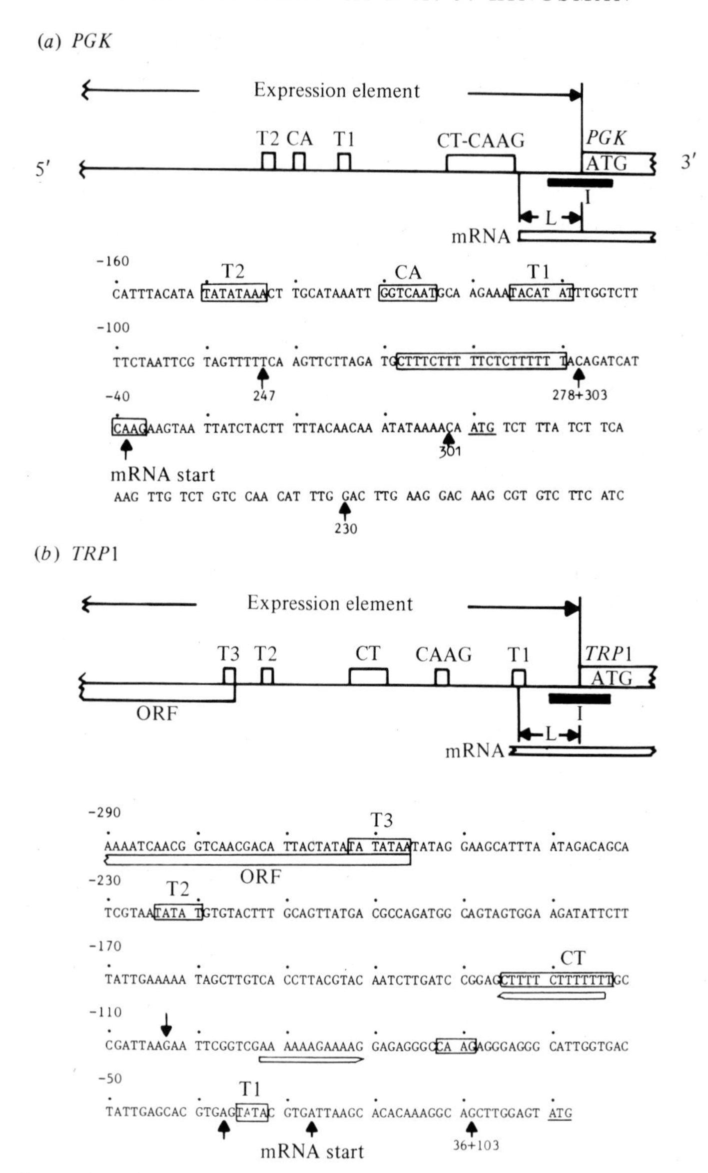

Fig. 3. Expression elements of the *PGK* and *TRP*1 genes of yeast. The line drawings represent features of the sense strand of the DNA; they are not to scale. The nucleotide sequence of part of the 5′ region is given for the sense strand, nucleotides are numbered from the A of the ATG as + 1. Arrows and numbers on the nucleotide sequence refer to the 3′ limit of the expression elements used in the *PGK* and *TRP*1 expression vectors, the arrow at nucleotide 103 in the *TRP*1 sequence indicates the 5′ limit of the expression element used in the vector pMA103. L, untranslated leader sequence; T, TATA box, *PGK* has two possible TATA boxes, *TRP*1 has three possible TATA boxes, T1 being least like the consensus sequence; CA, CAAT box; CT, a pyrimidine-rich tract (discussed in text); ORF, open reading frame; I, translation initiation environment; open arrows represent inverted repeats in the *TRP*1 sequence which can form a potential hairpin loop structure.

CAACAA or CAApyCAA is found about 10 bp upstream from the ATG of the *PGK* gene and in a number of other yeast genes (e.g. Dobson *et al.*, 1982a; Bennetzen & Hall, 1982a). Also the region upstream of the ATG tends to be AT rich although *TRP*1 seems to be an exception, being 43% GC (Dobson *et al.*, 1982a). An adenine is generally found 3 bp upstream from the ATG and a thymidine is often found at 6 bp downstream from the A of the ATG (Dobson *et al.*, 1982b). In addition it now seems likely that sequences in the 3′ flanking region may also be involved in determining efficient gene expression since deletions in this region of the *CYC*1 gene reduce the levels of mRNA (Zaret & Sherman, 1982).

There are therefore nucleotide sequences which may determine both the efficiency of transcription initiation and termination and the efficiency of translation in yeast but it is not yet clear which of these sequences is most significant and how they should be organized to direct maximum expression of interferon. However, it is likely that the principles developed with bacteria of using a highly efficient promoter and preserving correct transcription and translation signals will be effective for obtaining high levels of expression in yeast.

Several yeast genes that direct the synthesis of high levels of homologous mRNA and protein have been characterized (Gallwitz, Perrin & Siedel, 1981; Bennetzen & Hall, 1982a; Dobson *et al.*, 1982b) and it is likely that such genes contain the most efficient control signals for directing heterologous gene expression. The expression elements of two of these genes, *ADH*1 (Hitzeman *et al.*, 1981) and *PGK* (Tuite *et al.*, 1982) have been used to express interferon in yeast. In the first instance regions of the 5′ flanking sequence of the *ADH*1 gene terminating between 4 bp and 32 bp upstream from the ATG for *ADH* coding sequence were used to direct expression of a met-mature IFN-αD sequence. Overall yields from the cultures were low, 5.6×10^6 u/litre, in part because an unstable *ARS* – based plasmid was used as the vector and consequently the average plasmid copy number in the culture was low with less than 30% of the cells containing plasmid. More recently a 2 μm based vector has been used to direct very high levels of interferon production directed by the expression elements of the yeast *PGK* gene (Tuite *et al.*, 1982). The interferon sequence used in this study was a modified IFN-αA which had part of the signal sequence removed and replaced with a synthetic oligonucleotide to create a BamH1 restriction endonuclease site followed by an ATG translation initiation codon and then eight amino acids of the signal sequence followed by the mature sequence. A similar IFN-αA derivative with eight amino acids of signal sequence also gave very high yields in bacteria (Slocombe *et al.*, 1982). A yeast *PGK* expression element, in plasmid pMA230, comprising all the 5′ control sequences (extending for 1.5 kbp upstream from the ATG) and the first 11 N-terminal amino acids of *PGK*, was linked to this IFN-αA sequence to create an in-frame fusion protein with 11 *PGK* amino acids, five amino acids derived from the synthetic linker and the junction between the two sequences, eight amino acids from the signal sequence and the 165 amino acids from the mature sequence (Tuite *et al.*, 1982). This system produced f-IFN-αA as 1–5% of total yeast protein, i.e. at least 2×10^6 molecules per cell and the polypeptide was easily visible by polyacrylamide gel electrophoresis of total cellular protein as a distinct band of the molecular weight expected of the fusion protein (Tuite *et al.*, 1982; Fig. 4 lanes (*b*) and (*c*)). The maximum yield of interferon in this system is between 10 and 15 mg from a one-litre simple batch culture. An advantage with the *PGK* expression system is that levels of interferon can be regulated by the carbon source. The yeast glycolytic enzymes are expressed at low levels when yeast is grown on a non-fermentable carbon source such as acetate and induced about 100-fold when transferred to growth on glucose (Maitra & Lobo, 1971). A similar regulation is seen with *PGK* expression elements linked to the IFN-αA gene, for example after growth on acetate an efficient transcription/translation vector (pMA230) gave only 70 000 molecules of interferon per cell but eight hours after adding glucose to the culture this increases to about 10^6 molecules/cell (Tuite *et al.*, 1982). This ability to confine the high-level production of interferon to a precise stage in the growth of a culture may be important for optimizing industrial processes. In the construction described above both transcriptional and translational control signals of a highly expressed yeast gene are preserved which could explain the high yields. In order to determine whether it is necessary to preserve homologous translation signals or whether it is possible to synthesize interferon as efficiently in a transcription fusion vector with hybrid translation signals a second expression vector (pMA301) was constructed (M. J. Dobson, M. F. Tuite, A. J. Kingsman & S. M. Kingsman, unpublished data). This has all the 5′ flanking regions of *PGK* up to one nucleotide before the ATG for *PGK* (Fig. 3) and was fused to the IFN-αA sequence via a synthetic oligonucleotide

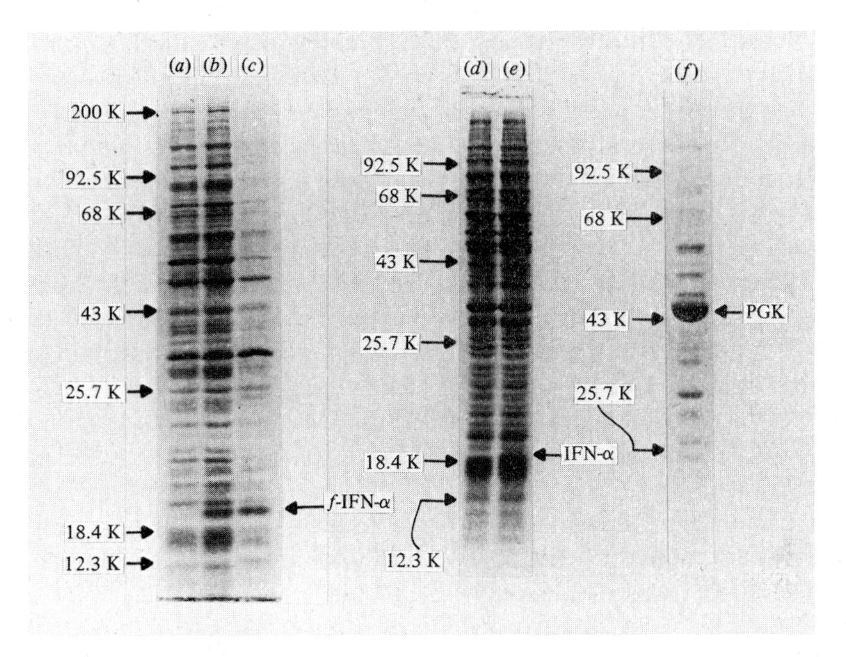

Fig. 4. Interferon production in yeast directed by *PGK* expression vectors. Coomassie stained SDS-PAGE protein profiles. (*a*) Total protein from yeast containing the transcription and translation fusion expression vector pMA230 alone. (*b*) Total protein from yeast containing expression vector pMA230 with the IFN-αA coding sequence; the arrow indicates the fusion interferon polypeptide, f-IFN-αA. (*c*) Protein from (*b*) after partial purification on a specific monoclonal antibody column. (*d*) Total protein from yeast containing the transcription fusion expression vector pMA301 alone. (*e*) Total protein from yeast containing expression vector pMA301 with the IFN-αA coding sequence, the arrow indicates the interferon polypeptide, IFN-αA. (f) Total protein from yeast containing plasmid pMA3a-*PGK* containing the whole of the yeast *PGK* gene (Dobson *et al.* 1982), the arrow indicates the *PGK* polypeptide. Gel conditions and molecular weight markers are given in Tuite *et al.* (1982).

linker which was A rich to mimic the natural translation initiation environment (Table 3, linker B). This construction gave the same levels of IFN-αA as obtained with the transcription/translation fusion vector in terms of biological activity and amount of protein observed on polyacrylamide gels, about 1–5% of total yeast protein was IFN-αA (Fig. 4, lane (*e*)) and interferon levels are also regulated by glucose (E. J. Mellor, M. J. Dobson, M. F. Tuite, A. J. Kingsman & S. M. Kingsman, unpublished data). This result indicates that coupling normal transcription control signals with an artificial ATG environment will result in high yields of interferon.

In order to analyse how much of the 5′ flanking region of a yeast gene is sufficient to direct high-level expression of interferon, different *PGK* expression fragments have been constructed which terminate at increasing distances from the ATG (Fig. 3). Results with the *ADH*1 system suggested that the extent of the mRNA leader sequence used in the expression fragment did not markedly affect interferon yields over a 28 bp sequence (Hitzeman *et al.*, 1981). However studies with the *PGK* system indicate that sequences upstream from the mRNA leader are critical for high-level expression. An expression element, present in plasmid pMA303, which lacks 47 bp upstream from the ATG initiation codon for *PGK* gives about 10-fold lower levels of interferon than the element on plasmid pMA301 which terminates one bp before the ATG (Table 3). It may be significant that the low-expression element terminates between the CT block and the CAAG sequences and the disruption of this sequence may be responsible for the low yields. If the entire CT-CAAG is removed as in plasmid pMA247, then yields are 100-fold lower (M. F. Tuite, E. J. Mellor, M. J. Dobson, A. J. Kingsman & S. M. Kingsman, unpublished data). All the promoter fragments studied by Hitzeman *et al.*, (1981) had retained an intact CT-CAAG structure. It seems likely therefore that some sequences in the 5′ untranscribed region are important for determining high-level gene expression and that the different expression efficiencies of yeast genes may be a function of these regions. This is further supported by studies with the expression elements from the *TRP*1 gene. A fragment extending from 1.5 kbp upstream from the ATG of the *TRP*1 gene and terminating eight bp upstream from the ATG was linked to the IFN-αA sequence via an oligonucleotide linker (Table 3, linker A). IFN-αA was expressed poorly in this vector, producing only 2000 molecules per cell (Dobson *et al.*, 1982*a*) which is 1000-fold lower than the most efficient *PGK* trancription fusion vector, pMA301, and 10-fold lower than obtained with the least efficient *PGK* vector, pMA247, where the same oligonucleotide linker was used. It is perhaps significant that pMA247 can still direct the synthesis of significant levels of interferon, 2×10^4 molecules per cell, despite lacking the CT-CAAG structure and the favoured mRNA initiation point. This may imply that, as is the case for other eukaryotic systems (for a review see Yaniv, 1982), upstream sequences are involved in determining expression levels. In this respect it is interesting that when most of the upstream region of the *TRP*1 gene was removed to leave a 95 bp expression element in plasmid pMA103 levels of interferon expression dropped to about 60 molecules per cell (Dobson *et al.*, 1982a). This comparison of the *PGK* and *TRP*1 systems suggests that to maximize interferon production in yeast it is essential to exploit the expression signals

Table 3. *Production of interferon-alpha in yeast*

	Vector	Expression element[a]	Linker[b]	Molecules/cell[c]
PGK				
	pMA230	−1500 to +37	A	6×10^6
	pMA301	−1500 to −2	B	2×10^6
	pMA278	−1500 to −48	A	5×10^4
	pMA303	−1500 to −48	B	2×10^5
	pMA247	−1500 to −83	B	2×10^4
*TRP*1	pMA103	−103 to −8	A	66[d]
	pMA36	−1500 to −8	A	2×10^3

Linker A: [CCGGATCCATGGG] CTGCAAG

Linker B: [CAAAATGATCCATGGG] CTGCAAG

[a] The 5′ end of the expression element in each plasmid is indicated by the arrow and plasmid number on the nucleotide sequence in Fig. 3.
[b] The expression elements are joined via synthetic oligonucleotides either linker A or linker B. The boxed nucleotides are the linker, the ATG initiation codon is underlined, nucleotides to the right of the box are from the IFN-αA fragment starting at nucleotide 44 in the signal sequence (Tuite *et al.*, 1982).
[c] All cultures are grown on glucose and the precise conditions, the extraction procedure and assay for interferon are described in Tuite *et al.*, 1982. The assay is a viral c.p.e. reduction assay on Hep-2 cells using EMC as the challenge virus and titres are adjusted to a reference interferon. Molecules per cell are calculated assuming a molecular weight of 20K and a specific activity of 2×10^8 units/mg in this assay procedure.
[d] This material is confirmed as interferon because it is neutralized by specific monoclonal antibody to IFN-αA.

from a highly expressed yeast gene such as *PGK*. It has been suggested that the immediate ATG environment may have little importance in determining efficient expression (Hitzeman *et al.*, 1981). However, some variations in expression efficiency have been observed when different synthetic oligonucleotides have been used to link *PGK* expression elements to the IFN-αA fragment. For example, the vector pMA303 which terminates between the CT block and the CAAG sequences consistently gives 4–6-fold higher levels of IFN-αA when an A rich linker is used than an identical fragment in vector pMA278 where a relatively GC rich linker is used (Table 3). It is therefore likely that yield improvements of at least five-fold can be achieved by optimizing the ATG environment but that this has a relatively minor effect compared with the importance of the 5′ non-transcribed region. Table 3 gives a comparison of interferon yields produced in yeast with expression elements containing different 5′ sequences and generating different ATG environments. Direct comparison between the *ADH*1 and the

PGK system is complex because different interferon sequences were used and different parameters used to calculate yields. However, when interferon is assayed on bovine cells and a specific activity of 2×10^8 u/mg is used the yields of IFN-αA directed by *PGK* are 2×10^7 molecules per cell (Tuite *et al.*, 1982; E. J. Mellor, M. J. Dobson, M. F. Tuite, A. J. Kingsman & S. M. Kingsman, unpublished data) while those for IFN-αD directed by *ADH* are 1.8×10^5 molecules per cell (Hitzeman *et al.*, 1981). Maximum levels reported to date in yeast have been with the *PGK* system yielding interferon as 1–5% of total cell protein (Tuite *et al.*, 1982). However, these yields fall short of the maximum yields expected from an expression element which directs synthesis of 1–5% of total cell protein as PGK on a plasmid which is present at about 50 copies per cell. In fact when the entire *PGK* gene is put in a 2μm plasmid, pMA3a-*PGK*(Dobson *et al.*, 1982b), the PGK polypeptide is made as 80% of total cell protein (Fig. 4, lane (*f*)). There may be several reasons for failing to achieve these levels with interferon constructions. The first concerns codon usage; it has been established that highly expressed yeast genes have a marked bias in their codon usage (Bennetzen & Hall, 1982b; Tuite *et al.*, 1982), which reflects the abundance of the relevant tRNA species in the cell. Poorly expressed genes such as *TRP*1 do not display such a marked bias. The codon usage in the interferon gene is different from the abundant yeast gene profile (Tuite *et al.*, 1982), in particular the most favoured codon for leucine in yeast is TTG whereas in interferon it is CTG, and the codon AGG is used for arginine 50% of the time in interferon but is never used in *PGK* (Perkins *et al.*, 1983; M. F. Tuite, M. J. Dobson, A. J. Kingsman & S. M. Kingsman, unpublished data). The translatability of interferon may therefore be reduced because of the requirement for rare tRNA species. A second possible reason for lower than theoretical yields may be that the small amount of signal sequence may affect either mRNA or protein stability, although no significant degradation of the interferon polypeptide has been observed in pulse chase experiments (E. J. Mellor, A. J. Kingsman & S. M. Kingsman, unpublished data). Finally it is possible that correct transcription termination may be important for determining maximum mRNA levels (Zaret & Sherman, 1982). The interferon sequence used in the present studies contains several hundred nucleotides of 3′ untranslated region of the interferon gene and the mRNA seems to terminate in this region in yeast. It is possible that higher levels of mRNA would be obtained if transcription terminated in a yeast sequence. To fulfill a possible requirement for both 5′ and 3′ yeast sequences for maximum heterologous gene expression we have constructed a new series of expression vectors referred to as 'sandwich expression vectors', a typical example is plasmid pMA91 shown in Fig. 5. This plasmid has the standard features of a yeast expression plasmid, that is, a 2μm origin and a *LEU2* gene for replication and selection in yeast and the pBR322 origin and ampicillin resistance gene for replication and selection in *E.coli*. The plasmid contains the 5′ flanking region of the yeast *PGK* gene from –1500 to –2. This region includes all upstream expression elements and sequences necessary for regulation of expression by glucose. The plasmid also contains the 3′ flanking region of the *PGK* gene (M. F. Tuite, M. J. Dobson, A. J. Kingsman & S. M. Kingsman, unpublished data) which contains the consensus yeast transcription termination sequence (Zaret & Sherman, 1982) and a possible polyadenylation signal. A synthetic A rich BglII linker is inserted between the *PGK* 5′ and 3′ regions to create an expression site for an interferon sequence which is then sandwiched between yeast 5′ and 3′ expression elements.

Clearly it is possible to synthesize high levels of IFN-α in yeast, comparable with levels produced in *E.coli*. Levels may eventually be increased by more precise manipulation of the various yeast gene expression signals in a way analogous to the manipulation of hybrid ribosome-binding sites in *E.coli*. To date there is little information on the expression of other interferons in yeast although expression of IFN-γ has been reported (quoted in Friedman, Epstein & Merigan, 1982). Future studies will indicate whether yeast is preferred over *E.coli* for the synthesis of IFN-β and IFN-γ which are not expressed maximally in *E.coli*, and whether secretion and processing of preinterferons occurs efficiently in yeast. A limited amount of secretion of preIFN-γ in yeast has been reported (quoted in Friedman, Epstein & Merigan, 1982) but it is not yet established whether the correct processing occurs and if yields of secreted processed preinterferon approach those of intracellular met-interferon. It will be essential to define conditions for complete fidelity of processing if the interferon is to be used pharmaceutically because of possible problems of antigenicity of any contaminating signal peptide. The clinical efficacy of microbially derived interferons is currently being assessed. At present there is no information on the clinical use of interferon produced in yeast but bacterial

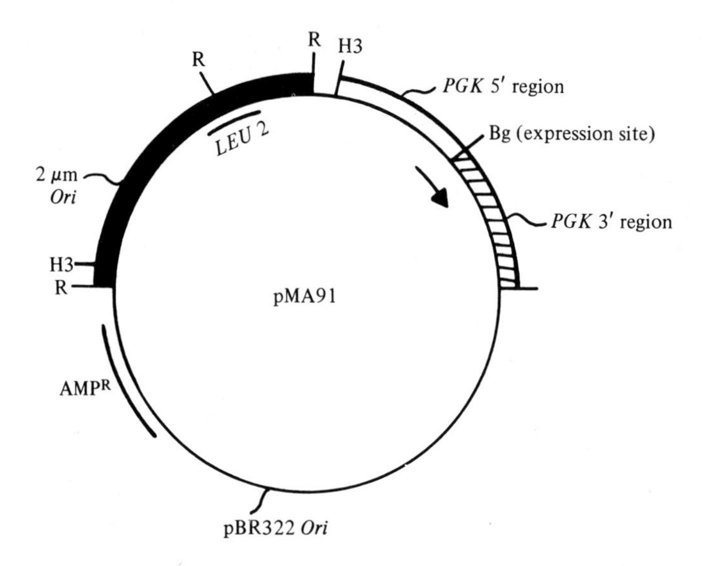

Fig. 5. A typical yeast expression vector, pMA91. Drawing is not to scale. Thin line, pBR322 sequences; thick line, yeast DNA region containing the 2μm origin of replication and the *LEU*2 gene; open box, 5′ region of *PGK*; hatched box, 3′ region of *PGK*; arrow marks direction of transcription; R, Eco RI; H3, Hind III; Bg Bgl II, the site at which interferon coding sequences are inserted.

interferons are showing some antiviral and antitumour activity (Friedman *et al.*, 1982; Newmark, 1982; refer to articles by A. Billiau, G. Scott and J. S. Malpas in this volume for a detailed discussion of the clinical properties of interferons).

THE BIOLOGICAL ACTIVITY OF INTERFERONS PRODUCED IN MICROORGANISMS

The broad spectrum of biological activities of natural interferons (Stewart, 1979) may be due partly to the heterogeneity of the preparations, particularly IFN-α preparations, and/or to multifunctional properties of single polypeptides. The ability to express each of the polypeptides in microorganisms allows each interferon to be analysed separately and in controlled combinations. It is also possible to mix different parts of each polypeptide by creating hybrid genes *in vitro* and specifically to alter single amino acids or regions in the polypeptide by targeted mutagenesis of the nucleotide sequence or by chemically synthesizing stretches of nucleotide sequence. These studies are helping to identify functional features of the interferons and to assign biological properties to individual polypeptides.

Many studies have now shown that interferons derived from *E.coli* (EcoIFN) and from yeast (YeIFN) have the general properties of natural interferons, that is, EcoIFN-α and EcoIFN-β are both stable at pH 2.0 whereas EcoIFN-γ is acid labile, they are all neutralized by broad specificity antisera to the corresponding natural interferon and they all induce an antiviral state in mammalian cells (e.g. Derynck *et al.*, 1980b; Goeddel *et al.*, 1980b; Nagata *et al.*, 1980b; Gray *et al.*, 1982). In addition EcoIFN-αD and EcoIFN-β enhance natural killer cell activity and antibody-dependent cytotoxicity *in vitro* (Masucci *et al.*, 1980; McCullagh *et al.*, 1982) and the proliferation of Daudi cells is inhibited by EcoIFNαD, EcoIFN-αA and EcoIFN-β (Masucci *et al.*, 1980; Stebbing *et al.*, 1981; McCullagh *et al.*, 1982). EcoIFN-αD also reduces macrophage migration inhibition mediated by phytohaemagglutinin-activated lymphocytes (Masucci *et al.*, 1980) and both natural and synthetic EcoIFN-αD and EcoIFN-β induce 2′5′ oligoisoadenylate synthetase activity in human cells (Nagata *et al.*, 1980*b*; McCullagh *et al.*, 1982; De Maeyer *et al.*, 1982). EcoIFN-αA is as effective as buffy coat interferon in protecting squirrel monkeys against lethal encaphalomyocarditis virus (EMC) infection (Goeddel *et al.*, 1980b) and against vaccinia virus infection of rhesus monkeys (Schellekens *et al.*, 1981). The serum clearance rates of EcoIFN-αA in rhesus monkeys are also similar to those of buffy coat interferon (Stebbing *et al.*, 1981). Clearly the absence of post-translational modifications, e.g. glycosylation of the interferon polypeptides produced in bacteria, does not affect their gross biological properties. On the whole EcoIFN-β has very similar biological properties to natural IFN-β although it may be more potent against HSV in certain human cell lines (McCullagh *et al.*, 1982). At present there is little published data on the comparative biological properties of microbial IFN-γ and natural IFN-γ.

The analysis of the biological activity of the different IFN-α molecular subtypes produced as individual polypeptides in microorganisms is revealing some significant differences between the polypeptides and identifying some activities which are not displayed by naturally synthesized heterogeneous interferon preparations. In particular some of the individual IFN-α polypeptides are more active on heterologous cells than natural heterogeneous preparations. For example, several microbial IFN-α polypeptides have marked activity on mouse cells whereas natural human IFN-α has little (or no) activity (Stewart, 1979) unless very high concentrations are used (Samuel & Farris, 1977). The antiviral activity of EcoIFN-αD on mouse L_{929} fibroblasts is comparable to that of natural mouse IFN-α, and both induce phosphokinase activity in these cells with identical kinetics (Samuel & Knutson, 1982a,b). Synthetic EcoIFN-αD also has significant activity on mouse cells (De Maeyer *et al.*, 1982). EcoIFN-αD also activates mouse natural killer cells, reduces mouse macrophage migration inhibition (Masucci *et al.*, 1980) and increases the survival of mice inoculated with L1210 leukaemia cells (Stebbing *et al.*, 1981) whereas natural human IFN-α has no effect in these systems. IFN-αA shows variable activity on mouse cells; in some studies it failed to protect mouse L_{929} cells significantly against a VSV challenge (Weck *et al.*, 1981a) or to induce significant phosphokinase activity in these cells (Samuel & Knutson, 1982*a*) whereas in other studies it has had some activity in the same system (Yelverton *et al.*, 1981; Slocombe *et al.*, 1982; Streuli *et al.*, 1981; Weck *et al.*, 1981b) and a YefIFN-αA also shows some activity on mouse cells (Tuite *et al.*, 1982). The variation in activity of microbial IFN-αA on mouse cells may be a function of the relative purity of the preparations as *E.coli* proteins are known to potentiate the antiviral effect of EcoIFN-αA on mouse cells (Stebbing *et al.*, 1981). EcoIFN-αD also has a higher activity on guinea pig cells than either natural IFN-α or EcoIFN-αA (Stewart *et al.*, 1980; Streuli *et al.*, 1981) and the microbial IFN-α polypeptides often show a higher relative titre on bovine versus human cells as compared to the natural IFN-α mixture (Stewart *et al.*, 1980; Weck *et al.*, 1981a; Slocombe *et al.*, 1982; Tuite *et al.*, 1982). Although EcoIFN-αD has a high activity on many heterologous cells, the activity on a wide range of human cell lines is consistently lower than that shown by either the natural IFN-α mixture or by EcoIFN-αA (Samuel & Knutson, 1981; Streuli *et al.*, 1981). Finally, EcoIFN-αB, EcoIFN-αC, EcoIFN-αD and buffy coat IFN are all active on a rabbit cell line but only EcoIFN-αD and buffy coat IFN protected primary rabbit cells. In addition to variations in cell specificity the different EcoIFN-α polypeptides have different potencies against different viruses in the same cell. Five EcoIFN-α polypeptides gave similar titres on Vero cells challenged with EMC virus but buffy coat, EcoIFN-αA and EcoIFN-αD were ten- to twenty-fold more potent against vesicular stomatitis virus (VSV) (Weck *et al.*, 1981a).

Similarly EcoIFN-αB was less effective against Sendai virus than against VSV in WISH cells (Yelverton *et al.*, 1981).

The specific activities of the different interferons may vary in different extracts and in some studies fusion polypeptides have been used. However, some general features of the specific antiviral properties of the different IFN-α polypeptides are emerging. On the whole EcoIFN-αA has a far higher activity on human cells than EcoIFN-αD. Both EcoIFN-αA and EcoIFN-αD have higher activity on mouse cells than the natural interferon-α mixture but EcoIFN-αD is significantly more active than EcoIFN-αA and approaches the activity of natural mouse interferon (Streuli *et al.*, 1981; Weck *et al.*, 1981b; Samuel & Knutson, 1982*a*). EcoIFN-αF seems to have the least activity on heterologous cells (Weck *et al.*, 1981*a*). It seems likely that the biological activities of the different microbial IFN-α polypeptides reflect inherent properties of the proteins rather than their microbial origin. However, this has not yet been confirmed by detailed analysis of the biological properties of the different molecular subtypes when purified from a naturally synthesized heterogeneous interferon preparation.

To begin analysing the basis for the different biological properties of microbial IFN-α polypeptides, in particular the high activity of EcoIFN-αA on human cells and the high activity of EcoIFN-αD on mouse cells, various hybrid interferon genes have been constructed *in vitro* which produce hybrid polypeptides in *E.coli.* IFN-αD and IFN-αA share restriction endonuclease sites for BglII at the codon for amino acid 61(αA) or 62(αD) and for PvuII at the codon for amino acid 91(αA) or 92(αD) and it is therefore possible, for example, to join the N-terminal 61 amino acids of the IFN-αA coding sequence to the 104 carboxy-terminal (C-terminal) amino acids of the IFN-αD coding sequence via their common BglII site in the nucleotide sequence and *vice versa.* These hybrids are referred to as D/A where the N-terminus is derived from IFN-αD and A/D if the N-terminus is derived from IFN-αA. No difference in properties is observed if the join is made at the BglII site or the PvuII site so no distinction is made between the constructions in the following discussion. Hybrids have been made between met-mature interferons (Weck *et al.*, 1981b) and between fusion interferons which have an additional 16 amino acids derived from *E.coli* β-galactosidase and part of the interferon signal sequence (Streuli *et al.*, 1981). Essentially similar results have been obtained. Hybrids with the N-terminus derived from IFN-αD retained the low activity on human cells characteristic of the IFN-αD parent molecule and this was correlated with the failure of the D/A hybrid to induce phosphokinase activity in human amnion U cells (Samuel & Knutson, 1981). The A/D hybrids, however, had a high activity on human cells characteristic of the parental IFN-αA and this was correlated with the ability to induce phosphokinase activity in U cells (Samuel & Knutson, 1981). The A/D hybrids were also highly active on mouse cells showing the characteristics of the IFN-αD parental molecule while the D/A hybrids had little or no activity on mouse cells and failed to induce phosphokinase activity in these cells (Samuel & Knutson, 1982a, b). In addition A/D hybrids reduced the incidence of tumour formation in mice injected with L1210 leukaemia cells (Stebbing *et al.*, 1981). In some cases the hybrid was even more active than the most active parental molecule, for example an A/D hybrid was 1000 times more active against VSV on mouse L_{929} cells than EcoIFN-αA, EcoIFN-αD or a D/A hybrid (Weck *et al.*, 1981b) and in inducing phosphokinase activity (Samuel & Knutson, 1982a). Conversely a D/A hybrid was less active than either EcoIFN-αA or EcoIFN-αD in inducing an antiviral state and phosphokinase activity in human U cells (Samuel & Knutson, 1981). There are also differences between the hybrids and the parental molecules in antiviral specificity against different viruses in the same cell line. On mouse L_{929} cells the relative potencies against EMC virus were A/D>D>D/A>A but the D/A hybrid was less active against VSV making the relative potencies against this virus A/D>D>A>D/A. However, on monkey cells (Vero) the relative potencies against VSV were A/D = A>D>>D/A but the D/A hybrid was more active against HSV making the relative potencies A/D = D/A>A = D (Weck *et al.*, 1981b).

These studies with hybrid interferons suggest that there are at least two biologically active components of the interferon polypeptides which are involved in inducing an antiviral state. Because there is no difference in properties between the hybrids constructed with BglII or with PvuII these particular components are clearly not located between amino acids 61 and 92. One component must be located in the first 62 amino acids and be involved in determining induction of the antiviral state in human cells and acts efficiently in the IFN-αA polypeptide and less efficiently in the IFN-αD polypeptide. The second component must be located in the C-terminal 74 amino acids and be involved in determining the induction of the antiviral state in mouse cells and acts efficiently in the IFN-αD polypeptide but less efficiently in the IFN-αA polypeptide. The two components may be spatially independent or form a single active domain in the folded polypeptide. The cases where hybrids have much higher or lower activities than either of the parental molecules might be explained by new conformational constraints in the hybrid molecule which affect these active sites to make them more or less effective.

It is not yet clear whether the different potencies of EcoIFN-αD and EcoIFN-αA on human and mouse cells reflect differences in the binding of the polypeptides to specific cell receptors or some other difference in activating the biochemical effectors of the antiviral state. It is possible that the conserved domains I and II represent the major binding components of human interferon for human and mouse cells respectively (Streuli *et al.*, 1981). However, these are highly conserved between IFN-αA and IFN-αD showing only nine amino acid differences out of 42 in domain I and three amino acids out of 31 in domain II (Goeddel *et al.*, 1981) and there may be insufficient variability in these regions to generate the observed differences in biological activity. However, a single amino acid change can have a significant effect on the biological properties of an interferon if it affects the molecular configuration. For example a single transition from cysteine to tyrosine at position 141 destroys the biological activity of EcoIFN-β presumably by disrupting the intramolecular disulphide bridge (Shepard *et al.*, 1981). It is likely that there are also conformational constraints on the biological activity of IFN-α because chemically synthesized fragments while retaining antigenicity lacked biological activity (Smith, Komoriya & Anfinsen, 1981). A more precise definition of the biologically significant regions of the interferon polypeptides is now possible by introducing smaller pieces of each polypeptide into hybrid molecules, by replacing parts of a molecule with a chemically synthesized fragment, and by site-directed mutagenesis (e.g. Shortle, DiMaio & Nathans, 1981) to alter specific amino acids. To date most of the comparisons between different interferons have related to their antiviral activities but it will also be interesting to identify differences in other properties, for example antiproliferative activity. However, although some of the biological properties of interferon may be due to distinct biologically active regions of the polypeptide it is evident that there are significant interactions between different interferons and with non-interferon proteins. The activity of natural IFN-α is not an average of the activities of individual polypeptides;

IFN-αA and IFN-αD are the predominant molecular subtypes in buffy coat and KG-1 cell interferons yet the natural interferon shows little or no activity on mouse cells and generally lower activity on bovine cells than either molecular subtype (Weck *et al.*, 1981a; Streuli *et al.*, 1981). Similarly a mixture of EcoIFN-αA and EcoIFN-αD does not display the mean activity of the individual components (Stebbing *et al.*, 1981; Weck *et al.*, 1981a); for example the mixture has higher activity on hamster BHK, mouse L_{929} and primary rabbit kidney cells than either of the parent molecules suggesting a synergistic effect, but a lower activity on Vero and HeLa cells suggesting some inhibitory effect. Similarly IFN-γ potentiates the antiviral (Fleischmann, Georgiades, Osborne & Johnson, 1979) and antiproliferative properties (Fleischmann, Kleyn & Baron, 1980) of both IFN-α and IFN-β. Non-interferon polypeptides also affect interferon activity, for example *E.coli* proteins potentiate the effect of EcoIFN-αA on mouse L_{929} cells whereas purified EcoIFN-αA has little effect on these cells (Stebbing *et al.*, 1981). Similarly EcoIFN-αA has little activity against rabbit cells *in vitro* (Weck *et al.*, 1981a) but is effective in preventing HSV infection of the rabbit eye (Smolin *et al.*, 1982) suggesting an interaction with host factors. It seems likely therefore that, for example, the induction by natural interferon of an antiviral state in a particular cell will depend on a subtle combination of the relative concentrations of different interferons in the preparation and the suppression or activation of their biological properties by other interferon polypeptides, by non-interferon proteins and by host factors. Undoubtedly a single interferon polypeptide is capable of showing all of the biological activities which define an interferon (Stewart, 1979) because, for example, IFN-β preparations are predominantly a single species and highly purified individual interferons display all the basic properties which define an interferon (Rubinstein *et al.*, 1981). However, on the whole, natural IFN-β has a narrower host range than IFN-α (Stewart, 1979) and bacterial IFN-β shows no detectable antiviral properties on monkey, feline, rabbit and mouse cells (Derynck *et al.*, 1980b; McCullagh *et al.*, 1982). It seems reasonable to suppose that the biological activities of natural IFN-α may be more versatile because of the spectrum of different polypeptides. A simple model might propose that each interferon polypeptide has the ability to interact efficiently with a different cell receptor structure and that different cells have quantitatively and qualitatively different receptors. Efficient induction of an antiviral state in a particular cell might depend on the concentration of the relevant interferon in the preparation and whether its activity is enhanced or suppressed by other proteins or host factors. It is also clear that there must be more than one biochemical mechanism which mediates the biological effects of the interferons. With regard to the antiviral state two different enzyme activities have been identified in interferon-treated cells, a phosphokinase which phosphorylates a 68K dalton ribosome-associated protein and the α subunit of initiation factor eIF2 (Farrell *et al.*, 1978; Samuel, 1979), and a 2′–5′ oligoisoadenylate synthetase, the product of which activates an endogenous ribonuclease, RNase F (Wreschner, McCauley, Skehel & Kerr, 1981). However, the phosphokinase and 2′–5′ oligoisoadenylate synthetase are not always both required to protect a cell from a particular virus and different viruses are affected by different biochemical effectors of the antiviral state. For example when mouse embryonal carcinoma (EC) cells are treated with mouse interferon there is no detectable phosphokinase activity yet the cells are still protected from picornavirus infection, while they are still susceptible to infection by VSV, Sindbis and influenza virus (Nilson, Ward & Baglioni, 1980). Similarly interferon-treated human amnion U cells are protected from VSV infection but are sensitive to reovirus despite a marked induction of phosphokinase (Samuel & Knutson, 1981). Interferon-treated NIH3T3 cells fail to synthesize active RNase F (Epstein *et al.*, 1981) yet are protected against murine leukaemia virus and are sensitive to EMC and VSV (Czarniecki, Sreevalsan, Friedman & Panet, 1981). Furthermore both the phosphokinase and 2′–5′-oligoisoadenylate synthetase inhibit translation of viral mRNA yet several viruses are inhibited by interferon at stages in their life cycle other than at translation. For example, SV40 is predominantly inhibited at the stage of mRNA synthesis (Metz, Levin & Oxman, 1976; Kingsman & Samuel, 1980) and some retroviruses are inhibited at the stage of virus particle maturation (Friedman, Chang, Ramseur & Myers, 1975), implying that there may be at least two more antiviral mechanisms. This idea of separate biochemical effectors of the antiviral state active against different viruses is supported further by studies showing that the translation of SV40 mRNA is unaffected while the translation of reovirus mRNA is markedly reduced in BSC-1 cells coinfected with these viruses and then treated with interferon (Daher & Samuel, 1982).

It is possible therefore that different interferon polypeptides are more or less efficient at triggering different effector pathways which inhibit different viruses. The low activity of a particular interferon polypeptide against one virus but not another in the same cell type might therefore be due to a failure to trigger the appropriate antiviral mechanism since presumably the ability to affect one virus implies that the interferon is bound to the cell. It will be interesting to see how the host range and the antiviral range of single microbially derived IFN polypeptides correlate with biochemical parameters of the antiviral state. For example, are any particular IFN polypeptides more effective against viruses blocked at transcription or maturation rather than at translation? Will any individual polypeptides render mouse EC cells resistant to Sindbis, VSV or Influenza virus and if so, is this correlated with the appearance of phosphokinase activity or some new biochemical activity? Is the failure of IFN-αD/A hybrids to induce phosphokinase activity in human amnion U cells due to a failure to activate the biochemical pathway or due to a failure to bind appropriate receptors? In answering these sorts of questions with microbially derived interferons it may be possible to match a specific interferon polypeptide with a cell receptor type or with a specific biochemical response and possibly to identify new biochemical effectors of the antiviral state. Moreover it may be possible to accentuate the antiviral state by using hybrid interferons, altered interferons, or specific mixtures of different interferons. This sort of analysis can be applied equally to all the biological properties of the interferons and it may soon be possible to identify regions of each polypeptide which have specific biological functions. In the future it may be possible to construct new interferons which combine active sites for different biological functions, to mix different interferon and non-interferon polypeptides in precise combinations, and to combine interferons to generate interferon preparations with greater specific clinical effectiveness than natural interferons.

CONCLUSIONS

Significant levels of both IFN-α and IFN-β can be produced in microorganisms, of the order of 10 mg from one litre of batch culture. This is a marked improvement on systems for producing natural interferon where for example it takes two litres of blood to make 1 ug of IFN-α (Rubinstein *et al.*, 1979). Although improvements to large-scale tissue culture production are currently being made it seems likely that the large-scale production of interferons in microorganisms will provide a realistic alternative. There may, of course, be some situations, for example using IFN-γ, where high-level production in microorganisms may not be possible and the gene may have to be reintroduced into mammalian cells for synthesis (Gray *et al.*, 1982; Devos *et al.*, 1982). However, the synthesis of interferons in microorganisms is not just important for making massive amounts of product because the ability to manipulate the individual genes in microorganisms is also opening up a new approach to the analysis of their structural and functional properties. The fundamental studies of interferon expression outlined in this review will now develop towards restructuring the interferon genes to generate novel polypeptides which combine desirable features from different interferons and which may have new and as yet unpredicted biological properties.

We would like to thank Drs K. Atherton, M. Houghton, and C. E. Samuel and Professor D. C. Burke for communicating their results before publication. We are indebted to our colleagues for continuing to generate the data for this review while we took time off to write it. In particular we thank Drs Melanie Dobson and Mick Tuite for critical reading of the manuscript. We would also like to thank Celltech for financial support for the research. S.M.K. is an E.P.A. Cephalosporin Fund Research Fellow of the Royal Society of London.

REFERENCES

ALLEN, G. & FANTES, K. H. (1980). A family of structural genes for human lymphoblastoid (leukocyte-type) interferon. *Nature*, **287**, 408–11.

BACKMAN, K. & PTASHNE, M. (1978). Maximising gene expression on a plasmid using recombination *in vitro*. *Cell*, **13**, 65–75.

BEGGS, J. D. (1978). Transformation of yeast by a replicating hybrid plasmid. *Nature*, **275**, 104–9.

BEGGS, J. D., VAN DEN BERG J., VAN OOYEN, A. & WEISSMAN, C. (1980). Abnormal expression of chromosomal rabbit β-globin gene in *Saccharomyces cerevisiae*. *Nature*, **283**, 835–40.

BENNETZEN, J. L. & HALL, B. D. (1982*a*). The primary structure of the *Saccharomyces cerevisiae* gene for alcohol dehydrogenase I. *Journal of Biological Chemistry*, **257**, 3018–26.

BENNETZEN, J. L. & HALL, B. D. (1982*b*). Codon selection in yeast. *Journal of Biological Chemistry*, **257**, 3026–31.

BENOIST, C., O'HARE, K., BREATHNACH, B. & CHAMBON, P. (1980). The ovalbumin gene – sequence of putative control regions. *Nucleic Acids Research*, **8**, 127–42.

BOLIVAR, F., RODRIGUEZ, R., GREENE R. J., BETLACH, M., HEYNEKER, H. L., BOYER, H. W., CROSA, J. & FALKOW, S. (1977). Construction and characterization of new cloning vehicles. *Gene*, **2**, 95–113.

BOLLEN, G. H. P. M., MOLENAAR, C. M. T., COHEN, L. H., VAN RAAMSDONK-DUIN, M. M. C., MAGER, W. H. & PLANTA, R. J. (1982). Ribosomal protein genes of yeast contain intervening sequences. *Gene*, **18,** 29–37.

BOSE, S., GURARI-ROTMAN, D., RUEGG, V. T., CORLEY, L. & ANFINSEN, C. B. (1976). Apparent dispensability of the carbohydrate moiety of human interferon for antiviral activity. *Journal of Biological Chemistry*, **251,** 1659–62.

BRACK, C., NAGATA, S., MANTEI, N. & WEISSMAN, C. (1981). Molecular analysis of the human interferon-α gene family. *Gene*, **15,** 379–94.

BRIDGEN, P. J., ANFINSEN, C. B., CORLEY, L., BOSE, S., ZOON, K. C., RUEGG, V. T. & BUCKLER, C. E. (1977). Human lymphoblastoid interferon. Large scale production and partial purification. *Journal of Biological Chemistry*, **252,** 6585–7.

BROACH J. R. (1982). The yeast plasmid 2μm circle. *Cell*, **28,** 203–4.

CANTELL, K. & HIRVONEN, S. (1977). Preparation of human leucocyte interferon for clinical use. In *The Interferon System: A Current Review*, ed. S. Baron and F. Dianzani, *Texas Reports in Biology and Medicine*, **35,** 138–44.

CLARKE, L. & CARBON, J. (1980). Isolation of a yeast centromere and construction of functional small circular chromosomes. *Nature*, **287,** 504–9.

CZARNIECKI, C. W., SREEVALSAN, T., FRIEDMAN, R. M. & PANET, A. (1981). Dissociation of interferon effects on murine leukaemia virus and encephalomyocarditis virus replication in mouse cells. *Journal of Virology*, **37,** 827–31.

DAHER, K. A. & SAMUEL, C. E. (1982). Mechanism of interferon action: differential effect of interferon on the synthesis of simian virus 40 and reovirus polypeptides in monkey kidney cells. *Virology*, **117,** 379–90.

DE MAEYER, E., SKUP, D., PRASAD, K-S. N., DE MAEYER-GUIGNARD, J., WILLIAMS, B., MEACOCK, P., SHARPE, G., PIOLI, D., HENNAM, J., SCHUCH, W. & ATHERTON, K. (1982). Chemically synthesised human α1 interferon gene. *Proceedings of the National Academy of Sciences of the USA*, **79,** 4256–9.

DERYNCK, R., CONTENT, J., DE CLERCQ, E., VOLCKAERT, G., TAVERNIER, J., DEVOS, R. & FIERS, W. (1980a). Isolation and structure of a human fibroblast interferon gene. *Nature*, **285,** 542–7.

DERYNCK, R., REMAUT, E., SAMAN, E., STANSSENS, P., DE CLERCQ, E., CONTENT, J. & FIERS, W. (1980b). Expression of human fibroblast interferon gene in *Escherichia coli. Nature*, **287,** 193–7.

DEVOS, R., CHEROUTRE, H., TAYA, Y., DEGRAVE, W., VAN HEUVERSWYN, H. & FIERS, W. (1982). Molecular cloning of human immune interferon cDNA and its expression in eukaryotic cells. *Nucleic Acids Research*, **10,** 2487–503.

DOBSON, M. J., TUITE, M. F., MELLOR, E. J., ROBERTS, N. A., KING R. M., BURKE, D. C., KINGSMAN, A. J. & KINGSMAN, S. M. (1982a). Expression in *Saccharomyces cerevisiae* of human interferon-alpha directed by the *TRP1* 5′ region. *Nucleic Acids Research*, **11,** 2287–302.

DOBSON, M. J., TUITE, M. F., ROBERTS, N. A., KINGSMAN, A. J. KINGSMAN, S. M., PERKINS, R. E., CONROY, S. C., DUNBAR, B. & FOTHERGILL, L. A. (1982b). Conservation of high efficiency promoter sequences in *Saccharomyces cerevisiae. Nucleic Acids Research*, **10,** 2625–37.

DONAHUE, T. F., FARABOUGH, P. J. & FINK, G. R. (1982). The nucleotide sequence of the HIS4 region of yeast. *Gene*, **18,** 47–59.

EDGE, M. D., GREEN, A. R., HETHCLIFFE, G. R., MEACOCK, P. A., SCHUCH, W., SCANLON, D. B., ATKINSON, T. C., NEWTON, C. R. & MARKHAM, A. F. (1981). Total synthesis of a human leucocyte interferon gene. *Nature*, **292,** 756–62.

EPSTEIN, D. A., CZARNIECKI, C. W., JACOBSEN, H., FRIEDMAN, R. H. & PANET, A. (1981). A mouse cell line which is unprotected by interferon against lytic virus infection lacks ribonuclease F activity. *European Journal of Biochemistry*, **118,** 9–15.

EPSTEIN, L. B. (1982). Interferon-gamma: success, structure and speculation. *Nature*, **295,** 453–4.

FARRELL, P. J., SEN, G. C., DUBOIS, M. F., RATNER, L., SLATTERY, E. & LENGYEL, P. (1978). Interferon action: two distinct pathways for inhibition of protein synthesis by double-stranded RNA. *Proceedings of the National Academy of Sciences of the USA*, **75,** 5893–7.

FINTER, N. B. & FANTES, K. H. (1980). The purity and safety of interferon prepared for clinical use: the case for lymphoblastoid interferon. In *Interferon*, ed. I. Gresser. London & New York: Academic Press.

FLEISHMANN, JR. W. R., GEORGIADES, J. A., OSBORNE, L. C. & JOHNSON, M. W. (1979). Potentiation of interferon activity by mixed preparations of fibroblast and immune interferon. *Infection and Immunity*, **26,** 248–53.

FLEISCHMANN, JR. W. R., KLEYN, K. M. & BARON, S. (1980). Potentiation of antitumour effect of virus induced interferon by mouse immune interferon preparations. *Journal of the National Cancer Institute*, **65,** 963–6.

FRIEDMAN, R. M., CHANG, E. A., RAMSEUR, J. M. & MYERS, M. W. (1975). Interferon directed inhibition of chronic murine leukaemia virus production in cell cultures: lack of effect on intracellular viral markers. *Journal of Virology*, **16,** 569–74.

FRIEDMAN, R. M., EPSTEIN, L. B., & MERIGAN, T. (1982). Interferon redux. *Nature*, **269,** 704–5.

GALLWITZ, D., PERRIN, F. & SEIDEL, R. (1981). The actin gene in yeast *Saccharomyces cerevisiae*: 5′ and 3′ end mapping, flanking and putative regulatory sequences. *Nucleic Acids Research*, **9,** 6339–50.

GALLWITZ, D. & SURES, I. (1980). Structure of a split yeast gene: complete nucleotide sequence of the actin gene in *Saccharomyces cerevisiae. Proceedings of the National Academy of Sciences of the USA*, **77,** 2546–50.

GANNON, F., O'HARE, K., PERRIN, F., LE PENNEC, J. B., BENOIST, C., COCHET, M., BREATHNACH, R., ROYAL, A., GARAPEN, A., CAMIC, B. & CHAMBON, P. (1979). Organization and sequences at the 5′ end of a cloned complete ovalbumin gene. *Nature*, **278,** 428–34.

GANOZA, M. C., FRASER, A. R. & NEILSON, T. (1978). Nucleotides contiguous to AUG affect translation initiation. *Biochemistry*, **17,** 2769–75.

GOEDDEL, D. V., LEUNG, D. W., DULL, T. J. GROSS, M., LAWN, R. M., MCCANDLISS, R., SEEBURG, P. H., ULLRICH, A., YELVERTON, E. & GRAY, P. W. (1981). The structure of eight distinct cloned human leukocyte interferon cDNAs. *Nature*, **290,** 20–6.

GOEDDEL, D. V., SHEPARD, H. M., YELVERTON, E., LEUNG D., CREA, R., SLOMA, A. & PESTKA, S. (1980a). Synthesis of human fibroblast interferon by *E. coli*. *Nucleic Acids Research*, **8,** 4057–4.

GOEDDEL, D. V., YELVERTON, E., ULLRICH, A., HEYNEKER, H. L., MIOZZARI, G., HOLMES, W., SEEBURG, P. H., DULL, T., MAY L., STEBBING, N., CREA, R., MAEDA, S., MCCARDLISS, R., SLOMA, A., TABOR, J. M., GROSS, M., FARRILLETTI, P. C. & PESTKA, S. (1980b). Human leukocyte interferon produced by *E. coli* is biologically active. *Nature*, **287,** 411–16.

GOLD, L., PRIBNOW, D., SCHREIDER, T., SHINEDLING, S., SINGER, B. S. & STORMO, G. (1981). Translational initiation in prokaryotes. *Annual Review of Microbiology*, **35,** 365–403.

GOTTESMAN, M., OPPENHEIM, A. & COURT, D. (1982). Retroregulation: control of gene expression from sites distal to the gene. *Cell*, **29,** 727–8.

GRAY, P. W. & GOEDDEL, D. W. (1982). Structure of the human immune interferon gene. *Nature*, **298,** 859–63.

GRAY, P. W., LEUNG, D. W., PENNICA, D., YELVERTON, E., NAJARIAN, R., SIMONSEN, C., C., DERYNCK, R., SHERWOOD, P. J., WALLACE, D. M., BERGER,

S. L., LEVINSON, A. D. & GOEDDEL, D. W. (1982). Expression of human immune interferon cDNA in *E. coli* and monkey cells. *Nature*, **295,** 503–8.

GROSVELD, G. C., DE BOER, E., SHWMAKER, C. K. & FLAVELL, R. A. (1982). DNA sequences necessary for transcription of the rabbit β-globin gene *in vivo. Nature*, **295,** 120–6.

HARDY, K., STAHL, S. & KÜPPER, H. (1981). Production in *B.subtilis* of Hepatitis B core antigen and of major antigen of foot and mouth disease virus. *Nature*, **293,** 481–3.

HAYES, T. G., YIP, Y. K. & VILCEK, J. (1979). *Le* interferon production by human fibroblasts. *Virology*, **98,** 351–63.

HAVELL, E. A., BERMAN, B., OGBURN, C. A., BERG, K., PAUCKER, K. & VILCEK, J. (1975). Two antigenically distinct species of human interferon. *Proceedings of the National Academy of Sciences of the USA*, **72,** 2185–7.

HAVELL, E. A., YIP, Y. K. & VILCEK, J. (1978). Characteristics of human lymphoblastoid (Namalva) interferon. *Journal of General Virology*, **38,** 51–60.

HINNEN, A., HICKS, J. B. & FINK, G. R. (1978). Transformation of yeast. *Proceedings of the National Academy of Sciences of the USA*, **75,** 1929–33.

HITZEMAN, R. A., HAGIE, F. E., LEVINE, H. L., GOEDDEL, D. V., AMMERER, G. & HALL, B. D. (1981). Expression of a human gene for interferon in yeast. *Nature*, **293,** 717–22.

HOUGHTON, M., DOEL, S. M., CATLIN, G. H., STEWART, A. G., PORTER, A. G., TACON, W. C., A., EATON, M. A. W., EMTAGE, J. S. & CAREY, N. H. (1981). The cloning and expression of a human fibroblast interferon gene in bacteria. In *Proceedings of the International Genetic Engineering Conference*, ed. Battelle Institute. Virginia, USA.

JIMENEZ, A. & DAVIES, J. (1980). Expression of a transposable antibiotic resistance element in *Saccharomyces. Nature*, **287,** 869–71.

KINGSMAN, A. J., CLARKE, L., MORTIMER, R. K. & CARBON, J. (1979). Replication in *Saccharomyces cerevisiae* of plasmid pBR313 carrying DNA from the yeast *TRP1* region. *Gene*, **7,** 141–52.

KINGSMAN, S. M. & SAMUEL, C. E. (1980). Mechanism of interferon action: interferon-mediated inhibition of Siman virus-40 early RNA accumulation. *Virology*, **101,** 458–65.

KISS, G. B., PEARLMAN, R. E., CORNISH, K. V., FRIESEN, J. D. & CHAN, V. L. (1982). The Herpes simplex virus thymidine kinase gene is not transcribed in *Saccharomyces cerevisiae. Journal of Bacteriology*, **149,** 542–7.

KOZAK, M. (1981). Possible role of flanking nucleotides in recognition of the AUG initiator codon by eukaryotic ribosomes. *Nucleic Acids Research*, **9,** 5233–52.

MAITRA, P. K. & LOBO, Z. (1971). A kinetic study of glycolytic enzyme synthesis in yeast. *Journal of Biological Chemistry*, **246,** 475–88.

MANTEI, N., SCHWARZSTEIN, M., STREULI, M., PANEM, S., NAGATA, S. & WEISSMAN, C., (1980). The nucleotide sequence of a cloned human leukocyte interferon cDNA. *Gene*, **10,** 1–10.

MASCUCCI, M. G., SZIGETI, R., KLEIN, E., KLEIN, G., GRUEST, J., MONTAGNIER, L., TAIRA, H., HALL, A., NAGATA, S. & WEISSMAN, C. (1980). Effect of interferon $\alpha 1$ from *E.coli* on some cell functions. *Science*, **209,** 1431–5.

MCCULLAGH, K. G., DAVIES, J. A., SIMM, I., DAWSON, K. M., O'NEIL, G. J., DOEL, S. M., CATLIN, G. H. & HOUGHTON, M. (1982). Biological properties of human IFN-β1 synthesised in recombinant bacteria. *Journal of Interferon Research* (in press).

MCNEIL, J. B. & FRIESEN, J. (1981). Expression of the *Herpes simplex* virus thymidine kinase gene in *Saccharomyces cervisiae. Molecular and General Genetics*, **184,** 386–93.

MESSING, J., CREA, R. & SEEBERG, P. H. (1981). A system for shotgun DNA sequencing. *Nucleic Acids Research*, **9**, 309–21.

METZ, D. H., LEVIN, M. J. & OXMAN, M. N. (1976). Mechanism of interferon action: further evidence for transcription as the primary site of action in simian virus 40 infection. *Journal of General Virology*, **32**, 227–40.

MILLER, J. H. & REZNIKOFF, W. S. (1980). *The Operon*. New York: Cold Spring Harbour Laboratories.

NAGATA, S., MANTEI, N. & WEISSMAN, C. (1980a). The structure of one of the eight or more distinct chromosomal genes for human interferon-α. *Nature*, **287**, 401–8.

NAGATA, S., TAIRA, H., HALL, A., JOHNSRUD, L., STREULI, M., ECSÖDI, J., BOLL, W., CANTELL, K. & WEISSMAN, C. (1980b). Synthesis in *E.coli* of a polypeptide with human leukocyte interferon activity. *Nature*, **284**, 316–20.

NEWMARK, P. (1982). Interferon: decline and stall. *Nature*, **291**, 105–6.

NILSON, T. W., WARD, D. L. & BAGLIONI, C. (1980). Virus specific effects of interferon in embryonal carcinoma cells. *Nature*, **286**, 178–80.

NOVICK, P., FIELD, C. & SCHEKMAN, R. (1980). Identification of 23 complementation groups required for post-translational events in the yeast secretory pathway. *Cell*, **21**, 205–17.

OLD, R. W. & PRIMROSE, S. B. (1981). *Principles of Gene Manipulation*. Oxford: Blackwell Scientific Publications.

PERKINS, R. E., CONROY, S. C., DUNBAR, B., FOTHERGILL, L. A., TUITE, M. F., DOBSON, M. J., KINGSMAN, S. M. & KINGSMAN, A. J. (1983). The complete amino acid sequence of yeast phosphoglycerate kinase. *Biochemical Journal*, **211**, 199–218.

PLATT, T. (1980). Regulation of gene expression in the tryptophan operon of *E.coli*. In *The Operon*, ed. J. H. Miller & W. S. Reznikoff. New York: Cold Spring Harbour Laboratory.

PRIBNOW, D. (1975). Nucleotide sequence of an RNA polymerase binding site at an early T7 promoter. *Proceedings of the National Academy of Sciences of the USA*, **72**, 784–8.

REMAUT, E., STANSSENS, P. & FIERS, W. (1981). Plasmid vectors for high-efficiency expression controlled by the P_L promoter of coliphage lambda. *Gene*, **15**, 81–93.

REZNIKOFF, W. S. & ABELSON, J. N. (1980). The *lac* promoter. In *The Operon*, ed. J. H. Miller & W. S. Reznikoff. New York: Cold Spring Harbour Laboratory.

ROBERTS, T. M., KACICH, R. & PTASHNE, W. (1979). A general method for maximizing the expression of a cloned gene. *Proceedings of the National Academy of Sciences of the USA*, **76**, 760–4.

RUBINSTEIN, M., LEVY, W. P., MOSCHERA, J. A., LAI, C-Y., HERSCHBERY, R. D., BERTELL, R. T. & PESTKA, S. (1981). Human leukocyte interferon: isolation and characterization of several molecular forms. *Archives of Biochemistry and Biophysics*, **210**, 307–18.

RUBINSTEIN, M., RUBINSTEIN, S., FAMILLETI, P. C., MILLER, R. S., WALDMAN, A. A. & PESTKA, S. (1979). Human leucocyte interferon. Production, purification to homogeneity and initial characterization. *Proceedings of the National Academy of Sciences of the USA*, **76**, 640–4.

SAMUEL, C. E. & FARRIS, D. A. (1977). Species specificity of interferon and of the interferon mediated inhibitor of translation from mouse, monkey and human cells. *Virology*, **77**, 556–65.

SAMUEL, C. E. (1979). Mechanism of interferon action. Kinetics of interferon action in mouse L_{929} cells: phosphorylation of protein synthesis initiation factor eIf-2 and ribosome associated protein P_1. *Virology*, **93**, 281–5.

SAMUEL, C. E. & KNUTSON, G. S. (1981). Mechanism of interferon action: cloned human leukocyte interferons induce protein kinase and inhibit vescular stomatitis virus but not reovirus replication in human amnion cells. *Virology*, **114,** 302–6.

SAMUEL C. E. & KNUTSON, G. (1982a). Mechanism of interferon action: kinetics of induction of the antiviral state and protein phosphorylation in mouse fibroblasts treated with natural and cloned interferons. *Journal of Biological Chemistry*, **257,** 1791–5.

SAMUEL, C. E. & KNUTSON, G. (1982b). Mechanism of interferon action: kinetics of decay of the antiviral state and protein phosphorylation in mouse fibroblasts treated with natural and cloned interferons. *Journal of Biological Chemistry*, **257,** 1796–1801.

SCHELLEKENS, H., DEHEUS, A., BAHIUS, R., FOUNTOULAKIS, M., SCHEIN, C., ESCODI, J., NAGATA, S. & WEISSMAN, C. (1981). Comparative antiviral efficiency of leukocyte and bacterially produced human α-interferon in rhesus monkeys. *Nature*, **292,** 775–6.

SHEPARD, H. M., LEUNG, D., STEBBING, N. & GOEDDEL, D. V. (1981). A single amino acid change in IFN-β1 abolishes its antiviral activity. *Nature*, **294,** 563–4.

SHEPARD, H. M., YELVERTON, E. & GOEDDEL, J. V. (1982). Increased synthesis in *E.coli* of fibroblast and leukocyte interferon through alterations in ribosome binding sites. *DNA*, **1,** 125–31.

SHINE, J. & DALGARNO, L. (1975). Determinant of cistron specificity in bacterial ribosomes. *Nature*, **254,** 34–8.

SHORTLE, D., DIMAIO, D. & NATHANS, D. (1981). Directed mutagenesis. *Annual Review of Genetics*, **15,** 265–94.

SIEBENLIST, U. (1979). Nucleotide sequence of the three major early promoters of bacteriophage T7. *Nucleic Acids Research*, **6,** 1895–907.

SIEBENLIST, U., SIMPSON, R. B. & GILBERT, W. (1980). *E.coli* RNA polymerase interacts homologously with two different promoters. *Cell*, **20,** 269–81.

SLOCOMBE, P., EASTON, A., BOSELEY, P. A. & BURKE, D. C. (1982). High level expression of an interferon-α2 gene cloned in phage M13mp7 and subsequent purification with a monoclonal antibody. *Proceedings of the National Academy of Sciences of the USA*, **79,** 5455–9.

SMITH, M. E., KOMORYA, A. & ANFINSEN, C. B. (1981). Chemical synthesis and properties of fragments of human leukocyte interferon. In *The Biology of the Interferon System*, ed. E. De Maeyer, G. Galasso & H. Schellekens. Amsterdam: North-Holland/Elsevier.

SMOLIN, G., STEBBING, N., FIEDLAENDER, M., FRIEDLAENDER, R. & OKUMOTO, M. (1982). Natural and cloned human leukocyte interferon in Herpes virus infections of rabbit eyes. *Archives of Opthalmology*, **100,** 481–3.

STEBBING, N., WECK, P. K., FENNO, J. T., APPERSON, S. & LEE, S. H. (1981). Comparison of the biological properties of natural and recombinant DNA derived human interferons. In *The Biology of the Interferon System*, ed. E. De Maeyer, G. Galasso and H. Schellekens. Amsterdam: North-Holland/Elsevier.

STEWART II, W. E. (1979). *The Interferon System.* Vienna: Springer-Verlag.

STEWART II, W. E., SARKAR, F. H., TAIRA, H., HALL, A., NAGATA, S. & WEISSMAN, C. (1980). Comparisons of several biological and physicochemical properties of human leukocyte interferons produced by human leukocytes and by *E.coli. Gene*, **11,** 181–6.

STREULI, M., HALL, A., BOLL, W., STEWART II, W. E., NAGATA, S. & WEISSMAN, C. (1981). Target cell specificity of two species of human interferon-α produced in *Escherichia coli* and of hybrid molecules derived from them. *Proceedings of the National Academy of Sciences of the USA*, **78,** 2848–52.

STREULI, M., NAGATA, S. & WEISSMAN, C. (1980). At least three human type α interferons: structure of α2. *Science*, **209,** 1343–8.

STRUHL, K. (1981). Deletion mapping a eukaryotic promoter. *Proceedings of the National Academy of Sciences of the USA*, **78,** 4461–5.

STRUHL, K., STINCHCOMB, D. T., SCHERER, S. & DAVIS, R. W. (1979). High frequency transformation of yeast: autonomous replication of hybrid DNA molecules. *Proceedings of the National Academy of Sciences of the USA*, **76,** 1035–9.

TALMADGE, K., KAUFMAN, J. & GILBERT, W. (1980). Bacteria mature preproinsulin to proinsulin. *Proceedings of the National Academy of Sciences of the USA*, **77,** 3988–92.

TAN, Y. A., BARAKAT, F., BERTHOLD, W., SMITH JOHANSSEN, H. & TAN C. (1979). The isolation and amino acid/sugar composition of human fibroblast interferon. *Journal of Biological Chemistry*, **254,** 8067–73.

TANIGUCHI, T., GUARENTE, L., ROBERTS, T. M., KIMDMAN, D., DOUHAN III, J. & PTASHNE, M. (1980a). Expression of the human fibroblast interferon gene in *Escherichia coli. Proceedings of the National Academy of Sciences of the USA*, **77,** 5230–3.

TANIGUCHI, T., MANTEI, N., SCHWARZSTEIN, M., NAGATA, S., MURAMATSU, M. & WEISSMAN, C. (1980b). Human leucocyte and fibroblast interferons are structurally related. *Nature*, **285,** 547–9.

TANIGUCHI, T., OHNO, S., FUJII-KURIYAMA, Y. & MURAMATSU, M. (1980c). The nucleotide sequence of human fibroblast interferon cDNA. *Gene*, **10,** 11–15.

TAVERNIER, J., DERYNCK, R. & FIERS, W. (1981). Evidence for a unique human fibroblast interferon (IFN-β1) chromosomal gene devoid of intervening sequences. *Nucleic Acids Research*, **9,** 461–71.

TUITE, M. F., DOBSON, M. J., ROBERTS, N. A., KING, R. M., BURKE, D. C., KINGSMAN, S. M. & KINGSMAN, A. J. (1982). Regulated high efficiency expression of human interferon alpha in *Saccharomyces cerevisiae. EMBO Journal*, **1,** 603–8.

VALENZUELA, P., MEDINA, A., RUTTER, W. J., AMMERER, G. & HALL, B. D. (1982). Synthesis and assembly of Hepatitis B virus surface antigen particles in yeast. *Nature*, **298,** 347–50.

WECK, P. K., APPERSON, S., MAY, L. & STEBBING N. (1981a). Comparison of the antiviral activities of various cloned human interferon-α subtypes in mammalian cell cultures. *Journal of General Virology*, **57,** 233–7.

WECK, P. K., APPERSON, S., STEBBINGS, N., GRAY, P. W., LEUNG, D., SHEPARD, H. M. & GOEDDEL, D. V. (1981b). Antiviral activities of two major human leucocyte interferons. *Nucleic Acids Research*, **9,** 6153–66.

WETZEL, R. (1981). Assignment of the disulphide bonds of leucocyte interferon. *Nature*, **289,** 606–7.

WETZEL, R., PERRY, L. J., ESTELL, D. A., LIN, N., LEVINE, H. L., SLINKER, B., FIELDS, F., ROSS, M. J. & SHIVELY, J. (1981). Properties of a human alpha-interferon purified from *E.coli* extracts. *Journal of Interferon Research*, **1,** 381–90.

WRESCHNER, D. H., MCCAULEY, J. W., SKEHEL, J. J. & KERR, I. M. (1981). Interferon action – sequence specificity of the ppp $(A2'p)_n$ A-dependent ribonuclease. *Nature*, **289,** 414–17.

YANOFSKY, C. (1981). Attenuation in the control of expression of bacterial operons. *Nature*, **289,** 751–8.

YANIV, M. (1982). Enhancing elements for activation of eukaryotic promoters. *Nature*, **297,** 17–18.

YELVERTON, E., LEUNG, D., WECK, P., GRAY, P. W. & GOEDDEL, D. V. (1981). Bacterial synthesis of a novel human leukocyte interferon. *Nucleic Acids Research*, **9,** 731–41.

YIP, Y. K., PANG, R. H. L., URBAN, C. & VILCEK, J. (1981). Partial purification and characterization of human γ (immune) interferon. *Proceedings of the National Academy of Sciences of the USA*, **78,** 1601–5.

ZALKIN, H. & YANOFSKY, C. (1982). Yeast gene *TRP5*: structure, function, regulation. *Journal of Biological Chemistry*, **257,** 1491–500.

ZARET, K. S. & SHERMAN, F. (1982). DNA sequence required for efficient transcription termination in yeast. *Cell*, **28,** 563–73.

ZOON, K. C., SMITH, M. E., BRIDGEN, P. J., ANFINSEN, C. B., HUNKAPILLER, M. W. & HOOD, L. E. (1980). Amino terminal sequence of the major component of human lymphoblastoid interferon. *Science*, **207,** 527.

THE PHARMACOKINETICS AND TOXICOLOGY OF INTERFERON

A. BILLIAU

Rega Institute, University of Leuven, B-3000 Leuven, Belgium

INTRODUCTION

There are, in essence, two ways of approaching the pharmacokinetics of interferons. The physiologist wants to know what happens to interferon molecules generated at a certain site in the organism, e.g. at the site of viral replication. In particular this question is pertinent to understanding protection of certain organs (e.g. the brain) by interferon produced at a distance (e.g. in the spleen). It should be noted here that it is not at all certain that interferon was designed to exert such protection at a distance. Interferons, like many other cytokines, seem rather to function as carriers of short-range signals. If this is so, the physiologist will be interested in the pharmacokinetics of interferon only in the context of its catabolism, inactivation and elimination from the body. These processes may be of extreme importance since it has appeared recently that chronic endogenous interferon production may cause extensive tissue damage.

The experimental therapist takes a view on interferon pharmacokinetics that is quite different from that of the physiologist. His main concern is to know what happens to interferon that is injected intravenously, intramuscularly or by any other route. The premise here is that the injected interferon will affect organs at a distance to the injection site, to the extent that it reaches the cells of these organs and stays in contact with them for a sufficient amount of time.

Scarce experimental data pertinent to both types of questions are available in the literature and these will be reviewed. An important reason for scarcity of data is that, until recently, interferons were available only in small quantities, so that, even after injection of the highest possible doses, barely detectable concentrations of interferon were present in tissues. In recent months this situation has changed and more data are now forthcoming. In particular, comparative studies with different pure interferons are now possible. Another improvement to be foreseen is the use of immunoassays for detection of interferons in tissues. These assays, while only slightly less sensitive than bioassays, are more specific and thus of higher utility for determination of interferon levels in sera and tissue extracts.

The toxicology of interferon has also been approached from different angles by physiologists and therapists. It is now clear that endogenous interferon is responsible for some of the symptoms of virus diseases. On the other hand various side-effects of interferon therapy have been studied in the context of phase I clinical trials. The second part of this review will deal with these aspects of interferon pharmacology.

PHARMACOKINETICS OF INTERFERON, AS INFERRED FROM ITS GLYCOPROTEIN NATURE

Some but not all of the interferons formed under natural conditions are glycoproteins (see evidence reviewed by Stewart, 1979). In the case of human interferons (HuIFN), it is now well established that HuIFN-β and HuIFN-γ are certainly glycosylated. In contrast, all but one of the HuIFN-α subtypes have no glycosylation site on their peptide chain. The extent of glycosylation, and, in particular, the number of terminal sialic acid residues are held responsible for charge heterogeneity amongst interferons of a given molecular weight class. In all probability the presence of sugar residues is not essential for the interferon molecules to be able to exert their antiviral action on cells. Extensively desialylated or deglycosylated interferon, as well as non-glycosylated interferon proteins elaborated by bacteria carrying cloned interferon genes, have antiviral activity comparable to that of naturally occurring, native interferons.

On the other hand, the presence of sugar residues seems to be of crucial importance for the pharmacokinetic behaviour of glycoproteins in general. In particular, studies with ceruloplasmin have led to the concept (Ashwell & Morell, 1974) that removal of at least two sialic acid residues, exposing the preterminal galactosyl residues, renders glycoproteins recognizable by a receptor molecule on liver cells. In accord with this observation, the vascular clearance of desialylated ceruloplasmin and certain other glycoproteins was found to be accelerated. A similar phenomenon was described by Bocci *et al.* (1977), who studied uptake of rabbit interferon by perfused rabbit liver. Also, lymphoblastoid interferon injected intravenously into rats was cleared more rapidly when it was desialylated prior to injection (Bose & Hickman, 1977).

After uptake by hepatocytes, desialylated glycoproteins are apparently engulfed into lysosomes and degraded (Ashwell & Morell, 1974). From this it seems reasonable to assume that the liver plays an important role in catabolizing injected interferons. One could also anticipate that the *in vivo* life span of an injected interferon molecule would decrease when it is less sialylated.

It is not known when and where in the body native interferon molecules are desialylated. Bocci *et al.* (1977) suggested that it occurs with circulating interferon in locations where the circulation is sluggish. They also postulated that interferon would be desialylated by membrane-bound sialidase at the time of transient attachment to target cells. Although these are attractive hypotheses, evidence for them is rather inferential.

PHARMACOKINETICS OF HUMAN INTERFERON AS APPARENT FROM CLINICAL STUDIES

Systemic interferon treatment in humans is usually performed by deep intramuscular injection. Sometimes subcutaneous bolus injections (Strander, Cantell, Carlström & Jokobsson, 1973) or intravenous infusions (Strander *et al.*, 1973; Jordan, Fried & Merigan, 1974; Emödi, Just, Hernandez & Hirt, 1975a; Treuner *et al.*, 1980) have been used. The main concern in designing the treatment regimens has been to obtain and maintain measurable titres of antiviral activity in the serum, with the least possible side-effects or risks of inadvertent reactions. With leucocyte interferon this has not been a major problem (Strander *et al.*, 1973; Ahstrom *et al.*, 1974; Cantell, Pyhälä & Strander, 1974; Jordan *et al.*, 1974; Emödi *et al.*, 1975a; Emödi, Rufli, Just & Hernandez, 1975b; Greenberg *et al.*, 1976; Arvin, Feldman & Merigan, 1978; Edy, Billiau & De Somer, 1978; Merigan *et al.*, 1978; Billiau *et al.*, 1979a; Cheeseman *et al.*, 1979; Pazin *et al.*, 1979; Greenberg, Harmon & Couch, 1980; Gutterman *et al.*, 1980; Priestman, 1980; Lucero *et al.*, 1982). Following a single intramuscular injection of about 3×10^6 units into adult patients interferon activity became detectable in the serum within three hours, reached peak levels at about six hours and remained detectable for over 24 h. In patients given daily injections of 3×10^6 units, interferon remained detectable at all times, with titres oscillating between 80 and 300 u/ml (Greenberg *et al.*, 1980). These findings have since been confirmed in various additional clinical trials (Borden *et al.*, 1982) including these done with HuIFN-α prepared from lymphoblastoid cell lines (Priestman, Johnston & Whiteman, 1982) and HuIFN-α prepared from genetically manipulated bacteria (Gutterman *et al.*, 1982).

Fibroblast interferon has also been given by intramuscular injection to numerous patients by several research groups (Desmyter *et al.*, 1976; Weimar *et al.*, 1977; Edy *et al.*, 1978; Horoszewicz *et al.*, 1978; Kingham *et al.*, 1978; Weimar *et al.*, 1978; Billiau *et al.*, 1979a; Dolen *et al.*, 1979; Weimar, Heijtink, Schalm & Schellekens, 1979; Billiau & De Somer, 1980; Lucero *et al.*, 1982). Low or undetectable levels of serum interferon were found despite the fact that the injected preparations were of similar potency to those of leucocyte interferon. A minimum of 10 to 30×10^6 units had to be given for interferon to become detectable in the serum (Edy *et al.*, 1978; Billiau *et al.*, 1979a; Billiau & De Somer, 1980; Lucero *et al.*, 1982). One possible explanation could be that fibroblast interferon may be destroyed before reaching the circulation. This consideration has led some investigators to use the intravenous administration route (Treuner *et al.*, 1980; Pape, Hadam, Eisenburg & Riethmüller, 1981; Ezaki *et al.*, 1982; Misset *et al.*, 1982). Although there is some evidence that spontaneous inactivation of interferon in body fluids occurs more rapidly with fibroblast than with leucocyte interferon (Cesario, 1977; Cesario, Vaziri, Slater & Tilles, 1979), there is no proof that this mechanism is sufficient to account for the differences in pharmacokinetics observed.

It may be that fibroblast interferon reaches the blood stream as effectively as leucocyte interferon, but is then removed more rapidly from the circulation to be absorbed by cells and organs. This possibility is favoured by pharmacokinetic studies in mice (to be discussed below) and by the observations on the *in vivo* activation of natural killer (NK) cells in patients. Specifically, Lucero *et al.* (1982) studied the activation of NK-cells in tumour patients given a single injection of 10×10^6 units of either leucocyte or fibroblast interferon. Despite low blood titres, the NK-cell system was activated by fibroblast interferon. Dose-response curves revealed that, for both interferons, 3×10^6 units intramuscularly was the threshold dose to obtain this effect. Thus, it seems that intramuscularly injected fibroblast interferon does reach certain internal organs and cells as effectively as leucocyte interferon.

PHARMACOKINETICS OF HUMAN INTERFERONS IN EXPERIMENTS ON ANIMALS

Various animal species have been used to study the pharmacokinetics of human interferons (Cantell & Pyhälä, 1973; Cantell *et al.*, 1974; Cantell & Pyhälä, 1976; Edy, Billiau & De Somer, 1976; Billiau *et al.*, 1979a; Vilček, Sulea, Zerebeckyj & Yip, 1980). One of the main issues which can be resolved by such experiments is the comparative pharmacokinetics of the different human interferon types. Early experiments performed in rabbits failed to reveal any difference in pharmacokinetic behaviour between the fibroblast and leukocyte interferons (Edy *et al.*, 1976). The fibroblast interferon used for these experiments was obtained by superinduction in diploid skin muscle cells and had been partially purified by fractional precipitation with ammonium sulphate, a method that yielded recoveries of around 10%. Later experiments (Billiau *et al.*, 1979a) were performed with interferon that was also prepared from diploid cells but had been partially purified by adsorption/elution from controlled pore glass (CPG) beads, a method that routinely yields recoveries of over 50% (Billiau *et al.*, 1979c). With this interferon a clear-cut difference from leucocyte interferon was observed in serum titres obtained after intramuscular injection in rabbits. Unexpectedly, the same preparations injected intramuscularly in monkeys failed to reveal a significant difference: high serum titres were obtained with either leucocyte or fibroblast interferon (Billiau *et al.*, 1981). Hence, the rabbit was considered to be a better model than the monkey for interferon pharmacokinetics in man. In an independent study, Vilček *et al.*, (1980) confirmed that intramuscular injection of fibroblast interferon into rabbits resulted in lower blood titres than those obtained with similar injections of leucocyte interferon. In this study the difference between the two interferons was, however, less pronounced. The fibroblast interferon used by these authors was of diploid cell origin and had not undergone any purification.

Intravenous injections of leucocyte and fibroblast interferons into rabbits failed to reveal significant differences in clearance rate (Billiau *et al.*, 1979a; Vilček *et al.*, 1980). It is doubtful, however, whether these experiments were accurate and sensitive enough to reveal small differences that were sufficiently important to account for different blood values after intramuscular injection. Thus, these experimental results, while failing to support the idea of a more rapid clearance of fibroblast type interferon, did not refute the possibility either.

In our laboratory, experiments were done in mice to determine whether fibroblast interferon would be taken up more rapidly by tissues as compared to leucocyte interferon (Billiau *et al.*, 1981). Leucocyte or fibroblast interferon (0.1 to 1×10^6 units) were injected intraperitoneally and antiviral activity was determined in serum, lungs and spleen. The fibroblast interferon used was of diploid cell origin and had been purified by CPG-adsorption. A 10- to 30-fold difference was apparent in the serum levels reached by leucocyte and fibroblast interferon. In contrast, antiviral activities measured in the spleen and the lungs were similar, regardless of which interferon was used. These data go along with the idea that lower blood titres obtained with fibroblast interferon are not due to destruction at the site of injection but rather to more rapid uptake by some organs or organ systems.

However, evidence to support the contrary was reported by Hanley, Wiranowska-Stewart & Stewart (1979). They report rapid destruction of human fibroblast interferon, and resistance to destruction of leucocyte interferon upon incubation with muscle homogenate. The finding was correlated with low blood titres of fibroblast interferon, as opposed to high titres of leucocyte interferon, after intraperitoneal or intramuscular injections into mice. The authors also found lower titres of fibroblast interferon in the tissues of injected mice.

Some of the discordant results in the literature may be accounted for by the molecular heterogeneity of fibroblast interferon. It has been known for quite some time that natural HuIFN-β, although consisting of a single type of peptide, is heterogeneous with respect to glycosylation (Morser, Kabayo & Hutchinson, 1978). Evidence that this or other forms of molecular heterogeneity may affect the pharmacokinetics of fibroblast interferon was obtained in studies from our laboratory (Heine *et al.*, 1981; Heine, Billiau, Van Damme & De Somer, 1982). In particular it was found that fibroblast interferon, when chromatographed on zinc chelate columns, separates into two subpopulations eluting from the column at pH 5.9 and pH 5.2, respectively. Each of these was prepared in sufficient quantities to perform pharmacokinetic experiments in rabbits. Intramuscular injection of the pH 5.2-variant yielded 30-fold lower blood titres than a comparable dose of leucocyte interferon. The pH 5.9-variant, on the other hand, yielded a blood curve that was superimposable on that obtained with leucocyte interferon.

It was also found (Heine *et al.*, 1981) that fibroblast interferon prepared from most diploid cell strains contains more than 90% of the pH 5.2-variant. In contrast, fibroblast interferon from MG-63 cells or from certain high passage diploid cell strains was found to contain the two components in about equal proportions. These data provide a possible explanation for the discordance in results obtained in earlier pharmacokinetic studies with fibroblast interferons. Most probably, the preparations used by the different research groups contained different proportions of the two molecular variants of fibroblast interferon.

Sarkar (1982) compared natural leucocyte interferon and bacteria-derived HuIFN-α_1 for their abilities to distribute systemically in mice, following intramuscular and intraperitoneal injections. Both were found equally effective in entering the bloodstream. However, titres with natural leucocyte interferon lasted longer, indicating that different subtypes of HuIFN-α do indeed have different pharmacokinetic properties.

EXOGENOUS INTERFERON AND THE CENTRAL NERVOUS SYSTEM

Acute viral infections seldom cause death or life-long invalidity. If they do, the central nervous system is often the critical target organ involved. In order for interferon to be useful in preventing or curing these complications, it should be delivered to the meninges and/or the neural tissues in a suitably high concentration. Administration by intramuscular or intravenous injection offers little chance of achieving this goal. This can be inferred from studies in mice in which the concentrations of interferon in various organs were followed throughout the course of a fatal encephalitogenic virus infection (Heremans, Billiau & De Somer, 1980). During the early pre-encephalitic phase of the disease high levels of interferon were found in the spleen and serum, with very little if any detectable interferon in the brain. Brain interferon was detectable only during the later (encephalitic) phase, as a result of local viral replication. This type of experiment indicates that there is a strong blood-brain barrier for interferon. This view is also supported by studies in which interferon remained undetectable in the cerebrospinal fluid of patients receiving human interferons by intramuscular injections (Strander *et al.*, 1973; Ververken, Carton & Billiau, 1978; Priestman, 1980).

Experiments with exogenous leucocyte interferon administered intramuscularly or intravenously to monkeys showed that some of the interferon crossed the barrier between the vascular and cerebrospinal fluid compartments (Petralli, Merigan & Wilbur, 1965). In patients treated by intramuscular injections of lymphoblastoid cell interferon, no antiviral activity could be detected in the cerebrospinal fluid (Priestman *et al.*, 1982).

In view of the obvious difficulty in reaching the central nervous system by intramuscular administration, several investigators have attempted to obtain better penetration by delivering the interferon directly into the cerebrospinal fluid compartment. Proteins such as albumin and immunoglobulins are relatively excluded from the central nervous system under normal conditions by the anatomical barrier of tight junctions between capillary endothelial cells, arachnoid cells and choroid plexus epithelial cells. Within the central nervous system, free traffic between extracellular and cerebrospinal fluid is assumed because of the absence of tight junctions between ventricular ependymal cells. Continuity between cerebrospinal fluid and intrathecal space in the brain matter was documented by studies using horse-radish peroxidase (Brightman *et al.*, 1975).

Interferon injected into the cerebrospinal canal was found to reach the blood stream. In particular this was found to be the case in monkeys injected with human leucocyte interferon (Habif, Lipton & Cantell, 1975). Fibroblast interferon injected intrathecally into monkeys was found to diffuse rapidly throughout the cerebrospinal fluid: within an hour after injection in the lumbar region interferon could be recovered from fluid taken suboccipitally (Ververken *et al.*, 1978). Furthermore, the interferon injection did not bring about alterations in the electro-encephalogram, nor produce any anatomopathologic changes that would indicate local toxicity for the meninges. The question whether interferon delivered to the cerebrospinal fluid penetrates into the brain was studied by sacrificing the animals three hours after injection, in order to take brain samples for interferon detection (Billiau *et al.*, 1981). When great care was taken to remove all cerebrospinal fluid from the brain hemispheres before slicing, no interferon could be recovered from the white matter or cortex. Some activity could be recovered from the pia mater, scraped off together with superficial cortex. This interferon had probably penetrated into the tissue as it was not inactivated by washing the brain with a saline solution containing strong antibody to fibroblast interferon.

From these results it would appear that, in spite of high levels of residual interferon in the cerebrospinal fluid, very little if any can penetrate in the deep layers of brain matter. The data are in line with those obtained in studies using radioiodinated albumin (Paulson & Kapp, 1967), which was also found to rapidly diffuse from the cerebrospinal fluid into the blood and to penetrate barely, if at all into the brain matter.

In conclusion, it seems reasonable to assume that intrathecally administered fibroblast interferon can effectively reach the membranes surrounding the central nervous system but that there is little chance of its reaching the deep layers of white and grey matter.

Very few data are available on the pharmacokinetics of intrathecal administration of interferon in man. One child suffering from herpetic encephalitis received several such injections of leucocyte interferon: interferon was found to reach the blood thereby confirming the results obtained in monkeys (De Clercq *et al.*, 1975). An adult patient, suffering from Creutzfeldt Jacob disease (Ververken *et al.*, 1978), received three daily intrathecal injections of 10^6 units of fibroblast interferon. No serum interferon could be detected at any time, but a high concentration of interferon (3000 u/ml) was found in fluid taken from the cysterna magna five hours after lumbar injection, indicating efficient diffusion throughout the spinal canal.

In a recent study by Jacobs, O'Malley, Freeman & Ekes (1981), fibroblast interferon was given repeatedly by intrathecal injection to patients suffering from multiple sclerosis. From the discussions surrounding this widely publicized clinical trial (Abb *et al.*, 1982), it is not clear what the rationale was for preferring intrathecal to intramuscular or intravenous administration. The general premise of the study seemed to be that interferon was expected to correct some hypothetical immune dysfunction which is believed to be at the basis of the disease. There is, however, no reason to believe that this dysfunction is localized in/or restricted to the meningeal compartment. Thus, from a pharmacokinetic point of view, this study lacked a clear rationale.

RENAL AND HEPATIC CATABOLISM

Bocci and his coworkers have extensively studied (Bocci *et al.*, 1977, 1981a,b, 1982a,b) and reviewed (Bocci, 1977, 1981) renal and hepatic uptake and catabolism of human and rabbit interferon, especially by using organ-perfusion systems. In early studies these authors found that unmodified HuIFN-β as well as a rabbit interferon were partially taken up by liver tissue; desialylation caused an increase in uptake (Bocci *et al.*, 1977). In more recent studies the same research team compared hepatic catabolism of natural leucocyte and fibroblast interferons with HuIFN-α_2 obtained from genetically manipulated bacteria (Bocci *et al.*, 1982a). Very little if any uptake of leucocyte interferon and of HuIFN-α_2 occurred, but native and particularly desialylated fibroblast interferon underwent marked hepatic uptake, thereby confirming the importance of the liver as a catabolic site for glycosylated interferons.

The kidney is now recognized as an important organ for turnover of proteins with a molecular weight below 50 000. Bocci and coworkers have used several indirect and direct approaches to evaluate the catabolism of interferon in kidney. They found that in rabbits the disappearance of exogenous interferon plasma is significantly delayed by nephrectomy. Furthermore, tubular damage induced by maleate markedly reduced luminal uptake of interferon (measured by arteriovenous differences in interferon concentration) and also led to increased interferonuria (*loc. cit.* Bocci *et al.*, 1981b). In more recent work (Bocci *et al.*, 1981a, 1982b) the isolated and perfused kidney was used to evaluate renal handling of interferons. It was shown that the interferon was filtered by the kidney, largely reabsorbed by the tubuli and catabolized by the tubular epithelium, leaving small amounts for excretion in the urine.

THE TOXICOLOGY OF HUMAN INTERFERON

Interferon exerts a variety of effects on cells (as reviewed by Stewart, 1979) which suggest that treatment of intact animals or man with interferon may be accompanied by various side-effects. However, from studies in mice it would appear that therapeutically effective regimens of interferon treatment are fairly non-toxic. Only when large doses of interferon were given to very young animals could any outright toxicity be observed (Gresser, Tovey, Maury &

Chouroulinkow, 1975; Gresser *et al.*, 1976). Although such high doses have never been used in man, distinct undesirable side-effects of interferon therapy have been noted.

Fever

Most authors who have used leucocyte interferon to treat patients mention mild to severe fever reactions (Strander *et al.*, 1973; Ahstrom *et al.*, 1974; Jordan *et al.*, 1974; Emödi *et al.*, 1975a,b, 1976; Greenberg & Mosny, 1977; Merigan *et al.*, 1978; Pollard & Merigan, 1978; Cheeseman *et al.*, 1979; Idestrom *et al.*, 1979; Pazin *et al.*, 1979; Weimar *et al.*, 1979; Billiau *et al.*, 1980; Gutterman *et al.*, 1980; Priestman, 1980; Weimar *et al.*, 1980). They are most prominent after the first injections and tend to become less severe after prolonged treatment. The fever is usually accompanied by chills and malaise. In immunosuppressed renal transplant patients fever reactions were not seen with leucocyte interferon (Cheeseman *et al.*, 1979).

Fever reactions were also seen in patients treated with fibroblast interferon (Desmyter *et al.*, 1976; Weimar *et al.*, 1978, 1979; Billiau *et al.*, 1979a; Billiau & De Somer, 1980) injected intramuscularly or intravenously (Billiau *et al.*, 1980; Treuner *et al.*, 1980). Fibroblast interferon prepared by alternative purification schedules was reported to be non-pyrogenic (Kingham *et al.*, 1978; Dolen *et al.*, 1979).

The key question is whether these fever reactions are an effect of the interferons or of impurities in the preparations. There seems to be little doubt that much of the pyrogenicity of relatively impure interferon preparations can be removed by extensive and careful purifications. However, it is still not excluded that the interferons themselves have a certain pyrogenic activity. Highly pure HuIFN-α produced by genetically manipulated bacteria was found to be pyrogenic for man when injected intramuscularly (Gutterman *et al.*, 1982). Natural HuIFN-α produced by lymphoblastoid cells and purified by affinity chromatography on monoclonal antibodies was also found to have retained its pyrogenic effect (Scott *et al.*, 1982). From the data obtained in man, one would thus tend to conclude that HuIFN-α is indeed pyrogenic.

One problem in studying pyrogenicity of interferon is that there are no good animal models. Schellekens *et al.* (1981, and personal communication) found that rhesus monkeys developed fever in response to some but not all batches of leucocyte interferon, but did not develop fever in response to HuIFN-α_2 from a bacterial source. However, in man and in chimpanzees this bacterial interferon, as well as natural leucocyte interferon, did induce fever.

The pyrogenicity of HuIFN-β also continues to be debated. Extensive purification of fibroblast interferon using affinity chromatography on zinc-chelate columns removed all pyrogenicity detectable after intramuscular injection (Billiau *et al.*, 1980). Yet, intravenous infusion of this interferon which had a specific activity in excess of 10^9 u/mg protein, still caused some fever. Thus, in all probability, fever after intramuscular injection of fibroblast interferon is largely but not completely due to impurities. We also found that HuIFN-β prepared from an osteosarcoma cell line (MG-63) rather than from diploid cells was completely devoid of pyrogenic activity in patients (Pouillart *et al.*, 1982) treated with intramuscular injections, indicating that the pyrogenic impurities were not produced by MG-63 cells.

Fever is a constant symptom occurring during interferon induction by viruses, endotoxins or synthetic inducers (Petralli *et al.*, 1965; Siegert, Shu & Kohlhage, 1967; Lindsay, Trown, Brandt & Forbes, 1969). This accounts for speculations that interferon may be responsible for fever and malaise during the acute phase of virus infection. The data obtained with pure fibroblast interferon would rather indicate that fever and malaise are due to factors co-induced with interferon. It may therefore be important to define and characterize the pyrogenic impurites in current interferon preparations.

Skin reactivity

Intracutaneous injection of fibroblast interferon preparations ($\geq$ 20 000 units in 0.1 ml) was reported to provoke painless erythema and induration at the site of injection (Billiau *et al.*, 1979a). This reaction occurred in patients who had never received any interferon before, but also in patients under continuous interferon therapy. The size of the reaction (average diameter with 20 000 units: 1 to 2 cm) varied from one test person to another but was remarkably constant when tested at different time intervals in a single individual. Patients with spontaneous or iatrogenic immunosuppression reacted as strongly as patients with a normal immune system. Histopathologically the reaction resembled a delayed-type hypersensitivity response, being characterized by perivascular mononuclear infiltrate, few polynuclear cells, very little oedema and absence of vascular lesions. Some authors working with fibroblast interferon purified by different techniques found either very mild skin reactivity or none at all (Scott *et al.*, 1977; Horoszewicz *et al.*, 1978; Scott, Cartwright, Ledu & Dicker, 1978; Carter *et al.*, 1979).

From experiments with completely pure fibroblast interferon (Billiau *et al.*, 1980), it is now evident that this erythematous reaction is not caused by interferon, but by substances which, in all probability, are co-produced with interferon and not removed by earlier partial purification methods. Again, it may eventually be important to define and characterize these skin-reactive factors in fibroblast interferon preparations as they may throw light on inflammatory responses seen during virus infections and on the antitumour effect of interferon injected directly into tumour nodules (Horoszewicz *et al.*, 1978).

Cytostatic effects

It has been known for a long time that cell division is inhibited by *in vitro* exposure to interferon preparations, and that this effect is largely to be attributed to the action of the interferon molecules present in these preparations (Gresser & Tovey, 1978). In most clinical studies with patients receiving human leucocyte or fibroblast interferon no severe cytostatic effects, such as those seen with chemotherapy, have been noted. However, in a recent study, Gutterman *et al.* (1980) reported hair-loss in six out of 38 cancer patients receiving daily doses of 3 to 9×10^6 units of leucocyte interferon. It cannot at present be determined whether this loss was due to interferon itself, to impurities in the injected preparations or to other peculiarities in the clinical situation of the patients.

Suppression of bone marrow function is another common complication of many forms of cytostatic chemotherapy, which has been observed rather frequently in patients treated with interferon. *In vitro*, interferon preparations were found to exert a suppressive effect on the outgrowth of colonies from bone marrow cells (Petralli *et al.*, 1965; McNeill & Gresser, 1973; Greenberg & Mosny, 1977; Nissen, Speck, Emödi & Iscove, 1977). The specificity of this *in vitro* myelosuppressive effect is still insufficiently documented, especially since so far all studies have been done with impure preparations of interferon. Therefore, disagreement between authors (Greenberg & Mosny, 1977; van't Hulle, Schellekens, Löwenberg & de Vries, 1978) on species specificity as well as comparisons between relative sensitivities to different types of interferon (Weimar *et al.*, 1980) are difficult to interpret. Yet, it is worthwhile to mention that human leucocyte interferon was reported to be more active than fibroblast interferon is suppressing myeloid colonies (van't Hulle *et al.*, 1978). Recently, it was shown that maturation of human peripheral blood monocytes into macrophages is reversibly inhibited by interferon (Lee & Epstein, 1980). Again, the interferons used in this study were impure and the antisera used to characterize the effect were polyspecific.

In most clinical trials with leukocyte interferon mild signs of myelosuppression have been noted. Specifically, leukopenia (Greenberg *et al.*, 1976; Cheeseman *et al.*, 1979; Weimar *et al.*, 1979, 1980; Gutterman *et al.*, 1980), granulocytopenia (Pazin *et al.*, 1979; Weimar *et al.*, 1979; Gutterman *et al.*, 1980; Priestman, 1980) and thrombocytopenia (Greenberg *et al.*, 1976; Cheeseman *et al.*, 1979; Gutterman *et al.*, 1980) were reported in patients treated for several days or weeks with daily doses in excess of 10^6 units. In general these haematological changes were not precluding continuation of treatment. Lack of bone marrow engraftment, as evidence for myelosuppression, was not seen in a study in which leucocyte interferon was administered to bone-marrow transplant patients in order to prevent viral infections (O'Reilly *et al.*, 1976). A decrease in the white blood cell count from 5000 to 3000 was found in patients given a single injection of 1.2 to 3.0×10^6 units of lymphoblastoid interferon (Priestman, 1980); this leukopenia was maximal 12 h post injection and was not yet completely restored to normal after 48 h. In patients treated with fibroblast interferon there was no evidence for myelosuppression (Weimar *et al.*, 1978; Carter *et al.*, 1979) except in a single patient (Billiau, Edy & De Somer, 1979b; Weimar *et al.*, 1979) who showed mild leukopenia during the first days of treatment. This patient reacted to leucocyte interferon with severe leukopenia necessitating temporary interruptions in therapy. A constant finding in patients receiving fibroblast interferon (purified by adsorption-elution from porous glass) was a relative lymphopenia (Billiau *et al.*, 1979b), maximal around three to four hours post injection and restored to normal within 24 h post injection. Most probably this effect was due to redistribution of cells rather than to myelosuppression.

Allergic reactions

Virtually all clinical studies up to the present date were done with preparations of leucocyte and fibroblast interferon that contained various proteins of nonhuman origin: Sendai virus and egg proteins in leucocyte interferon and calf serum proteins in fibroblast interferon preparations. Therefore, it is surprising to note how little mention has been made of allergic reactions in patients treated for several weeks with these preparations.

Patients treated with fibroblast interferon from our laboratory were monitored for the development of type I allergy by regular skin testing (intracutaneous injection of 20 000 units/0.1 ml of interferon). A minority of patients showed reactivity other than that also observed in virgin patients. Occasionally a patient developed severe reactivity as manifested by a large weal and flare reaction upon skin testing. When these patients were subsequently given skin tests with highly pure interferon ($> 10^9$ units/mg protein) no reaction ensued, indicating that sensitization was not directed to interferon and that the purification method (adsorption to controlled porous glass + zinc-chelate chromatography) effectively removed foreign proteins (Billiau *et al.*, 1980).

Interferon and tissue damage

Daily inoculation of suckling mice with approximately 10^6 units of mouse (Type I) interferon results in marked inhibition of growth, diffuse liver cell necrosis and death (Gresser *et al.*, 1975, 1976, 1981). When such treatment is stopped after about one week, some of the mice survive, but they develop fatal glomerulonephritis. These observations first done with grossly impure preparations of interferon, and criticized on that basis, were subsequently confirmed by experiments with electrophoretically pure mouse interferon (Gresser *et al.*, 1981). It is not known at present how these changes are brought about. The most important questions to be resolved are: why is the liver the sole site of extensive parenchymal necrosis? Are the early renal lesions (thickening of glomerular basement membrane) a primary effect of the treatment, or are they secondary to liver damage? Why is the phenomenon of renal and hepatic tissue damage restricted to treatment in the early postnatal period?

When interferon treatment is started at age one week the lesions do not develop, indicating that neonatal immaturity is an important component of the syndrome. The glomerulonephritis in surviving mice develops long after all interferon has been eliminated, and it resembles the disease induced in suckling mice by neonatal infection with lymphocytic choriomeningitis virus. Anti-interferon globulin inhibits the development of glomerulonephritis in such mice, indicating that the interferon formed during the first week of the neonatal infection is responsible (Gresser *et al.*, 1978).

Another system in which renal lesions have been linked to interferon is the lupus-like nephritis developing in F_1 hybrids of New Zealand Black and New Zealand White mice. Treatment of such mice with interferon was found to cause accelerated development of the lesions and symptoms (Heremans *et al.*, 1978). The mechanism of these tissue damages is still unclear. In general, the background of their pathogenesis seems to be immunological: (1) lupus is considered as an auto-immune disease; (2) neonatal immaturity of mice certainly encompasses immaturity of the immune system; and (3) interferons are known to have profound effects on the immune system.

Aside from the practical implications for interferon therapists, these observations seem to open a new road to understanding the physiology and physiopathology of the interferon system.

CONCLUDING REMARKS

From the studies reviewed here, it is clear that there remain large gaps in our knowledge of the pharmacokinetics of interferons. Questions such as: 'Where does interferon go when it is injected; and where is it catabolized?' have been answered unsatisfactorily. The main reason for this is that interferon preparations available up to now, although prepared and purified with the greatest care, were still heterogeneous and of low potency. With the advent of new methods of producing large amounts of pure interferons of a single molecular type, and of assaying interferons with more specific tests such as immunosorbent assays, rapid progress in answering these questions can now be foreseen. Perhaps the unravelling of the pharmacokinetics will allow us better to understand what the requirements for an adequate interferon therapy are.

With the increase in the number of clinical trials with interferon the problem of toxic side-effects has received more and more attention. It is now generally recognized that interferons are not innocuous substances. Very much like the lipopolysaccharides of Gram-negative bacteria, interferons seem to affect all organs and organ systems. Their toxicity is very aspecific. In fact many of the side effects resemble the symptoms of acute viral diseases such as influenza, and the question arises whether blocking of interferon production or interferon action might not be a useful form of therapy for such diseases. Along this line of thinking it is also possible that chronic production of interferon is responsible for irreparable tissue damage in diseases such as rheumatoid arthritis, lupus erythematosus, and possibly others.

Thus the areas for research at the level of the intact animal are multiple. Many of the questions which have remained unanswered in the decades between 1960 and 1980 can now be tackled with new techniques. The answers should be of utmost relevance for understanding the physiology and physiopathology of interferons and for designing adequate interferon therapeutic regimens.

REFERENCES

Abb, J., Deinhardt, F., Zander, H., Tenser, R. B., Rapp, F., Goust, J.-M., Fudenberg, H. H., Vilcek, J., Ho, M., Merigan, T. C., Oldstone, M. B. A. & Jackson, G. G. (1982). Trials of interferon therapy for multiple sclerosis. *The Journal of Infectious Diseases*, **146,** 109–15.

Ahstrom, L., Dohlwitz, A., Strander, H., Carlstrom, G. & Cantell, K. (1974). Interferon in acute leukemia in children. *Lancet*, **i,** 166–7.

Arvin, A. M., Feldman, S. & Merigan, T. C. (1978). Human leukocyte interferon in the treatment of varicella in children with cancer: a preliminary controlled trial. *Antimicrobial Agents and Chemotherapy*, **13,** 605–7.

Ashwell, G. & Morell, A. G. (1974). The role of surface carbohydrates in the hepatic recognition and transport of circulating glycoproteins. *Advances in Enzymology*, **41,** 99–128.

Billiau, A. & De Somer, P. (1980). Clinical use of interferons in viral infections. In *Interferon and Interferon Inducers, Clinical Applications*, ed. D. A. Stringfellow, pp. 113–44. New York, Basel: Marcel Dekker.

Billiau, A., De Somer, P., Edy, V. G., De Clercq, E. & Heremans, H. (1979a). Human fibroblast interferon for clinical trials: pharmacokinetics and tolerability in experimental animals and humans. *Antimicrobial Agents and Chemotherapy*, **16,** 53–63.

Billiau, A., Edy, V. G. & De Somer, P. (1979b). The clinical use of fibroblast interferon. In *Antiviral Mechanism in the Control of Neoplasia*, ed. P. Chandra, pp. 675–96. New York, London: Plenum Publishing Corp.

Billiau, A., Heine, J. W., Van Damme, J., Heremans, H. & De Somer, P. (1980). Tolerability of pure fibroblast interferon in man. *Annals of the New York Academy of Sciences*, **350,** 374–5.

Billiau, A., Heremans, H., Ververken, D., Van Damme, J., Carton, H. & De Somer, P. (1981). Tissue distribution of human interferons after exogenous administration in rabbits, monkeys, and mice. *Archives of Virology*, **68,** 19–25.

Billiau, A., Van Damme, J., Van Leuven, F., Edy, V. G., De Ley, M., Cassiman, J. J., Van den Berghe, H. & De Somer, P. (1979c). Human fibroblast interferon for clinical trials: production, partial purification, and characterization. *Antimicrobial Agents and Chemotherapy*, **16,** 49–55.

Bocci, V. (1977). Distribution of interferon in body fluids and tissue. *Texas Reports on Biology and Medicine*, **35,** 436–42.

Bocci, V. (1981). Pharmacokinetic studies of interferons. *Pharmacology and Therapeutics*, **13,** 421–40.

Bocci, V., Pacini, A., Bandinelli, J., Pessina, G. P., Muscettola, M. & Paulesu, L. (1982a). The role of the liver in the catabolism of human α- and β-interferon. *Journal of General Virology*, **60,** 397–400.

Bocci, V., Pacini, A., Muscettola, M., Paulesu, L. & Pessina, G. P. (1981a). Renal metabolism of rabbit serum interferon. *Journal of General Virology*, **55,** 297–304.

Bocci, V., Pacini, A., Muscettola, M., Paulesu, L., Pessina, G. P., Santiano, M. & Viano, I. (1981b). Renal filtration, absorption and catabolism of human alpha interferon. *Journal of Interferon Research*, **1,** 347–52.

Bocci, V., Pacini, A., Muscettola, M., Pessina, G. P., Paulesu, L. & Bandinelli, L. (1982b). The kidney is the main site of interferon catabolism. *Journal of Interferon Research*, **2,** 309–14.

Bocci, V., Pacini, A., Pessina, G. P., Barigli, V. & Russi, M. (1977). Metabolism of interferon: hepatic clearance of native and desialylated interferon. *Journal of General Virology*, **35,** 525–34.

Borden, E. C., Holland, J. F., Dao, T. L., Gutterman, J. U., Wiener, L., Chang, Y.-C. & Patel, J. (1982). Leukocyte-derived interferon (alpha) in human breast carcinoma. *Annals of Internal Medicine*, **97,** 1–6.

Bose, S. & Hickman, J. (1977). Role of carbohydrate moiety in determining the survival of interferon in the circulation. *Journal of Biological Chemistry*, **252,** 8336–7.

Brightman, R., Prescott, L. & Reese, T. S. (1975). Intracellular junctions of special ependyma. In *Brain Endocrine Interaction*, 2nd International Symposium, ed. K. M. Kniege, D. E. Scott, E. Kobayashi & P. Ischiis, pp. 146–65. Basel: Karger.

Cantell, K. & Pyhälä, L. (1973). Circulating interferon in rabbits after administration of human interferon by different routes. *Journal of General Virology*, **20,** 97–104.

Cantell, K. & Pyhälä, L. (1976). Pharmacokinetics of human leukocyte interferon. *Journal of Infectious Diseases*, **133,** A6–A12.

Cantell, K., Pyhälä, L. & Strander, H. (1974). Circulating human interferon after intramuscular injection into animals and man. *Journal of General Virology*, **22,** 453–5.

Carter, W. A., Dolen, J. G., Leong, S. S., Horoszewicz, J. S., Vladutiu, A. O., Leibowitz, A. I. & Nolan, J. P. (1979). Purified human fibroblast interferon in vivo: skin reactions and effect on bone marrow precursor cells. *Cancer Letters*, **7,** 243–9.

Cesario, T. C. (1977). The effect of body fluids on polynucleotide-induced fibroblast interferon and virus-induced leukocyte interferon. *Proceedings of the Society for Experimental Biology and Medicine*, **155,** 583–7.

Cesario, T., Vaziri, N., Slater, L. & Tilles, J. (1979). Inactivators of fibroblast interferon bound in human serum. *Infection and Immunity*, **24,** 851–5.

CHEESEMAN, S. H., RUBIN, R. H., STEWART, J. A., TOLKOFF-RUBIN, N. E., COSIMI, A. B., CANTELL, K., GILBERT, J., WINKLE, S. HERRIN, J. T., BLACK, P. H., RUSSELL, P. S. & HIRSCH, M. S. (1979). Controlled clinical trial of prophylactic human-leukocyte interferon in renal transplantation. *New England Journal of Medicine*, **300,** 1345–9.

DE CLERCQ, E., EDY, V. G., DEVLIEGER, H., EECKELS, R. & DESMYTER, J. (1975). Intrathecal administration of interferon in neonatal herpes. *Journal of Pediatrics*, **86,** 736–9.

DESMYTER, J., RAY, M. B., DE GROOTE, J., BRADBURNE, A. F., DESMET, V. J., EDY, V. G., BILLIAU, A., DE SOMER, P. & MORTELMANS, J. (1976). Administration of human fibroblast interferon in chronic hepatitis-B infection. *Lancet*, **ii,** 645–7.

DOLEN, J. G., CARTER, W. A., HOROSZEWICZ, J. S., VLADUTIU, A. O., LEIBOWITZ, A. I. & NOLAN, J. P. (1979). Fibroblast interferon treatment of a patient with chronic active hepatitis. *American Journal of Medicine*, **67,** 127–31.

EDY, V. G., BILLIAU, A. & DE SOMER, P. (1976). Comparison of rate of clearance of human fibroblast and leukocyte interferon from the circulatory system of rabbits. *Journal of Infectious Diseases*, **133,** A18–A21.

EDY, V. G., BILLIAU, A. & DE SOMER, P. (1978). Non-appearance of injected fibroblast interferon in the circulation. *Lancet*, **i,** 451–2.

EMÖDI, G., JUST, M., HERNANDEZ, R. & HIRT, H. R. (1975a). Circulating interferon in man after administration of exogenous human leukocyte interferon. *Journal of the National Cancer Institute*, **54,** 1045–9.

EMÖDI, G., O'REILLY, R., MÜLLER, A., EVERSON, L. K., BINSWANGER, U. & JUST, M. (1976). Effect of human exogenous leukocyte interferon in cytomegalovirus infections. *Journal of Infectious Diseases*, **133,** A199–A204.

EMÖDI, G., RUFLI, T., JUST, M. & HERNANDEZ, R. (1975b). Human interferon therapy for herpes zoster in adults. *Scandinavian Journal of Infectious Diseases*, **7,** 1–5.

EZAKI, K., OGAWA, M., OKABE, K., ABE, K., INOUE, K., HORIKOSHI, N. & INAGAKI, J. (1982). Clinical and immunological studies of human fibroblast interferon. *Cancer Chemotherapy and Pharmacology*, **8,** 47–55.

GREENBERG, H. B., POLLARD, R. B., LUTWICK, L. I., GREGORY, P. B., ROBINSON, W. S. & MERIGAN, T. C. (1976). Effect of human leukocyte interferon on hepatitis B virus infection in patients with chronic active hepatitis. *New England Journal of Medicine*, **295,** 517–22.

GREENBERG, P. L. & MOSNY, S. A. (1977). Cytotoxic effects of interferon *in vitro* on granulocytic progenitor cells. *Cancer Research*, **37,** 1794–9.

GREENBERG, S. B., HARMON, M. W. & COUCH, R. C. (1980). Exogenous interferon: stability and pharmacokinetics. In *Interferon and Interferon Inducers, Clinical Applications*, ed. D. A. Stringfellow, pp. 57–87. New York, Basel: Marcel Dekker.

GRESSER, I., AGUET, M., MOREL-MAROGER, L., WOODROW, D., PURRIEN-DUTILLEUL, F., GUILLON, J. C. & MAURY, C. (1981). Electrophoretically pure mouse interferon inhibits growth, induces liver and kidney lesions and kills suckling mice. *American Journal of Pathology*, **102,** 396–402.

GRESSER, I., MAURY, C., TOVEY, M., MOREL-MAROGER, L. & PONTILLON, F. (1976). Progressive glomerulonephritis in mice treated with interferon preparations at birth. *Nature*, **263,** 420–2.

GRESSER, I., MOREL-MAROGER, L., VERROUST, P., RIVIÈRE, Y. & GUILLON, J. C. (1978). Anti-interferon globulin inhibits the development of glomerulonephritis in mice infected at birth with lymphocytic choriomeningitis virus. *Proceedings of the National Academy of Sciences of the USA*, **75,** 3413–16.

GRESSER, I. & TOVEY, M. G. (1978). Antitumor effects of interferon. *Biochimica et Biophysica Acta*, **516,** 231–47.

GRESSER, I., TOVEY, M. G., MAURY, C. & CHOUROULINKOW, I. (1975). Lethality of interferon preparation for newborn mice. *Nature*, **258,** 76–8.

GUTTERMAN, J. U., BLUMENSCHEIN, G. R., ALEXANIAN R., YAP, H.-U., BUZDAR, A. U., CABANILLAS, F., HORTOBAGYI, G. N., HERSH, E. M., RASMUSSEN, S. L., HARMON, M., KRAMER, M. & PESTKA, S. (1980). Leucocyte interferon induced tumor regression in human metastatic breast cancer, multiple myeloma, and malignant lymphoma. *Annals of Internal Medicine*, **93,** 399–406.

GUTTERMAN, J. U., FINE, S., QUESDA, J., HORNING, S. J., LEVINE, J. F., ALEXANIAN, R., BERNHARDT, L., KRAMER, M., SPIEGEL, H., COLBURN, W., TROWN, P., MERIGAN, T. & DZIEWANOWSKI, Z. (1982). Recombinant leukocyte A interferon: pharmacokinetics, single-dose tolerance, and biologic effects in cancer patients. *Annals of Internal Medicine,* **96,** 549–56.

HABIF, D. V., LIPTON, R. & CANTELL, K. (1975). Interferon crosses blood-cerebrospinal fluid barrier in monkeys. *Proceedings of the Society for Experimental Biology and Medicine*, **149,** 287–9.

HANLEY, D. F., WIRANOWSKA-STEWART, M. & STEWART, W. E. II. (1979). Pharmacology of interferons. I. Pharmacologic distinctions between human leukocyte and fibroblast interferons. *International Journal of Immunopharmacology*, **1,** 219–26.

HEINE, J., BILLIAU, A., VAN DAMME, J. & DE SOMER, P. (1982). Human fibroblast interferon: subpopulation with different pharmacokinetic behavior. In *The Clinical Potential of Interferons*, ed. R. Kono & J. Vilcek, pp. 69–74. Tokyo: University of Tokyo Press.

HEINE, J., VAN DAMME, J., DE LEY, M., BILLIAU, A. & DE SOMER, P. (1981). Purification of human fibroblast interferon by zinc chelate chromatography. *Journal of General Virology*, **54,** 47–56.

HEREMANS, H., BILLIAU, A., COLOMBATTI, A., HILGERS, J. & DE SOMER, P. (1978). Interferon treatment of NZB mice: accelerated progression of autoimmune disease. *Infection and Immunity*, **21,** 925–30.

HEREMANS, H., BILLIAU, A. & DE SOMER, P. (1980). Interferon in experimental viral infections in mice: confrontation between tissue interferon levels resulting from the virus infection and from exogenous interferon therapy. *Infection and Immunity*, **30,** 513–22.

HOROSZEWICZ, J. S., LEONG, S. S., ITO, M., BUFFETT, R. F., KARAKOUSIS, C., HOLYOKE, E., JOB, L., DÖLEN, J. G. & CARTER, W. A. (1978). Human fibroblast interferon in human neoplasmia: clinical and laboratory study. *Cancer Treatment Reports*, **62,** 1899–1906.

IDESTROM, K., CANTELL, K., KILLANDER, D., NILSSON, K., STRANDER, H. & WILLEMS, J. (1979). Interferon therapy in myeloma. *Acta Medica Scandinavica*, **205,** 149–54.

JACOBS, L., O'MALLEY, J., FREEMAN, A. & EKES, R. (1981). Intrathecal interferon reduces exacerbations of multiple sclerosis. *Science*, **214,** 1026–8.

JORDAN, G. W., FRIED, R. P. & MERIGAN, T. C. (1974). Administration of human leukocyte interferon in herpes zoster. I. Safety, circulating antiviral activity and host responses to interferon. *Journal of Infectious Diseases*, **130,** 56–62.

KINGHAM, J. G. C., GANGULY, N. K., SHAARI, Z. D., MENDELSON, R., MCGUIRE, M. J., HOLGATE, S. J., CARTWRIGHT, T., SCOTT, G. M., RICHARDS, B. M. & WRIGHT, R. (1978). Treatment of HBsAg-positive chronic active hepatitis with human fibroblast interferon. *Gut*, **19,** 91–4.

LEE, S. H. & EPSTEIN, L. B. (1980). Reversible inhibition by interferon of the maturation of human peripheral blood monocytes to macrophages. *Cellular Immunology*, **50,** 177–90.

LINDSAY, H. L., TROWN, P. W., BRANDT, J. & FORBES, M. (1969). Pyrogenicity of poly I. poly C in rabbits. *Nature*, **223,** 717–18.

LUCERO, M. A., MAGDELENAT, H., FRIDMAN, W. H., POUILLART, P., BILLARDON, C., BILLIAU, A., CANTELL, K. & FALCOFF, E. (1982). Comparison of effects of leukocyte and fibroblast interferon on immunological parameters in cancer patients. *European Journal of Cancer and Clinical Oncology*, **18,** 243–51.

MCNEILL, T. A. & GRESSER, I. (1973). Inhibition of haemopoietic colony growth by interferon preparations from different sources. *Nature New Biology*, **244,** 173–4.

MERIGAN, T. C., RAND, K. H., POLLARD, R. B., ABDALLAH, P. S., JORDAN, G. W. & FRIED, R. P. (1978). Human leukocyte interferon for the treatment of herpes zoster in patients with cancer. *New England Journal of Medicine*, **298,** 981–7.

MISSET, J. L., MATHÉ, G., GASTIABURU, J., GOUTNER, A., DORVAL, T., GOUVEIA, J., HAYAT, M., JASMIN, C., SCHWARZENBERG, L., MACHOVER, D., RIBAUD, P., DE VASSAL, F. & HOROSZEWICZ, J. S. (1982) Treatment of lymphoid neoplasias with interferon. I. Human fibroblastic β-interferon in malignant gammapathies. Phase II trial. *Anticancer Research*, **2,** 63–6.

MORSER, J., KABAYO, J. P. & HUTCHINSON, D. W. (1978). Differences in sialic acid content of human interferons. *Journal of General Virology*, **41,** 175–8.

NISSEN, C., SPECK, B., EMÖDI, G. & ISCOVE, N. N. (1977). Toxicity of human interferon preparations for human bone marrow cultures. *Lancet*, ii, 203–4.

O'REILLY, R. J., EVERSON, L. K., EMÖDI, G., HANSEN, J., SMITHWICK, E. M., GRIMES, E., PAHWA, S., PAHWA, R., SCHWARTZ, S., ARMSTRONG, D., SHEGAL, F. P., GUPTA, S., DUPONT, B. & GOOD, R. A. (1976). Effects of exogenous interferon in cytomegalovirus infections complicating bone marrow transplantation. *Clinical Immunology and Immunopathology,* **6,** 51–61.

PAPE, G. R., HADAM, M. R., EISENBURG, J. & RIETHMÜLLER, G. (1981). Kinetics of natural cytotoxicity in patients treated with human fibroblast interferon. *Cancer Immunology and Immunotherapy*, **11,** 1–6.

PAULSON, G. & KAPP, J. P. (1967). Movement of sodium-22, radioiodinated protein and tritiated water from the cisterna magna into the cerebrovascular circulation. *Journal of Neurosurgery,* **27,** 138–41.

PAZIN, G. J., ARMSTRONG, J. A., LAM, M. P., TARR, G. C., JANNETTA, P. J. & HO, M. (1979). Prevention of reactivated herpes simplex infection by human leukocyte interferon after operation on the trigeminal root. *New England Journal of Medicine*, **301,** 225–30.

PETRALLI, J. K., MERIGAN, T. C. & WILBUR, J. R. (1965). Circulating interferon after measles vaccination. *New England Journal of Medicine*, **273,** 198–201.

POLLARD, R. B. & MERIGAN, T. C. (1978). Experience with clinical applications of interferon and interferon inducers. *Pharmacology and Therapeutics*, **2,** 783–811.

POUILLART, P., PALANGIE, T., JOUVE, M., GARCIA-GIRALT, E., FRIDMAN, H., MAGDELENAT, H., FALCOFF, E. & BILLIAU, A. (1982). Administration of fibroblast interferon to patients with advanced breast cancer: possible effects on skin metastasis and on hormone receptors. *European Journal of Cancer and Clinical Oncology* (in press).

PRIESTMAN, T. J. (1980). Initial evaluation of human lymphoblastoid interferon in patient with advanced malignant disease. *Lancet*, **ii,** 113–18.

PRIESTMAN, T. J., JOHNSTON, M. & WHITEMAN, P. D. (1982). Preliminary observations on the pharmacokinetics of human lymphoblastoid interferon given by intramuscular injection. *Clinical Oncology* (in press).

SARKAR, F. H. (1982). Pharmacokinetic comparison of leukocyte and *Escherichia coli*-derived human interferon type alpha. *Antiviral Research*, **2,** 103–6.

SCHELLEKENS, H., DE REUS, A., BOLHUIS, R., FOUNTOULAKIS, M., SCHEIN, C., EESÖDI, J., NAGATA, S. & WEISSMAN, C. (1981). Comparative antiviral efficiency of leukocyte and bacterially produced human α-interferon in rhesus monkeys. *Nature*, **292,** 775–6.

Scott, G. M., Butler, J. K., Cartwright, T., Richards, B. M., Kingham, J. G., Wright, R. & Tyrrell, D. A. J. (1977). Interferon skin reactivity and pyrexial reactions. *Lancet*, **ii,** 402–3.

Scott, G. M., Cartwright, T., Ledu, G. & Dicker, D. (1978). Effect of human fibroblast interferon on vaccination in volunteers. *Journal of Biological Standardization*, **6,** 73–6.

Scott, G. M., Wallace, J., Tyrrell, D. A. J., Cantell, K., Secher, D. S. & Stewart, W. E. (1982). Interim report on studies on 'toxic' effects of human leukocyte derived Interferon-alpha (HuIFN-α). *Journal of Interferon Research*, **2,** 127–30.

Siegert, R., Shu, H. L. & Kohlhage, H. (1967). Correlation between fever and interferon titer in rabbits after induction with myxoviruses. *Life Sciences*, **6,** 615–20.

Stewart, II. W. E. (1979). *The Interferon System.* Wien, New York: Springer.

Strander, H., Cantell, K., Carlström, G. & Jakobsson, P. A. (1973). Clinical and laboratory investigations on man: systemic administration of potent interferon to man. *Journal of the National Cancer Institute*, **51,** 733–42.

Treuner, J., Niethammer, D., Dannecker, G., Hagmann, R., Neef, V. & Hofschneider, P. H. (1980). Successful treatment of nasopharyngeal carcinoma with interferon. *Lancet*, **i,** 817–18.

van't Hulle, E., Schellekens, H., Löwenberg, B. & de Vries, M. J. (1978). The influence of interferon preparations on the proliferative capacity of human and mouse bone marrow cells in vitro. *Cancer Research*, **38,** 911–14.

Ververken, D., Carton, H. & Billiau, A. (1978). Intrathecal administration of interferon in MS patients? In *Human Immunity in Neurological Diseases. Nato Advanced Study Institutes Series*, **24,** 624–7.

Vilcek, J., Sulea, I. T., Zerebreckyj, I. L. & Yip, Y. K. (1980). Pharmacokinetic properties of human fibroblast and leukocyte interferons in rabbits. *Journal of Clinical Microbiology,* **11,** 102–5.

Weimar, W., Heijtink, R. A., Schalm, S. W. & Schellekens, H. (1979). Differential effects of fibroblast and leucocyte interferon in HBsAg positive chronic active hepatitis. *European Journal of Clinical Investigation*, **9,** 151–4.

Weimar, W., Heijtink, R. A., Schalm, S. W., van Blankenstein, M., Schellekens, H., Masurel, N., Edy, V. G., Billiau, A. & De Somer, P. (1977). Fibroblast interferon in HBsAg-positive chronic active hepatitis. *Lancet*, **ii,** 182.

Weimar, W., Heijtink, R. A., ten Kate, F. J. P., Schalm, S. W., Masurel, N. & Schellekens, H. (1980). Double-blind study of leucocyte interferon administration in chronic HBsAg-positive hepatitis. *Lancet*, **i,** 336–8.

Weimar, W., Schellekens, H., Lameijer, L. D. F., Masurel, N., Edy, V. G., Billiau, A. & De Somer, P. (1978). Double-blind study of interferon administration in renal transplant recipients. *European Journal of Clinical Investigation*, **8,** 255–8.

THE ANTIVIRAL EFFECTS OF INTERFERON

G. M. SCOTT

Clinical Research Centre, Harrow, Middlesex HA1 3UJ, UK

INTRODUCTION

Interferon was discovered because of its antiviral properties. Although it is now clear that 'interferon' is a family of proteins or glycoproteins with actions which extend beyond the mere rendering of cells resistant to virus replication, it has always been hoped that interferon would be a broad spectrum antiviral agent in man.

The purpose of this chapter is to summarize briefly the studies using exogenous interferons against virus infections of man with reference to animal models where appropriate and to make some predictions for the future use of interferon. Our failure, at the present time, to be able to define adequately any one single indication for interferon in the clinic (if one exists) has been entirely due to the difficulty in making sufficient interferon from eukaryocytes. A paucity of material has meant that relatively small numbers of patients or volunteers with a particular infection have been treated and that the data collected, although for many indications encouraging, are simply not yet adequate to allow the application for permission to market interferon. Furthermore, until genetic engineering made mass production feasible, there would not have been sufficient material to supply a demand for the treatment of patients.

Apart from the demonstration of efficacy of interferon in any single indication (and this includes demonstration of clear superiority over other antiviral agents either alone or in combination), one major problem has to be overcome. This is the adequate evaluation of the 'unwanted' effects of interferons, a family of natural proteins with remarkably diverse effects on cell growth and metabolism and on many aspects of learned and innate immune responses. Perhaps the endogenous interferon response to acute virus infections is concerned with recovery from the infection, not only by a direct antiviral action on cells but also by different effects on the immune system. The administration of exogenous interferon over short periods would mimic these effects if the interferon levels achieved in tissues were comparable with those achieved in acute infections.

Over prolonged periods, for example in the prophylaxis of herpes virus infections after renal transplantation or against respiratory virus infections in susceptible individuals, the acceptability of interferon would be considerably reduced if there were unwanted effects on cell growth or on the immune system. These effects are as yet hypothetical and ill-defined although recent reports of neutralizing antibody formation in response to exogenous interferons give rise for concern.

Unfortunately, it is not possible to rely on the results of extensive animal toxicity experiments to detect unwanted effects of interferons in man. This is partly because it is not only the antiviral effects which are relatively species selective (Finter, Woodrooffe & Priestman, 1982) but also because animals would be expected to raise neutralizing antibodies to human interferons as they would to repeated injections of any foreign protein. The paucity of material supply meant that interferon was administered to man without the preamble of extensive animal toxicity experiments which are ethically and legally required for any new drug. Regulatory authorities showed an unusual willingness to allow the early testing of interferons in man, in part perhaps because it was felt unlikely that a natural human product would have significant unwanted effects. This unreasonable assumption proved not to be the case (as it did thirty years ago with corticosteroids). Dr Billiau has dealt with the toxicity of interferon in detail and it is not the purpose of this chapter to repeat his account.

The possibility that interferon may become available as an antiviral or antitumour agent is now a real one, simply because of the cloning of interferon genes in prokaryocytes. The production of individual molecular species of interferon in sufficient quantities allows a high degree of chemical purification. The techniques pose a series of novel problems of quality control and toxicity and one of immediate interest to the clinician. If there are many different interferon proteins available, will it be possible to tell whether one is better than another for any particular indication? Animal models for comparing interferons have been devised but it will be difficult to demonstrate subtle differences between preparations in man. It is important that prokaryocyte products are sufficiently like natural interferons to be active without enhanced immunogenicity and toxicity. Clinical trials with cloned interferons have already begun and the results of some of these will be available by the time this book is published.

As will become apparent from the discussion below, the experimental and natural virus infections in which interferon has been shown to have an effect, have been mainly those in which a response to treatment could be demonstrated with ease, either clinically or virologically. It is not surprising, therefore, that the only established acute infections which are recorded to have been affected beneficially are those caused by herpes viruses. Not only do they have classical clinical features, but they may (in the situations tested) be associated with a failure of endogenous interferon production. Several uncontrolled experiments have been performed in patients with other acute virus infections but it is not possible to evaluate the effects of interferon in these. The discussion below, therefore, focuses on the results of adequately controlled clinical trials. These have been divided into situations where interferon was used prophylactically, in the treatment of acute conditions and in the treatment of chronic viral illnesses. It is these results which lay the groundwork for the development of rational study designs and treatment regimens.

PROPHYLAXIS

(1) Vaccinia

In 1804, Jenner reported that if a patient had cold sores, vaccination with cowpox might not 'take'. This may be the first description of virus interference. Vaccinia is sensitive to interferon *in vitro* and also induces a local interferon response, thought to be a possible explanation for the relative benignity of variolation smallpox (Wheelock, 1964). Inoculation of vaccinia virus has been used in the treatment of warts (McGee, 1967).

Lesions caused by vaccinia virus may easily be evaluated quantitatively but vary with the inoculum, method of inoculation and host immunity. Before interferon was discovered, it was shown that intradermal inoculations of heat-killed inactivated influenza virus into rabbit skin could protect against subsequent vaccination at the same site (Depoux & Isaacs, 1954) and it was rational to try cell supernatants containing interferon in the same model. Chick embryo cell interferon did not protect rabbit skin sites very well (Lindenmann, Burke & Isaacs, 1957) but rabbit kidney cell interferon did protect when inoculated one day before, but not at the same time as or later than, vaccination (Isaacs & Westwood, 1959).

Monkey kidney cell interferon was active against vaccinia *in vitro* (Sutton & Tyrrell, 1961) and in monkey skin (Andrews, 1961) and reduced the infectivity of vaccinia in monkey eyes (Cantell & Tommila, 1960).

This material was used in the first successful experiment with interferon in man. Thirty eight volunteers were inoculated intradermally at separate skin sites with crude monkey kidney cell interferon and placebo. There was complete protection against vaccinia inoculated 24 h later, at the interferon site in 24 volunteers and a reduction in lesion size in a further eight (Scientific Committee on Interferon, 1962). The doses of interferon used in these preliminary studies were not known because no appropriate standards were available.

Petralli, Merigan & Wilbur (1965) found that protection against vaccinia could be achieved at a certain interval after live measles vaccination, coinciding with the presence of circulating antiviral activity and generalized systemic reactions.

The Scientific Committee on Interferon (1970) showed reduced but not completely abolished vaccinia lesions in monkey skin treated with 100 u human leucocyte interferon intradermally. In monkeys immunosuppressed with horse anti-lymphocyte serum, primary vaccinia pustules were surrounded by satellite lesions which could be prevented by 5×10^5 u leucocyte interferon daily, even when started as late as the day after vaccination (Neumann-Haefelin, Shrestha & Manthey, 1976). In this model, intramuscular interferon was considerably less effective than the same dose given locally. Intramuscular fibroblast and leucocyte interferons were compared against vaccinia in normal monkeys (Weimar *et al.*, 1980b). Interferon was started the day before vaccination and continued for a week. Lower serum levels of antiviral activity were found after intramuscular fibroblast interferon compared with serum levels after the same dose of leucocyte interferon, and there was a slight (though not significant) reduction in the degree of protection seen at 5×10^5 u/kg/day. When the same antiviral activity was achieved in the serum, little difference was seen between the activity of the preparations. Because certain monkey cells are not sensitive to human interferons *in vitro*, it was proposed that intramuscular interferon protected rhesus monkeys against vaccinia by enhancing alternative host defence mechanisms (Schellekens, Weimar, Cantell & Stitz, 1979). The same model was used to compare natural leucocyte and cloned $\alpha 2$ interferons (Schellekens *et al.*, 1981). After intramuscular injection of 5×10^5 u/kg/day, slightly lower antiviral titres were obtained in the serum after $\alpha 2$ interferon and the protection against vaccinia was slightly (though again not significantly) less than that afforded by natural interferon.

Partially-purified natural fibroblast and leucocyte interferons were given intradermally at multiple skin sites on rhesus monkey backs in a dose-ranging comparative study against vaccinia given 24 h later. There was a suggestion of a reduction in vaccinia lesions at control sites when a large total dose of interferon had been given. No significant differences in the degree of protection afforded by fibroblast or leucocyte interferon could be shown (Scott *et al.*, 1977). In a double-blind placebo-controlled study, significant protection of skin sites against vaccinia was demonstrated by 5×10^5 u fibroblast interferon, confirming the results of the previous clinical study using monkey kidney interferon and proving that fibroblast interferon had antiviral activity in man (Scott, Cartwright, Le Du & Dicker, 1978).

Models of experimental infection with vaccinia in man and animals have proved extremely useful for demonstrating *in vivo* antiviral activity of interferons and for comparing different preparations. The relevance of these observations to the clinical use of interferon is doubtful. Variola infections are no longer a risk to man and vaccination is perhaps no longer ethical in volunteers. It is unlikely, therefore, that the model will be pursued further although it would be interesting to test interferon in molluscum contagiosum.

(2) Respiratory virus infections

The first demonstration of interferon induction was made using inactivated influenza virus (Isaacs & Lindenmann, 1957). Isaacs & Hitchcock (1960) showed the possible role of interferon in the recovery from experimental influenza in mice and similar observations have been made in man (Richman, Murphy, Baron & Uhlendorf, 1976). After experimental virulent influenza virus infection of volunteers, a rapid rise in the titres of virus in the nasopharyngeal secretions is seen to peak at 26 h. There is then a rapid fall of virus titre coinciding with the peak interferon activity in secretions and symptoms, some days before the appearance of specific antibody. This would suggest that interferon may be involved in the rapid recovery from this infection but it is interesting to note that Gresser, Tovey, Maury & Bandu (1976) found that a potent antibody to interferon in mice did not enhance pathogenicity of experimental influenza.

Takano, Jensen & Warren (1963) showed that tiny doses of chick embryo cell interferon given to mice exerted a protective effect against a low infective challenge with influenza, providing it was started the day before virus inoculation. The Scientific Committee on Interferon (1965) published the negative results of a study using low-titred monkey kidney cell interferon (similar to that used against vaccinia), given intranasally against parainfluenza 1, rhinovirus M or coxsackie A21 in small numbers of volunteers. This study was salutory and indicated that if the failure had been due simply to insufficient interferon (which is almost certainly the case), then the capacity at that time to produce and purify it was simply not sufficient for the needs of any single well-conducted trial, let alone mass therapy. It was natural, therefore, that experiments with interferon inducers would be performed during the late sixties and early seventies. The results of these were, on the whole, disappointing (Hill *et al.*, 1971a; Gatmaitan, Stanley & Jackson, 1973), so that a welcome return to exogenous leucocyte interferon was made possible by the efforts of Cantell. A series of controlled experiments were performed at the MRC Common Cold Unit in the early seventies. Initially, rhinovirus 2 infection was not affected by 1.8×10^5 u leucocyte interferon given intranasally in divided doses, from 20 h before until 43 h after virus infection (Tyrrell & Reed, 1973). Neither did a total of 8×10^5 u interferon given as 16 doses from the day before and on the day of influenza B virus challenge reduce virus shedding, seroconversion or symptoms (Merigan, Reed, Hall & Tyrrell, 1973). A delay in the onset of symptoms in the interferon group, together with previous avian influenza work (Portnoy & Merigan, 1971) suggest that treatment might be more effective if continued for longer. In a further experiment, therefore, 14×10^6 u leucocyte interferon were given as 39 doses (in groups of three doses within one hour, three times per day) by nasal spray, starting one day before rhinovirus 4 challenge and continuing for a total of four days (Merigan *et al.*, 1973). Treated volunteers were compared again with a matched control group who received placebo, in which there was a low rate of clinical colds. Nevertheless, there were significant reductions in the frequency of severe colds, in virus shedding and the serologic response to infection.

It was elating to be able to show that interferon could protect against rhinovirus but disappointing that the effect required so much interferon in terms of the output from Cantell's laboratory. It is interesting to note that only 50–150 μg interferon protein was given to each volunteer.

A considerable amount of research has gone into improving ways of delivering interferon in the nose. In rhesus monkeys and normal human subjects, there is an extremely short half-life of recovered interferon from the inferior turbinate of the nose, compared with a baseline recovery from the opposite side, after the local application of 10^4 u (Johnson *et al.*, 1976). Furthermore, human nasal secretions contain factors which partly inactivate human interferons *in vitro*, fibroblast interferon being considerably more susceptible to inactivation than leucocyte interferon (Harmon, Greenberg & Couch, 1976). Turbinate scrapings could be rendered resistant to vesicular stomatitis virus infection by 100 or 1000 u/ml of either leucocyte or fibroblast interferon. In order to induce an antiviral state consistently, turbinate cells required at least 4 h incubation with 1000 u/ml interferon (Harmon, Greenberg, Johnson & Couch, 1977; Greenberg, Harmon, Johnson & Couch, 1978). The mucus content of the nose is cleared completely in 20–30 min (Proctor, Andersen & Lundquist, 1973) and the initial half-life of intranasal radioactive albumin is about 20 min (Aoki & Crawley, 1976). An initial rapid decline in radioactivity after dosing was followed by a slower decline caused by the slow removal of material mainly from the anterior part of the nose. This was probably also observed in some volunteers given interferon, antiviral activity sometimes being recoverable even 24 h after dosing (Merigan *et al.*, 1973; Scott *et al.*, 1982b).

In an attempt to find a way to protect the nose against rhinovirus with a smaller dose than that used by Merigan *et al.*, Greenberg and coworkers (1982) applied interferon soaked in a cotton wool pledget in the nose for one hour or by continuous aerosol spray over one hour. All the volunteers received antihistamines which reduce mucociliary activity. Using single or multiple doses of leucocyte interferon (from 10^6 to 4×10^6 u in total), an overall reduction in the frequency of colds due to rhinovirus 13 was observed. At the Common Cold Unit, we failed to influence rhinovirus 2 infections using a prophylactic schedule of 6×10^5 u fibroblast interferon given in divided doses over four days. Although the intended dose was considerably higher than this, there was a marked loss of antiviral activity of this preparation during handling (Scott, Reed, Cartwright & Tyrrell, 1980).

When fresh supplies of highly purified natural leucocyte interferon (prepared by Cantell and purified by Secher using a monoclonal antibody), and cloned $\alpha2$ interferon became available in 1981, it was thought essential to repeat the previous successful study (Merigan *et al.*, 1973) to establish baseline protective activity of interferon against rhinoviruses before moving on to answer the two major unknown questions: first, could interferon therapy affect infection once it had become established and secondly, could the frequency of dosing be traded for a higher concentration of interferon?

In two recent studies (Scott *et al.*, 1982a,b), a high degree of protection against rhinovirus 9 infection was achieved using a total of 9×10^7 u of both highly purified natural and cloned leucocyte interferons. The dose schedule was almost exactly the same as that used in the earlier successful trial. Using a small volunteer-activated nasal spray pump, we have examined schedules of interferon given as prophylaxis against rhinovirus 9 and rhinovirus 14. Two viruses were used in order to increase the frequency of colds in the placebo group. Nevertheless, only about half of the volunteers receiving both viruses had definite colds (graded mild, moderate or severe). Using high-titred cloned $\alpha2$ interferon (10^7 u/ml), dose schedules of three, two or one applications per day appeared to offer protection against the double virus challenge, but failures in protection were seen if the concentration was reduced (10^6 or 10^5 u/ml) or when there was a long interval between interferon dose and virus inoculation. This is somewhat surprising in view of the supposed mechanism of action of interferon but the results suggest that dosing three times per day is necessary to protect against high-dose virus challenge at any time of day (Phillpotts, Higgins, Scott, Wallace and Tyrrell, in preparation). Now that material is freely available by genetic engineering, such high dose schedules are feasible for widescale prophylaxis in susceptible individuals but theoretically it may be possible to protect against natural colds using considerably less interferon. This is because the pathogenesis of experimental (where very large virus inocula are used) and natural infections (where the infectious dose is not known) may be different. Certainly in influenza, quite different illnesses may result (Little, Douglas, Hall & Roth, 1979).

Bearing this in mind, it is interesting to observe the protection claimed by Russian scientists using 10^4 units crude leucocyte interferon intranasally against influenza during an epidemic (Solov'ev, 1969). Similar experiments have been performed in China against rhinovirus infections using low titre interferon, but were not reported in sufficient detail to allow critical evaluation (Yunde *et al.*, 1981). Imanishi and coworkers (1980) studied a heterogeneous group of student volunteers in Japan given leucocyte interferon nose drops (5×10^3 u/day) or placebo for four months. A reduction in the number of volunteers complaining of upper respiratory and of general symptoms (particularly fever) was seen in the interferon group. From the examination of paired sera, it appears that there was a small parainfluenza outbreak during the trial but no virus isolations were reported. It is difficult to reconcile these results from Russia, China and Japan with the results of experiments in volunteers under highly controlled conditions. If high doses of topical interferon turn out to cause local symptoms due to an inflammatory effect in the nose (Scott, 1982), then lower doses than we have shown to be necessary to protect against experimental colds will need to be tested in field trials anyway. It is only in large numbers of volunteers exposed to natural colds that this sort of problem can be sorted out. Trials are presently under way to determine effective prophylactic schedules against influenza and coronaviruses. Field trials designed to test protection of large groups of volunteers against wild viruses have already started in the USA and will shortly begin in the UK (Scott, Phillpotts & Tyrrell, 1982).

(3) Herpes simplex labialis

When the trigeminal nerve is severed in the course of surgery for *tic dolouroux*, an outbreak of herpes labialis is very likely to occur, more so if the patient is seropositive (Ellison, Canton & Rose, 1959). This may be aggravated by the administration of large doses of dexamethasone designed to reduce local post-operative oedema. In a double-blind trial, leucocyte interferon (7×10^4 u/kg/day) or placebo were given intramuscularly once daily from one day before operation on the trigeminal root, for a total of five days (Pazin *et al.*, 1979). Of 18 patients on placebo, 10 had herpes labialis and 15 shed herpes virus in the oropharyngeal secretions; of 19 on interferon, only five had clinical herpes and eight excreted virus. Interferon treatment itself caused some fever but clearly reduced the incidence of reactivation herpes. Not surprisingly, this treatment did not affect subsequent reactivation herpes in these patients (Haverkos, Pazin, Armstrong & Ho, 1980) and we have seen outbreaks of herpes labialis in association with administration of exogenous leucocyte interferons (Scott, personal observations).

(4) Renal transplant recipients

Two groups have studied the effects of interferon prophylaxis on the reactivation of viruses and on the outcome of grafts after transplantation.

The first study reported was performed in 16 renal transplant recipients allocated at random to receive 3×10^6 u fibroblast interferon or placebo intramuscularly twice a week (Weimar *et al.*, 1978). Treatment was started just before the transplantation operation and carried on for three months. Cytomegalovirus infection occurred at the same rate in the two groups and volunteers in the interferon group but not the placebo group, had clinical influenza A (one) and rubella (one). More volunteers in the placebo group had evidence of herpes simplex (four) than in the interferon group (one). It was concluded that fibroblast interferon in this schedule had had no demonstrable effect. Perhaps it is not surprising if one considers the small numbers of patients treated and the expected frequency of virus infections over this period. Whatever else may be drawn from this study, overt virus infections did occur in the face of this dose schedule of interferon.

In the second, larger-scale study, 41 renal transplant recipients had either 3×10^6 u leucocyte interferon or placebo twice a week for a total of 15 doses (Cheeseman *et al.*, 1979, 1980a,b). These volunteers were also randomized to receive horse antilymphocyte globulin (ATG) or not during the month after transplant. In the interferon group there was a significant delay in cytomegalovirus (CMV) excretion in the urine and a reduction in frequency of viraemia. ATG appeared to enhance the risk of CMV viraemia. There were no differences between groups in the shedding of herpes simplex virus, but there was a non-significant reduction and delay in Epstein-Barr virus shedding particularly in those receiving ATG plus interferon. ATG was associated with a rise in titre of anti-EBV VCA. One patient on ATG plus placebo died of disseminated varicella infection. There was no evidence for an effect of interferon on the excretion of papova viruses.

The results of these trials are of very great interest and they are presently being extended so that patients will receive higher doses of interferon for longer periods (Hirsch *et al.*, 1981).

(5) Rubella vaccine and other togavirus infections

Several rubella virus strains were shown to be sensitive to the antiviral action of interferon in culture but strains varied markedly in their ability to stimulate an interferon response in human placental cell culture (Potter, Banatvala & Best, 1973). Best & Banatvala (1975) gave leucocyte interferon (6.5×10^5 u as two doses before virus) followed by live attenuated rubella vaccine to four sero-negative volunteers. Four other volunteers were vaccinated but did not receive interferon and acted as controls. A delay in virus excretion and lower post-infection HAI titres were seen in the volunteers on active treatment.

Draper (Scientific Committee on Interferon, 1970) showed that human leucocyte interferon could delay viraemia due to FN strain attenuated yellow fever vaccine in rhesus monkeys but did not alter antibody production. Stephen *et al.* (1975) gave Poly I:C conjugated with poly-L-lysine and carboxymethyl-cellulose (Poly I:C-L/C) to rhesus monkeys before and for 17 days after inoculation of a lethal dose of virulent yellow fever virus. Untreated animals died within four to six days but treated animals survived. Both the inducer and virus infection induced detectable serum interferon. No work has been reported with interferon in arbovirus infections of man.

TREATMENT OF ACUTE INFECTIONS

(1) Vaccinia eye disease

The first attempts to treat an established virus infection were made by Jones, Galbraith & Al-Hussaini (1962), who reported the results of treating seven patients with autoinoculation vaccinial keratitis with monkey kidney cell interferon. No toxicity of this preparation given as eye drops was seen. The study was not controlled, but no worsening of the granular opaque corneal epithelial lesions was seen after treatment had begun. Healing of superficial lesions occurred within four to 24 h after starting treatment but the impression was gained that deep stromal infection and extensive ulceration were not affected. The trial was more confused because topical vaccinia convalescent gamma-globulin was given to all patients and in four patients who received this treatment alone, the superficial lesions healed within one to four days. It is therefore difficult to know whether interferon had any effect at all on this disease.

(2) Herpes simplex (HSV) eye disease

Following extensive animal testing of Poly I : C against experimental herpes simplex infections in eyes (Kaufman, Ellison & Waltman, 1969; Pollikoff, Cannavale, Dixon & Di Puppo, 1970), and preliminary studies in man (Guerra *et al.*, 1970; Oosterhuis *et al.*, 1976), an extensive double-blind study conducted over several years concluded that Poly I : C was as effective as IDU in acute keratitis (Galin, Chowchuvech & Kronenberg, 1976). Human, rabbit and monkey interferons were tested in rabbit and money eye models. In 1960, Cantell & Tommila failed to demonstrate any protective effect of unconcentrated rabbit kidney cell interferon against HSV in rabbit eyes. Sugar, Kaufman & Varnell (1973) also failed to reduce the infectivity of HSV using rabbit kidney interferon eye drops at 1.9×10^5 u/ml. In the animals that survived, continuing interferon eye drops at the same dose did not alter the frequency of recurrences. There was a slight reduction in the severity of infection due to herpes simplex in monkey eyes by leucocyte interferon at 1.2×10^4 or 5×10^4 u/ml. One to 1.5×10^5 u/ml leucocyte interferon drops successfully inhibited HSV in African Green Monkey eyes, when given 15 h before or simultaneously with virus (Neumann-Haefelin *et al.*, 1975). The same workers compared human leucocyte with fibroblast interferons in rhesus monkey eyes and the 50% effective concentration of fibroblast interferon was found to be 1.9×10^5 u/ml and at this concentration, leucocyte interferon was slightly though not significantly better (Neumann-Haefelin, Sundmacher, Skoda & Cantell, 1977).

Early uncontrolled studies with interferon in clinical HSV keratitis are impossible to evaluate (Tommila, 1963) but comparative trials suggested that crude low titre interferon might be slightly less effective than IDU (Patterson *et al.*, 1963).

Sundmacher, Neumann-Haefelin, Manthey & Muller, (1976) described the use of human leucocyte interferon (6.25×10^4 u/ml, three drops three times per day), either alone or in combination with thermocautery in simple dendritic ulceration. Thermocautery was more rapidly effective than interferon alone in promoting healing. At this concentration, interferon did not accelerate healing after thermocautery. A higher concentration of interferon (3×10^6

u/ml) in combination with thermocautery, did accelerate healing and inhibit virus shedding compared with thermocautery alone (Sundmacher, Neumann-Haefelin & Cantell, 1976).

Jones, Coster, Falcon & Cantell (1976) adopted a slightly different study design. Fifty-five HSV dendritic ulcers were treated with minimal wiping debridement which is regarded as effective therapy in about 50% of superficial ulcers. The other half would be expected to develop fresh dendrites at the edge of the ulcer within a week. Daily eye drops of 11–31 $\times$ 10^6 u/ml leucocyte interferon significantly reduced the infection rate but did not completely prevent recurrences. This result is in contrast to the effect of acyclovir in the same model which completely prevented recurrences (Jones *et al.*, 1979). Restriction of interferon to single daily dosing was determined by the results of rabbit eye HSV studies which indicated that single daily doses of high potency interferon were just as effective as multiple daily doses of lower potency material (McGill *et al.*, 1976).

In a small uncontrolled study of fibroblast interferon (approximately 10^7 u/ml given three times a day) in otherwise untreated dendritic ulcers (McGill *et al.*, 1977), four out of five ulcers healed within three to eight days but one failed to heal after eight days, although the ulcer remained stationary and subsequently failed to heal on adenine arabinoside. Sundmacher and coworkers (1978b) treated 38 patients with acute dendritic keratitis with either leucocyte or fibroblast interferon at 10^6 u/ml after thermocautery, and could demonstrate no difference in the healing rate between the two treatment groups. No placebo group was included so it is difficult to be sure that either interferon had any effect.

Several groups have compared interferon with chemical antiherpetic agents *in vitro*: Schacter *et al.* (1970) demonstrated a relative insensitivity of HSV to interferon compared with IDU *in vitro*. Lerner & Bailey (1976) studied the different sensitivities of HSV types I and II to interferon and adenine arabinoside in human fibroblasts. Several strains of HSV II were more resistant to the action of added interferon than HSV I. Interferon and adenine arabinoside seemed synergistic against HSV I but only additive against HSV II. This assay system used the antivirals incorporated in the overlay after the virus had been added. When interferon was applied in the medium 24 h before virus, the inhibitory concentrations required were several log dilutions lower, although the differences in the sensitivities of different HSV strains were still observed. This experiment showed that relatively large amounts of interferon were needed to modify plaque development when added to tissue culture with virus. Bryson & Kronenberg (1977) showed that the dynamics of the plaque inhibitory effect of interferon in fibroblast cultures were quite different from and did not interact with those of adenine arabinoside or its metabolites, suggesting that, *in vivo*, the drugs should be at least additive in effect.

Sundmacher, Cantell & Neumann-Haefelin (1978a) felt from their previous studies that interferon was not sufficiently effective as single therapy of herpes keratitis. They therefore performed a study in 37 new patients comparing the effect of adding albumin or leucocyte interferon (10^6 to 30×10^6 u/ml) eye drops to standard therapy with trifluorothymidine drops. The lower concentration of interferon did not improve the mean time to healing (5.3d) over albumin alone (5.7d) but the higher dose caused a significant reduction (2.9d). By increasing the numbers in the studies a clear gradation of the improvement achieved over trifluorothymidine by adding different concentrations of interferon could be shown (Sundmacher, Cantell & Neumann-Haefelin, 1981). The results were confirmed by De Koning, Van Bijsterveld & Cantell (1981).

There have been single case reports of interferon treatment of severe herpetic stromal disease. In an infant with severe indolent ocular infection following primary gingivostomatitis, there had been failure to heal with IDU alone or with corticosteroids; leucocyte interferon eye drops were substituted for IDU and the disease rapidly improved (Pallin, Lundmark & Brege, 1976). Kobza *et al.*, (1975) reported a similar case. Their patient was a transplant recipient who had primary anogenital herpes followed by an eye infection and then a chronic recurrent generalized rash. Gradual deterioration ensued despite courses of IDU and cytosine arabinoside, but small doses of leucocyte interferon given intramuscularly (5×10^5 u daily for seven days) caused rapid resolution of skin and genital lesions. When the skin lesions recurred in association with a rejection episode treated with methylprednisolone, they healed again after a second course of interferon.

Single case reports are often of interest but are not useful in trying to make a case for the therapeutic use of interferon. Nevertheless, it is clear from controlled clinical trials that interferon is active in the treatment of established herpetic eye disease. Perhaps the optimum therapy for simple superficial ulcers is debridement plus a powerful nucleoside analogue plus interferon (at a titre over 10^6 u/ml) but this needs to be firmly established. The treatment of stromal disease is a more important goal but because of the chronicity and difficulties in showing clear clinical responses, it will be considerably more difficult to prove an effect of any schedule. Prevention of recurrent *herpes corneae* is also a prime target for antiviral therapy. Kaufman *et al.* (1976) were unable to show a protective effect of a low-titred interferon preparation used regularly, but improved material supply will presumably now allow the testing of better regimens in susceptible patients over long periods, provided that higher concentrations of interferon are not toxic.

(3) Varicella zoster

Herpes zoster is a common complication in patients who are immunosuppressed, particularly those with lymphoma, and the disease is complicated in such patients by a prolonged course with dissemination of the virus beyond the primary dermatome to skin and viscera (Armstrong, Gurwith, Waddell & Merigan, 1970; Stevens & Merigan, 1972). In patients with disseminated disease, the appearance of virus specific complement-fixing (CF) antibody in the blood and the appearance of vesicle interferon were delayed. The resolution of disease (cessation of new lesion formation and crusting of established lesions) followed closely on the peak interferon titres in vesicle fluid. In half of 15 patients studied closely, CF antibody had not appeared in the serum by this time. This suggested that a selective delay in the production of interferon at the lesion site could favour a prolonged illness and dissemination. Varicella zoster appeared sensitive to interferon *in vitro* (Rasmussen, Holmes, Hofmeister & Merigan, 1977), so it seemed natural to try exogenous interferon against the disease, initially in cancer patients.

Feldman, Hughes, Darlington & Kim (1976), in a well conducted study of topical Poly I : C or placebo applied wet under occlusive dressings, were unable to demonstrate any significant benefit of active treatment. Strander, Cantell, Carlström & Jakobsson (1973) were impressed by the effect of high titre leucocyte interferon in three patients with herpes zoster complicating malignancy. Relief of pain and crusting of lesions occurred within one week of beginning therapy in all three.

In a partly controlled trial, in which only six out of 28 patients with herpes zoster had cancer, all the cases were given 10^6 u leucocyte interferon intramuscularly daily for eight days. Nine normal patients with herpes zoster were given placebo instead of interferon but it is not clear whether the patients were blind – the investigators were not. The only clinical criterion recorded in detail was the persistence of neuralgia and there was some evidence that the use of interferon was associated with a reduced proportion of patients with persistent pain (Emödi, Rufli, Just & Hernandez, 1975). In a larger study, 36 patients received interferon (10^6–3×10^6 u/day) and 26 placebo. It is still not clear whether the study was single- or double-blind or whether the results from the first study were incorporated in these data (Emödi & Rufli, 1977). Details of patients were not given and although the interferon recipients were pain free on average more rapidly than the placebo group, more rapid negative vesicle virus cultures were observed on placebo. There were no differences in the incidence of post-herpetic neuralgia.

Following a preliminary tolerance study (Jordan, Fried & Merigan, 1974), larger groups of cancer or lymphoma patients were entered into three phases of a double-blind placebo-controlled trial (Merigan *et al.*, 1978). Patients with cancer or reticulo-endothelial disorders were warned that they might develop shingles and were asked to report as soon as possible after the onset of the disease. This ensured that patients were treated very quickly, but they were excluded anyway if distal cutaneous spread had occurred before presentation. As the trial progressed, patients were treated earlier and earlier.

In the first phase, 4.2×10^4 ($\times$ 1), in the second, 1.7×10^5 ($\times$ 1), and in the third, 2.55×10^5 u/kg ($\times$ 4), were given initially followed by half the dose twelve-hourly. Treatment was continued for a week or longer if new lesions continued to occur. The results showed that only the highest dose could effectively prevent new lesions in the primary dermatome or distal spread. This dose, however, caused fever more commonly. Overall, the interferon recipients had a reduction in the rate of serious complications, a reduction in the amount of pain during acute treatment and a significant reduction of post-herpetic neuralgia.

This study represents the most comprehensive and convincing evaluation of interferon against established acute virus infection. It showed some activity of interferon against early herpes zoster at the two higher dose schedules tested and it is important to test higher doses still. The same group retested the highest dose given for only two days at the start of herpes zoster (Merigan, Gallagher, Pollard & Arvin, 1981). A small significant effect was seen on cutaneous dissemination and post-herpetic neuralgia but the results were not so striking as in the original trial. Prolonged treatment appears to be necessary.

Similar schedules were used to treat chicken-pox in children with malignancies (Arvin *et al.*, 1982a). There were no differences between placebo or interferon, at dose schedules of 4.2×10^4–2.55×10^5 u/kg/day, in the time to cessation of new lesion formation. Four out of nine interferon recipients continued to excrete virus and have new vesicle formation for 7–11 days after starting treatment. A reduction in serious complications was noted in the interferon group and one patient died with severe disseminated varicella on the fifth day of placebo treatment. Two others died of unrelated pneumonias some weeks after new lesions had ceased, one on placebo and one on interferon. A higher dose schedule was tried (3.5×10^5 u/kg/day for 48 h followed by half the dose daily for three days). At this dose the inhibition of new lesion formation was clearer but one further patient who had received interferon in the high dose group died from varicella pneumonia two weeks after starting treatment. This would suggest that interferon treatment needs to be continued for longer, perhaps at high dose, but this requires further testing. With the demonstration that various nucleoside analogues such as adenine arabinoside and acyclovir have activity against varicella infections in immunocompromised patients, it may no longer be ethical to give placebo treatment, and combination therapy with interferon needs to be evaluated in new large-scale trials.

(4) Adenovirus eye disease

An encouraging report has emerged of the use of fibroblast interferon in the double-blind treatment of adenovirus conjunctivitis during an epidemic in Israel (Romano *et al.*, 1980). Interferon treatment at a dose of 1–2×10^5 u/day given by repeated eye drops reduced the mean time to healing from 27 to 6.5 days and reduced the incidence of ocular complications of disease. Control patients were treated with dexamethasone eye drops and the experiment was not double-blind. It is surprising that the effect of fibroblast interferon should have been so dramatic, particularly at this low concentration. The successful experiments with interferon against herpes simplex in the eye indicate that endogenous interferon responses may not play much of a role in the local resolution of corneal disease. However, the administration of steroids may have led to prolonged illness in the control group. Preliminary experiments by Sundmacher *et al.* (1981) have failed to confirm an effect of leucocyte interferon against adenovirus infection. More detailed results are awaited with interest.

(5) Other acute viral infections

(a) Fulminant hepatitis

Three groups were unable to find circulating interferon activity in acute hepatitis B (Taylor & Zuckerman, 1968; Wheelock, Schenker & Combes, 1968; Hill, Walsh & Purcell, 1971b). Hill and coworkers examined serum samples taken prospectively during the pre-icteric phase of transfusion hepatitis at a time when active virus replication would be expected. It was surprising therefore that Levin & Hahn (1982) should find circulating antiviral activity at considerable titres in acute hepatitis but they used very sensitive bovine cells in their assay. Five out of six patients with fulminant hepatitis (caused by hepatitis B virus, hepatitis A virus, herpes virus or non-A, non-B hepatitis) had no circulating interferon on presentation and leucocytes from all six had a reduced capacity to make α or γ interferon responses compared with patients with 'normal' hepatitis and healthy controls. Treatment of five patients with small doses of leucocyte interferon led to a recovery in the ability of leucocytes to resist infection with vesicular stomatitis virus *in vitro*. Three of these five patients survived the infection and two of four patients with fulminant hepatitis-B infection also survived after treatment with fibroblast interferon (P. DeSomer, personal communication). Because of the variable outcome of the disease, it is not possible to say whether interferon was actually beneficial, but observations made by Levin and Hahn on the state of the interferon system (which perhaps should also include studies of the 2′5′A-synthetase system) in acutely ill patients are of great interest.

(b) Acute cytomegalovirus pneumonia

Meyers and coworkers (1982) reported the administration of interferon combined with adenine arabinoside to transplant recipients who had documented cytomegalovirus pneumonia. Treatment was associated with a significant reduction in titre of infective virus between lung biopsy and autopsy specimens and a reduction in titre of virus isolated from other sites. Despite this treatment, which had to be stopped in four out of seven patients because of severe bone marrow depression and neurotoxicity, all but one of the patients died of this disease. The treatment was neither effective nor safe.

(c) Other conditions where interferon production is suppressed

There are many diseases like lymphoproliferative disorders where the ability of white cells to mount an interferon response is depressed. This is illustrated well for herpes zoster in tumour patients (see above). There are perhaps more surprising examples, such as the observations of a prolonged deficiency in leucocyte interferon response in children with recurrent upper respiratory virus infections (Isaacs *et al.*, 1981).

In addition, some viruses such as respiratory syncytial virus do not induce a very good interferon response either *in vitro* or *in vivo* (Hall, Douglas, Simons & Geiman, 1978; Hall, Douglas & Simons, 1981; Moehring & Forsyth, 1971). This may be one reason for the protracted illnesses caused by this virus, and would suggest that exogenous interferon treatment may be useful in the treatment of this condition. However, interesting questions about early diagnosis of childhood pneumonia and how the interferon should be administered are raised. Finter (1968) showed that high doses of interferon given intraperitoneally did not get into the respiratory secretions of mice. If this is so for man, then interferon would perhaps best be given by aerosol but only clinical trials can show whether aerosolized interferon will be free of inflammatory effect in the lung at the doses needed to treat viral pneumonia. Identifying defects in the endogenous immune response is of great importance in determining a role for interferon.

CHRONIC VIRUS INFECTIONS

(1) Hepatitis B virus associated chronic hepatitis

The discovery of circulating markers of whole Hepatitis B virus particles (Dane particles, HBcAg, HBeAg and HBV DNA polymerase) made it possible to identify patients with chronic hepatitis who had measurable ongoing virus replication in the liver. Changes in these markers in response to treatment could be measured objectively so that it was unnecessary, at least initially, to perform double-blind placebo-controlled trials. When interpreting the results, therefore, it is necessary to take into account the occasional spontaneous regression of chronic hepatitis and, in particular, spontaneous changes from aggressive to persistent forms.

Various schedules of leucocyte interferon were given to four patients with chronic active hepatitis (CAH), three of whom had persistently elevated circulating Hepatitis B virus DNA polymerase (HBV DNAP) (Greenberg *et al.*, 1976). The interferon was given at test doses ranging from 17×10^4 to 0.12×10^4 u/kg/day. In the first patient, HBV DNAP fell dramatically on three separate occasions at progressively reducing dose schedules of interferon. With further reduction in dose of interferon given daily over three months, HBV DNAP remained undetectable and titres of HBcAg and HBsAg fell gradually to undetectable levels.

Starting with a lower dose in the second patient (1.5×10^4 u/kg/day), no effect on HBV DNAP was seen. When the dose was doubled, HBV DNAP fell, and at 6×10^4 u/kg/day given over five weeks, HBcAg fell gradually to low levels and HBsAg fell from a titre of 256 to 64. In the third patient, only transient effects on HBV DNAP and HBcAg were seen on one week's treatment at 1.2×10^4 u/kg/day. HBeAg was cleared in the two patients who had prolonged treatment.

The fourth patient had no detectable HBV DNAP and no effects of treatment were observed. Very few studies have been performed on such patients although Kingham and coworkers (1978) showed that 10^7 u/day fibroblast interferon in two HBV associated CAH patients who were HBV DNAP and HBeAg negative, caused a dramatic fall in HBcAb and anti-DNA antibody titres. In one of these a persistently raised AsT fell gradually to normal after treatment.

Another group of seven patients with CAH and one with chronic persistent hepatitis were treated by Scullard and coworkers (1979). The dose of leucocyte interferon given was 5×10^4 u/kg/day or at a reduced dose because of unacceptable falls in neutrophil and platelet counts. Only one patient appeared to respond with a definite fall in HBV DNAP in response to treatment so the course was repeated with similar results. HBeAg became negative after the second course. The changes in circulating markers in the other patients are difficult to interpret.

A further four patients with CAH and serum HBeAg and elevated HBV DNAP were treated with 10^6 u/day leucocyte interferon for several months. There were consistent falls in HBV DNAP in all four with clearance of HBeAg and a change in liver biopsy histological appearances from 'aggressive' to 'persistent' type in two. In the other two, liver function tests improved on treatment and returned to pretreatment levels one month after stopping treatment (Kato, Kobayashi, Suyama & Hattori, 1979).

These promising results led Weimar and coworkers (1980a) to do some preliminary open studies and then enter a double-blind placebo-controlled study of 16 patients with well-documented HBV CAH; half were allocated to leucocyte interferon 12×10^6 u daily for one week, then halving doses weekly for five weeks. All eight on active treatment had a fall in HBV DNAP soon after starting treatment but so did one on placebo. No changes in HBsAg, HBeAg or HBcAb were observed. These authors were understandably pessimistic about the transient and superficial nature of the effect of interferon on HBV replication. Merigan, Robinson & Gregory (1980) countered by suggesting that the dose of interferon given had been neither high enough nor prolonged enough to achieve a persistent fall in HBV DNAP and a fall in the other markers of infection. A more likely explanation of these observations is that a more permanent response to interferon treatment alone depends on factors independent of liver biopsy histology and the presence of chronic virus replication; they may have something to do with cell-mediated immunity, and may account for occasional spontaneous regression of disease. Treatment with interferon may not only reduce virus replication but may also have an appropriate effect on immune responses in certain patients, but this has yet to be established. Identifying the patients who respond well to interferon is a thorny problem and has not been achieved using simple tests of immunologic competence.

Several groups have tried fibroblast interferon in HBV CAH (Desmyter *et al.*, 1976; Weimar *et al.*, 1977; Dolen *et al.*, 1979; Weimar *et al.*, 1979; Kingham *et al.*, 1978). The reports describe the treatment of 12 patients altogether. In only one patient was there a dramatic fall in HBV DNAP with interferon treatment (Dolen *et al.*, 1979). In the rest, the changes were less dramatic and difficult to relate to treatment. In one study fibroblast interferon was given for two weeks to three patients and after an interval of two months, leucocyte interferon was then given for two weeks (Weimar *et al.*, 1979); one became HBeAg negative six weeks after the fibroblast interferon. Liver 'function' tests improved gradually in all three but the fibroblast interferon did not cause the dramatic falls in HBV DNAP which were seen later with leucocyte interferon. Perhaps fibroblast interferon has more effect on the immune status of the patient (Kingham *et al.*, 1978) and encourages gradual resolution of the disease without so dramatically altering virus replication. If so, a combination of leucocyte and fibroblast interferon might be synergistic, but this is a rather tenuous hypothesis.

Merigan *et al.* (1980) suggested that leucocyte interferon alone was insufficient therapy to treat most patients with chronic hepatitis. There were reports that adenine arabinoside (AraA) (Pollard *et al.*, 1978) and its soluble derivative AraA monophosphate (Ara AMP) (Weller *et al.*, 1980) had a similar effect to interferon on HBV replication so combination therapy with these drugs was tried. One of the problems has been the toxicity of these compounds, AraA causing dose-related neurotoxicity, aggravated by interferon (Sacks *et al.*, 1982) and Ara AMP causing thrombocytopenia (Weller *et al.*, 1980) or muscular pains (Smith *et al.*, 1982).

In open studies, Scullard *et al.* (1981a,b) showed that appropriate dose schedules of leucocyte interferon alone caused complete clearance of HBV DNAP in four out of 16 patients, of AraA in one out of six patients, whereas intermittent combination therapy was effective in seven out of 16 male patients. This is of particular interest because it is thought that the patients who respond best to interferon therapy have chronic active hepatitis, are females who have had recent hormone therapy. One important result of successful treatment is the rendering of patients' serum non-infectious (Scullard *et al.*, 1982). The demonstration of an effective schedule in males has encouraged Merigan's group to go into a five-year double-blind placebo-controlled study with this schedule in appropriate patients. This will be of great value, because it will clearly define the natural history of chronic active hepatitis in the placebo recipients and the need for therapy which may be prolonged and may have unpleasant side-effects.

(2) Cytomegalovirus infections

Clinically, infections with cytomegalovirus (CMV) are a problem when children are affected congenitally, and at any age during immunosuppression, particularly after transplantation.

The theoretical value of antiviral chemotherapy in congenitally-infected infants revolves around the hypothesis that continuing post-natal virus replication continues to cause pathological damage. This is not clearly proven, although the assumption is a reasonable one, particularly in children with hepatosplenomegaly with abnormal liver function tests or with thrombocytopenic purpura and neutropenia. Several antiherpetic nucleoside analogues have been given to patients with chronic CMV viruria in congenital or acquired infections. These have been associated principally with a transient reduction in viruria usually with no clinical benefit.

In 1966, Falcoff, Falcoff, Fournier & Chany described a method for manufacture of crude leucocyte interferon for clinical use. The final preparation contained 162 u/ml (not standard units), and doses of 10–40 ml per day were given for between three and 400 days to patients with a wide variety of conditions. Three infants with congenital CMV were treated and 'survived' (the treatment as well as the condition). Live measles vaccine (Glasgow, Hanshaw, Merigan & Petralli, 1967) or pyrancopolymer (Plotkin & Stetler, 1970) used to induce an endogenous interferon response had no effect on CMV infection. However, Emödi & Just (1974) showed that cells from infants with congenital CMV have a reduced capacity to make interferon and it seemed reasonable to repeat Falcoff's experiment with higher doses as interferon became available.

Five infants with congenital CMV infection and with evidence of active pathological viral replication were given six courses of leucocyte interferon treatment (1.7×10^5 to 3.5×10^5 u/kg/day for 7–14 days) (Arvin, Yeager & Merigan, 1976). The one infant given two courses had 1.6×10^6 u/day on the second occasion but this caused unacceptable fever. A transient reduction in viruria was seen in this infant. Three courses of treatment caused a rise in liver enzymes and all the children showed a reduced weight gain. No effect on clinical manifestations was seen even when interferon was started very soon after the appearance of hepatosplenomegaly in one child.

Emödi *et al.* (1976) had better results. Two congenitally infected infants had complete and permanent resolution of viruria with courses of 10^6 u/day for 8–10 days. A third had transient suppression of viruria and a fourth had no response.

Interferon had an effect on CMV excretion in the urine in five out of nine infants with congenital infection, two clearing virus from the urine, and these results suggest that interferon can inhibit the replication of CMV *in vivo*, although no therapeutic benefit has been observed in these uncontrolled trials. Interferon must therefore rank with idoxuridine, fluoroxuridine, AraA and transfer factor. Further studies are needed to examine higher dose schedules of interferon and combination therapy.

(3) Papillomavirus infections (warts)

Body warts (verrucae) and genital warts are caused by different papillomaviruses (Coleman, 1980), have different pathological features and a quite different natural history. The virus of juvenile laryngeal papillomas is acquired at parturition and the lesions are similar to those of genital warts. They are not keratinized, are soft, fleshy, oedematous and infiltrated by mononuclear cells and virus is scarce. Verrucae are keratinized, anergic and virus is shed in profusion by desquamation.

McGee (1967) used heat-killed or virulent vaccinia virus to treat verrucae at various sites with some success. Allyn & Waldorf (1968) showed that the response to virus was proportional to the inflammatory lesion induced and because of morbidity and the risks of vaccination, the method never caught on. It was proposed that vaccinia worked by local viral interference, perhaps by inducing interferon, so it was reasonable to try interferon against warts.

In Russia, verrucae were treated in open trials with topical or intralesional crude low-titred leucocyte interferon (Borzov, Kuznetsov & Lobanovski, 1971). The warts were reputed to have disappeared with this treatment! In Yugoslavia, Ikić and coworkers (1975) performed a series of trials using leucocyte interferon ointment (4000 u/g) applied to vulval warts several times a day. A good response in open studies prompted a double-blind placebo-controlled trial where all of 10 patients showed complete regression of warts by 12 weeks after starting treatment whereas only three of ten placebo users had regression by that time. The placebo recipients were then cured using the active preparation. In a different study, penile warts appeared less responsive to interferon than vulval warts. Injections of single small doses of crude fibroblast interferon into penile warts appeared to stop them growing when compared with controls (Scott & Csonka, 1979). Ho and coworkers (1981) studied in detail the effects of different dose schedules of partially purified leucocyte interferon in one patient with multiple intractable warts for eight years. Despite the enhancement of natural killer cell activity, parenteral interferon (2.4×10^6 u twice a week for 55 doses) had no effect on the warts. Repeated local injections, on the other hand, did cause individual warts to resolve.

Seven children with juvenile laryngeal papillomas were given leucocyte interferon intramuscularly (3×10^6 u three times a week, then reducing) (Haglund, Lundquist, Cantell & Strander, 1981). All the children needed frequent surgery up to the time of interferon treatment. It seems that there were no major recurrences of warts after removal nor major progression of established disease by one month after starting treatment. Large masses of tumour did not resolve and it was considered necessary to continue with surgery until the warts were removed. After stopping treatment, the warts recurred within a month. If treatment was reduced (2×10^6 u twice a week or 10^6 u once a week), warts recurred but in four to five months. Chronic treatment was considered necessary to keep the warts under control but the dose needed is in the toxic range. Fibroblast interferon was less effective than leucocyte interferon in two other cases (Göbel *et al.*, 1981).

From these preliminary studies we can make the following hypotheses. Local but not systemic interferon induces the resolution of verrucae and high doses of systemic leucocyte interferon prevent recurrences of laryngeal papillomas, although fibroblast interferon may be less effective. Topical crude interferons are effective against genital warts.

Interferons could work by inhibiting papillomavirus replication, by suppressing the growth of proliferating epidermal cells or by enhancing non-specific immune responses (such as inflammation) or specific cell-mediated responses. The latter seems unlikely in the response of verrucae to local interferon, because the effect on distal warts is minimal when intralesional treatment is successful or when interferon is given systemically. The difference in response of juvenile papillomas and verrucae may reflect a variable sensitivity to interferons of papillomaviruses (which cannot be tested in tissue culture) or differences in the immune reactivity in these conditions. Perhaps laryngeal papillomas, in which virus is scarce and there is a mononuclear infiltrate, are more susceptible to NK cell or macrophage activation induced by systemic interferon. There are many unanswered questions here, some of which can be answered with the greater availability of material. The treatment of papillomas seems a clear goal for interferon therapy.

(4) Chronic rubella infections

Disappointing results were obtained by Arvin, Schmidt, Cantell & Merigan (1982b) in the treatment of three congenitally infected children with rubella. A transient fall in pharyngeal shedding of the virus was seen at a dose of 2×10^5–7×10^5 u/kg/day leucocyte interferon given for 10 days. No beneficial clinical effects were seen. However, in one other child with congenital rubella syndrome with persistent viraemia and an unusual cutaneous vasculitis, 3×10^6 u leucocyte interferon daily for two weeks was associated with resolution of viraemia and regression of skin lesions (Larsson *et al.*, 1976). Although skin lesions had disappeared two months after treatment, virus continued to be excreted in the urine and a high serum IgM rubella antibody persisted, suggesting ongoing virus replication. Whether interferon will eventually have a role in chronic rubella seems doubtful from these preliminary results.

SUMMARY

From this brief review of the important clinical trials which have been conducted with various interferons, I think it is clear that in several quite different experimental and natural infections in man, interferon may inhibit virus replication. Beneficial clinical effects have been shown in patients with herpes simplex eye disease, chronic active hepatitis, herpes zoster in immunocompromised patients, and papillomas. In the first two indications, the effects of interferon are enhanced by combination therapy with nucleoside analogues and a similar additive or synergistic effect would be expected in the third.

Progress towards bringing interferon to the clinic has, up to now, been very slow but this will accelerate rapidly because of the improved supply of human interferons from prokaryocytes. Commercial rather than purely scientific pressures will now be brought to bear to determine whether a role for interferon exists.

The eventual use of interferon will depend first on the demonstration of clear benefit (either given alone or in combination with other antiviral agents), in any chosen indication. Secondly, it may be necessary to show that one interferon type is better than another in a particular indication and this may prove exceedingly difficult. Thirdly, the risks of unwanted effects, either overt clinical side-effects or occult effects on the immune system and bone marrow must be acceptable, in proportion to the chosen indication. Finally, the eventual cost of interferon must be appropriate to the chosen indication.

There are many clinical questions which remain unanswered. For example, can we protect against wild upper respiratory tract infections, even influenza, using doses of interferon which have no unwanted effects? Is the benefit of interferon in varicella infections apparent in normal patients (as opposed to immunosuppressed ones) and is it better than, additive to, or even synergistic with, nucleoside analogues? Are the febrile reactions to interferon an important feature of the antiviral activity and, if so, will suppressing them with aspirin-like drugs be detrimental?

It is only by performing rational controlled trials in man that these questions will be answered. An optimistic prediction would be that interferon will be widely available in clinical use for the prevention of common infections such as colds, for the treatment of many acute virus infections such as respiratory syncytial pneumonia in children and for the treatment of many chronic viral illnesses. A pessimistic one would be that interferon will be shown to have interesting effects with little clinical relevance in a wide range of conditions, but that the unwanted effects and cost would make such treatment unacceptable.

REFERENCES

ALLYN, B. & WALDORF, D. S. (1968). Treatment of verruca with vaccinia. *Journal of the American Medical Association*, **203,** 807.

ANDREWS, R. D. (1961). Specificity of interferon. *British Medical Journal*, **i,** 1728–30

AOKI, F. Y. & CRAWLEY, J. C. W. (1976). Distribution and removal of human serum albumin – technetium 99M instilled intranasally by drops and spray. *British Journal of Clinical Pharmacology*, **3,** 869–78.

ARMSTRONG, R., GURWITH, M., WADDELL, D. & MERIGAN, T. (1970). Cutaneous interferon production in patients with Hodgkin's disease and other cancers infected with varicella or vaccinia. *New England Journal of Medicine*, **283,** 1182–7.

ARVIN, A. M., KUSHNER, J. H., FELDMAN, S., BAEHNER, L., HAMMOND, D. & MERIGAN, T. C. (1982a). Human leukocyte interferon for the treatment of varicella in children with cancer. *New England Journal of Medicine*, **306,** 761–5.

ARVIN, A. M., SCHMIDT, N. J., CANTELL, K. & MERIGAN, T. C. (1982b). Alpha interferon administration to infants with congenital rubella. *Antimicrobial Agents and Chemotherapy*, **21,** 259–61.

ARVIN, A., YEAGER, A. & MERIGAN, T. (1976). Effect of leukocyte interferon on urinary excretion of cytomegalovirus by infants. *Journal of Infectious Diseases*, **133,** A205–A212.

Best, J. M. & Banatvala, J. E. (1975). The effect of a human interferon preparation on vaccine induced rubella infection. *Journal of Biological Standardization*, **3,** 107–12.

Borzov, M. V., Kuznetsov, V. P. & Lobanovski, I. (1971). Relating to the use of interferon for the treatment and prevention of viral dermatoses. *Vestnik Dermatologii i Venerologii*, **45,** 14–17.

Bryson, Y. J. & Kronenberg, L. H. (1977). Combined antiviral effects of interferon, adenine arabinoside, hypoxanthine arabinoside and adenine arabinoside 5′ monophosphate in human fibroblast cultures. *Antimicrobial Agents and Chemotherapy*, **11,** 299–306.

Cantell, K. & Tommila, V. (1960). Effect of interferon on experimental vaccinia and herpes simplex virus infections in rabbits' eyes. *Lancet*, **ii,** 682–4.

Cheeseman, S. H., Black, P. H., Rubin, R. H., Cantell, K. & Hirsch, M. S. (1980a). Interferon and BK Papovavirus, clinical and laboratory studies. *Journal of Infectious Diseases*, **141,** 157–61.

Cheeseman, S. H., Henle, W., Rubin, R. H., Tolkoff-Rubin, N. E., Cosimi, B., Cantell, K., Winkle, S., Herrin, J. T., Black, P. H., Russell, P. S. & Hirsch, M. S. (1980b). Epstein-Barr virus infection in renal transplant recipients. *Annals of Internal Medicine*, **93,** 39–42.

Cheeseman, S. H., Rubin, R., Stewart, J., Tolkoff-Rubin, N., Cosimi, A., Cantell, K., Gilbert, J., Winkle, S., Herrin, J., Black, P. Russell, P. & Hirsch, M. (1979). Controlled clinical trial of prophylactic human leukocyte interferon in renal transplantation. *New England Journal of Medicine*, **300,** 1345–9.

Coleman, D. V. (1980). Recent developments in the papovaviruses: the human papillomaviruses. In *Recent Advances in Clinical Virology*, Vol. 2, ed. A. P. Waterson, pp. 79–88. Edinburgh: Churchill Livingstone.

De Koning, E. W. J., Van Bijsterveld, O. P. & Cantell, K. (1981). Human leukocyte interferon and trifluorothymidine versus albumin placebo and trifluorothymidine in the treatment of dendritic keratitis. In *The Biology of the Interferon System*, ed. E. De Maeyer, G. Galasso & H. Schellekens, pp. 351–4. London, New York, Amsterdam: Elsevier/North-Holland.

Depoux, R. & Isaacs, A. (1954). Interference betwen influenza and vaccinia viruses. *British Journal of Experimental Pathology*, **35,** 415–18.

Desmyter, J., DeGroote, J., Desmet, V. J., Billiau, A., Ray, M. B., Bradburne, A. F., Edy, V. G., Desomer, P. & Mortelmans, J. (1976). Administration of human fibroblast interferon in chronic hepatitis-B infection. *Lancet*, ii, 645–7.

Dolen, J. G., Carter, W. A., Horoszewicz, J. S., Vladutiu, A. O., Leibowitz, J. I. & Nolan, J. P. (1979). Fibroblast interferon treatment of a patient with chronic active hepatitis. *The American Journal of Medicine*, **67,** 127–31.

Ellison, S. A., Canton, S. A. & Rose, H. M. (1959). Studies of recurrent herpes simplex infections following section of the trigeminal nerve. *Journal of Infectious Diseases*, **105,** 161–7.

Emödi, G. & Just, M. (1974). Impaired interferon response of children with congenital cytomegalovirus disease. *Acta Paediatrica Scandinavica*, **63,** 183–7.

Emödi, G., O'Reilly, R., Muller, A., Everson, L. K., Binswanger, U. & Just, M. (1976). Effect of exogenous leukocyte interferon in CMV infections. *Journal of Infectious Diseases*, **133,** A199–A204.

Emödi, G. & Rufli, T. (1977). Antiviral action of interferon in man: use of interferon in varicella zoster infections in man. *Texas Reports of Biology and Medicine*, **35,** 511–15.

EMÖDI, G., RUFLI, T., JUST, M. & HERNANDEZ, R. (1975). Human interferon therapy for herpes zoster in adults. *Scandinavian Journal of Infectious Diseases*, **7,** 1–5.

FALCOFF, E., FALCOFF, R., FOURNIER, F. & CHANY, C. (1966). Production en masse, purification partielle et caractérisation d'un interféron destiné a des essais thérapeutiques humains. *Annales de l'Institut Pasteur*, **111,** 562–84.

FELDMAN, S., HUGHES, W. T., DARLINGTON, R. W. & KIM, H. K. (1976). Evaluation of topical polyinosinic acid-polycytidylic acid in treatment of localized herpes zoster in children with cancer: a randomized double-blind controlled study. *Antimicrobial Agents and Chemotherapy*, **8,** 289–94.

FINTER, N. B. (1968). Of mice and men: studies with interferon. In *Interferon*, ed. G. Wolstenholme & M. O'Connor, pp. 204–17. London: Churchill.

FINTER, N. B., WOODROOFFE, J. & PRIESTMAN, T. J. (1982). Monkeys are insensitive to pyrogenic effects of human alpha-interferons. *Nature*, **298,** 301.

GALIN, M. A., CHOWCHUVECH, E. & KRONENBERG, B. (1976). Therapeutic use of interferon inducers on herpes simplex keratitis in humans. *Annals of Ophthalmology*, **8,** 72–6.

GATMAITAN, B. G., STANLEY, E. G. & JACKSON, G. G. (1973). The limited effect of nasal interferon induced by rhinoviruses and a topical chemical inducer on the course of infection. *Journal of Infectious Diseases*, **127,** 401–7.

GLASGOW, L. A., HANSHAW, J. B., MERIGAN, T. C. & PETRALLI, J. K. (1967). Interferon and cytomegalovirus *in vivo* and *in vitro*. *Proceedings of the Society for Experimental Biology and Medicine*, **125,** 843–9.

GÖBEL, U., ARNOLD, W., WAHN, V., TREUNER, J., JÜRGENS, H. & CANTELL, K. (1981). Comparison of human fibroblast and leukocyte interferon in the treatment of severe laryngeal papillomatosis in children. *European Journal of Pediatrics*, **137,** 175–6.

GREENBERG, H. B., POLLARD, R., LUTWICK, L., GREGORY, P., ROBINSON, W. & MERIGAN, T. (1976). Effect of human leukocyte interferon on hepatitis B virus infection in patients with chronic active hepatitis. *New England Journal of Medicine*, **295,** 517–22.

GREENBERG, S. B., HARMON, M. W., COUCH, R. B., JOHNSON, P. E., WILSON, S. Z., DACSO, C. C., BLOOM, K. & QUARLES, J. (1982). Prophylactic effect of low doses of leukocyte interferon against infection with rhinovirus. *Journal of Infectious Diseases*, **145,** 542–6.

GREENBERG, S. B., HARMON, M. W., JOHNSON, P. E. & COUCH, R. B. (1978). Antiviral activity of intranasally applied human leukocyte interferon. *Antimicrobial Agents and Chemotherapy*, **14,** 596–600.

GRESSER, I., TOVEY, M. G., MAURY, C. & BANDU, M. T. (1976). Role of interferon in the pathogenesis of virus diseases in mice as demonstrated by the use of anti-interferon serum: II. Studies with herpes simplex, Moloney sarcoma, vesicular stomatitis, Newcastle disease and influenza viruses. *Journal of Experimental Medicine*, **144,** 1316–27.

GUERRA, R., FREZZOTTI, R., BONANNI, R., DIANZANI, F. & RITA, G. (1970). A preliminary study on the treatment of human herpes simplex keratitis with an interferon inducer. *Annals of the New York Academy of Sciences*, **173,** 823–9.

HAGLUND, S., LUNDQUIST, P. G., CANTELL, K. & STRANDER, H. (1981). Interferon therapy in juvenile laryngeal papillomatosis. *Archives of Otolaryngology*, **107,** 327–32.

HALL, C. B., DOUGLAS, R. G. & SIMONS, R. L. (1981). Interferon production in adults with respiratory syncytial viral infection. *Annals of Internal Medicine*, **94,** 53–5.

HALL, C. B., DOUGLAS, R. G., SIMONS, R. L. & GEIMAN, J. M. (1978). Interferon production in children with respiratory syncytical influenza and parainfluenza virus infections. *Journal of Pediatrics*, **93,** 28–32.

HARMON, M. W., GREENBERG, S. B. & COUCH, R. R. (1976). Effect of human nasal secretions on the antiviral activity of human fibroblast and leukocyte interferon. *Proceedings of the Society for Experimental Biology and Medicine*, **152,** 598–602.

HARMON, M. W., GREENBERG, S. B., JOHNSON, P. E. & COUCH, R. B. (1977). A human nasal epithelial cell culture system: evaluation of the response to human interferons. *Infection and Immunity*, **16,** 480–5.

HAVERKOS, H. W., PAZIN, G. J., ARMSTRONG, J. A. & HO, M. (1980). Follow up of interferon treatment of herpes simplex. *New England Journal of Medicine*, **303,** 699–700.

HILL, D. A., BARON, S., LEVY, H. B., BELLANTI, J., BUCKLER, C. E., CANNELLOS, G. & CARBONE, P. (1971a). Clinical studies of induction of interferon by polyinosinic-polycytidylic acid. In *Perspectives in Virology*, vol. 7, pp. 198–222. New York: Academic Press.

HILL, D. A., WALSH, J. H. & PURCELL, R. H. (1971b). Failure to demonstrate circulating interferon during incubation period and acute stage of transfusion-associated hepatitis. *Proceedings of the Society for Experimental Biology and Medicine*, **136,** 853–6.

HIRSCH, M. S., CHEESEMAN, S. H., RUBIN, R. H., SCHOOLEY, R. T., HENLE, W., LENNETTE, E. T., ANDREWS, C., SHAH, K. V. & CANTELL, K. (1981). Prophylactic interferon-alpha (IFNα) in renal transplant recipients. In *The Biology of the Interferon System*, ed. E. DeMaeyer, G. Galasso & H. Schellekens, pp. 339–42. London, New York, Amsterdam: Elsevier/North-Holland.

HO, M., PAZIN, G. J., WHITE, L. T., HAVERKOS, H., WECHSLER, R. L., BREINIG, M. K., CANTELL, K. & ARMSTRONG, J. A. (1981). Intralesional treatment of warts with interferon-α and its long term effect on NK cell activity. *In The Biology of the Interferon System*, ed. E. De Maeyer, G. Galasso & H. Schellekens, pp. 361–5. London, New York, Amsterdam: Elsevier/North-Holland.

IKIĆ, D., BOSNIĆ, N., SMERDEL, S., JUŠIĆ, D., ŠOOŠ, E. & DELIMAR, N. (1975). Double-blind study with human leukocyte interferon in the therapy of condylomata acuminata. In *Proceedings of a Symposium on Clinical Use of Interferon*, ed. D. Ikić, pp. 229–33. Zagreb: Yugoslav Academy of Sciences and Arts.

IMANISHI, J., KARAKI, T., SASAKI, O., MATSUO, A., OISHI, K., PAK, P. B., KISHIDA, T., TODA, S. & NAGATA, H. (1980). The preventive effect of human interferon-alpha preparations on upper respiratory diseases. *Journal of Interferon Research*, **1,** 169–79.

ISAACS, D., CLARKE, J. R., TYRRELL, D. A. J., WEBSTER, A. D. B. & VALMAN, H. B. (1981). Deficient production of leucocyte interferon (interferon-α) *in vitro* and *in vivo* in children with recurrent respiratory tract infections. *Lancet*, **ii,** 950–2..

ISAACS, A. & HITCHCOCK, G. (1980). Role of interferon in recovery from virus infections. *Lancet*, **ii,** 69–71.

ISAACS, A. & LINDENMANN, J. (1957). Virus interference 1. The interferon. *Proceedings of the Royal Society of London*, **B147**, 258–67.

ISAACS, A. & WESTWOOD, M. A. (1959). Inhibition by interferon of the growth of vaccinia virus in the rabbit skin. *Lancet*, **ii,** 324–5.

JENNER, E. (1804). Letter to the editor. *Medical and Physical Journal*, **12,** 97.

JOHNSON, P. E., GREENBERG, S. B., HARMON, M. W., ALFORD, B. R. & COUCH, R. B. (1976). Recovery of applied human leucocyte interferon from the nasal mucosa of chimpanzees and humans. *Journal of Clinical Microbiology*, **4,** 106–7.

JONES, B. R., COSTER, D. J., FALCON, M. G. & CANTELL, K. (1976). Topical therapy of ulcerative herpetic keratitis with human interferon. *Lancet*, **ii,** 128.

JONES, B. R., COSTER, D. J., FISON, P. N., THOMPSON, G. M., COBO, L. M. & FALCON, M. G. (1979). Efficacy of acycloguanosine (Wellcome 248u) against herpex simplex corneal ulcers. *Lancet*, **i,** 243–4.

Jones, B. R., Galbraith, J. E. K. & Al-Hussaini, M. K. (1962). Vaccinial keratitis treated with interferon. *Lancet*, **i,** 875–9.

Jordan, G. W., Fried, R. P. & Merigan, T. C. (1974). Administration of human leucocyte interferon in herpes zoster. 1. Safety, circulating antiviral activity and host response to infection. *Journal of Infectious Diseases*, **130,** 56–62.

Kato, Y., Kobayashi, K., Suyama, T. & Hattori, N. (1979). Effect of human leucocyte interferon on hepatitis B virus in patients with chronic active hepatitis. *Gastroenterology*, **77,** A21.

Kaufman, H. E., Ellison, E. D. & Waltman, S. R. (1969). Double-stranded RNA, an interferon inducer in herpes simplex keratitis. *American Journal of Ophthalmology*, **68,** 486–91.

Kaufman, H. E., Meyer, R. F., Laibson, P. R., Waltman, S. R., Nesburn, A. B. & Shuster, J. J. (1976). Human leukocyte interferon for the prevention of recurrences of herpetic keratitis. *Journal of Infectious Diseases*, **133,** A165–A168.

Kingham, J., Ganguly, N., Shaari, Z., Mendelson, R., McGuire, M., Holgate, S., Cartwright, T., Scott, G., Richards, B. & Wright, R. (1978). Treatment of HBsAg-positive chronic active hepatitis with fibroblast interferon. *Gut*, **19,** 91–4.

Kobza, K., Emödi, G., Just, M., Hilti, E., Leuenberger, A., Binswanger, U., Theil, G. & Brunner, K. P. (1975). Treatment of herpes infection with human exogenous interferon. *Lancet*, **i,** 1343–4.

Larsson, A., Forsgren, M., Hård, A. F., Segerstad, S., Strander, H. & Cantell, K. (1976). Administration of interferon to an infant with congenital rubella syndrome involving persistent viraemia and cutaneous vasculitis. *Acta Paediatrica Scandinavica*, **65,** 105–10.

Lerner, M. A. & Bailey, E. J. (1976). Differential sensitivity of herpes simplex virus Types I and II to human interferon: antiviral effects of interferon plus 9 β-D-arabinofuranosyladenine. *Journal of Infectious Diseases*, **134,** 400–4.

Levin, S. & Hahn, T. (1982). Interferon system in acute viral hepatitis. *Lancet*, **i,** 592–4.

Lindenmann, J., Burke, D. C. & Isaacs, A. (1957). Studies on the production, mode of action and properties of interferon. *British Journal of Experimental Pathology*, **38,** 551–62.

Little, J. W., Douglas, R. G., Hall, W. J. & Roth, F. K. (1979). Attenuated influenza produced by experimental intranasal inoculation. *Journal of Medical Virology*, **3,** 177–88.

McGee, A. R. (1967). Wart treatment by vaccination with smallpox vaccine: a preliminary report. *Canadian Medical Association Journal*, **99,** 119–21.

McGill, J. I., Cantell, K., Collins, P., Finter, N. B., Laird, R. & Jones, B. R. (1977). Optimal usage of exogenous human interferon for prevention therapy of herpetic keratitis. *Transactions of the Ophthalmological Society of the United Kingdom*, **97,** 324–6.

McGill, J., Collins, P., Cantell, K., Jones, B. R. & Finter, N. B. (1976). Optimal studies for use of interferon in the corneas of rabbits with herpes simplex keratitis. *Journal of Infectious Diseases*, **133,** A13–A19.

Merigan, T. C., Gallagher, J. G., Pollard, R. B. & Arvin, A. M. (1981). Short course human leukocyte interferon in treatment of herpes zoster in patients with cancer. *Antimicrobial Agents and Chemotherapy*, **19,** 193–5.

Merigan, T. C., Rand, K. H., Pollard, R. B., Abdallah, P. S., Jordan, G. W. & Fried, R. P. (1978). Human leucocyte interferon for the treatment of herpes zoster in patients with cancer. *New England Journal of Medicine*, **298,** 981–7.

Merigan, T. C., Reed, S. E., Hall, T. S. & Tyrrell, D. A. J. (1973). Inhibition of respiratory virus infection by locally applied interferon. *Lancet* **i,** 563–7.

MERIGAN, T. C., ROBINSON, W. S. & GREGORY, P. B. (1980). Interferon in chronic hepatitis B infection. *Lancet* **i,** 422–3.

MEYERS, J. D., MCGUFFIN, R. W., BRYSON, Y. J., CANTELL, K. & THOMAS, E. D. (1982). Treatment of cytomegalovirus pneumonia after marrow transplant with combined vidarabine and human leukocyte interferon. *Journal of Infectious Diseases*, **146,** 80–4.

MOEHRING, J. M. & FORSYTH, B. R. (1971). The role of interferon system in respiratory syncytial virus infection. *Proceedings of the Society for Experimental Biology and Medicine*, **138,** 1009–14.

NEUMANN-HAEFELIN, D., SHRESTHA, B. & MANTHEY, K. F. (1976). Effective antiviral prophylaxis and therapy by systemic application of human interferon in immunosuppressed monkeys. *Journal of Infectious Diseases*, **133,** A211–A216.

NEUMANN-HAEFELIN, D., SUNDMACHER, R., SAUTER, B., KARGES, H. E. & MANTHEY, K. F. (1975). Effect of human leukocyte interferon on vaccinia and herpes virus infected cell cultures and monkey corneas. *Infection and Immunity*, **12,** 148–55.

NEUMANN-HAEFELIN, D., SUNDMACHER, R., SKODA, R. & CANTELL, K. (1977). Comparative evaluation of human leucocyte and fibroblast interferon in the prevention of herpes simplex virus keratitis in a monkey model. *Infection and Immunity*, **17,** 468–70.

OOSTERHUIS, J. A., SIE, S. H., NOLEN, W. A., VERSTEEG, J. & JELTES, I. G. (1976). Therapy of herpetic keratitis with a combination of poly I : C and idoxuridine. *Ophthalmological Research*, **8,** 152–9.

PALLIN, O., LUNDMARK, K. M., & BREGE, K. G. (1976). Interferon in severe herpes simplex of the cornea. *Lancet* **i,** 1187.

PATTERSON, A., FOX, A. D., DAVIES, G., MAGUIRE, C. H., SELLORS, P. J. WRIGHT, P., RICE, N. S. C., COB, B. & JONES, B. R. (1963). Controlled studies of IDU in the treatment of herpetic keratitis. *Transactions of the Ophthalmological Society of the United Kingdom*, **83,** 583–91.

PAZIN, G. J., ARMSTRONG, J. A., LAM, M. T., TARR, G. C., JANNETTA, P. J. & HO, M. (1979). Prevention of reactivated herpes simplex infection by human leukocyte interferon. *New England Journal of Medicine*, **301,** 225–9.

PETRALLI, J. K., MERIGAN, T. C. & WILBUR, J. R. (1965). Action of endogenous interferon against vaccinia infection in children. *Lancet*, **ii,** 401–5.

PLOTKIN, S. A. & STETLER, H. (1970). Treatment of congenital cytomegalovirus inclusion disease with antiviral agents. *Antimicrobial Agents and Chemotherapy*, **10,** 372–9.

POLLARD, R. B., SMITH, J. L., NEAL, E. A., GREGORY, P. B., MERIGAN, T. C. & ROBINSON, W. S. (1978). Effect of vidarabine on chronic hepatitis B virus infection. *Journal of the American Medical Association*, **239**, 1648–50.

POLLIKOFF, R., CANNAVALE, P., DIXON, P. & DI PUPPO, A. (1970). Effect of complexed synthetic RNA analogues on herpes simplex virus infection in rabbit cornea. *American Journal of Ophthalmology*, **69,** 650–9.

PORTNOY, J. & MERIGAN, T. C. (1971). The effect of interferon and interferon inducers on avian influenza. *Journal of Infectious Diseases*, **124,** 545–52.

POTTER, J. E., BANATVALA, J. E. & BEST J. M. (1973). Interferon studies with Japanese and U.S. rubella virus strains. *British Medical Journal*, **i,** 197–9.

PROCTOR, D. F., ANDERSEN, I. & LUNDQUIST, G. (1973). Clearance of inhaled particles from the human nose. *Archives of Internal Medicine*, **131,** 132–9.

RASMUSSEN, L., HOLMES, A. R., HOFMEISTER, B. & MERIGAN, T. C. (1977). Multiplicity-dependent replication of varicella zoster virus in interferon treated cells. *Journal of General Virology*, **35,** 361–8.

RICHMAN, D. D., MURPHY, B. R., BARON, S. & UHLENDORF, C. (1976). Three strains of influenza-A virus (H_3N_3): interferon sensitivity *in vitro* and interferon production in volunteers. *Journal of Clinical Microbiology*, **3,** 223–6.

ROMANO, A., REVEL, M., GUARARI-ROTMAN, D., BLUMENTHAL, M. & STEIN, R. (1980). Use of human fibroblast derived (Beta) interferon in the treatment of epidemic adenovirus keratoconjunctivitis. *Journal of Interferon Research*, **1,** 95–100.

SACKS, S. L., SCULLARD, G. H., POLLARD, R. B., GREGORY, P. G., ROBINSON, W. S. & MERIGAN, T. C. (1982). Antiviral treatment of chronic hepatitis B virus infection IV. Pharmacokinetics and side effects of interferon and adenine arabinoside alone and in combination. *Antimicrobial Agents and Chemotherapy*, **21,** 95–100.

SCHACTER, N., GALLIN, M. A., WEISSENBACHER, M., BARON, S. & BILLIAU, A. (1970). Comparison of antiviral action of interferon, interferon inducers and IDU against herpes simplex and other viruses. *Annals of Ophthalmology*, **2,** 795–8.

SCHELLEKENS, H., DE REUS, A., BOLHUIS, R., FOUNTOULAKIS, M., SCHEIN, C., ECSÖDI, J., NAGATA, S. & WEISSMANN, C. (1981). Comparative antiviral efficiency of leukocyte and bacterially produced human alpha-interferon in rhesus monkeys. *Nature*, **292,** 775–6.

SCHELLEKENS, H., WEIMAR, W., CANTELL, K. & STITZ, L. (1979). Antiviral effect of interferon *in vivo* may be mediated by the host. *Nature*, **278,** 742–3.

SCIENTIFIC COMMITTEE ON INTERFERON (1962). Effect of interferon on vaccination in volunteers. *Lancet*, **i,** 873–5.

SCIENTIFIC COMMITTEE ON INTERFERON (1965). Experiments with interferon in man. *Lancet*, **i,** 505–6.

SCIENTIFIC COMMITTEE ON INTERFERON (1970). Progress towards trials of human interferon in man. *Annals of the New York Academy of Sciences*, **173,** 770–81.

SCOTT, G. M. (1982). Interferon: pharmacokinetics and toxicity. *Philosophical Transactions of the Royal Society of London*, **B299**, 91–107.

SCOTT, G. M., CARTWRIGHT, T., LE DU, G. & DICKER, D. (1978). Effect of human fibroblast interferon on vaccination in volunteers. *Journal of Biological Standardization*, **6,** 73–6.

SCOTT, G. M., CARTWRIGHT, T., TYRRELL, D. A. J., BUTLER, J. K., PORTEOUS, M. & STEVENS, R. M. (1977). Preliminary experience with fibroblast interferon against vaccinia virus in human and monkey skin. In *Proceeding of a Symposium on Preparation, Purification and Clinical use of Interferon*, ed. D. Ikić, pp. 135–42. Zagreb: Yugoslav Academy of Sciences and Arts.

SCOTT, G. M. & CSONKA, G. W. (1979). Effect of injections of small doses of human fibroblast interferon into genital warts: a pilot study. *British Journal of Venereal Disease*, **55,** 442–5.

SCOTT, G. M., PHILLPOTTS, R. J. & TYRRELL, D. A. J. (1982). Interferon and the common cold. *Lancet*, **ii,** 657.

SCOTT, G. M., PHILLPOTTS, R. J., WALLACE, J., GAUCI, C. L., GREINER, J. & TYRRELL, D. A. J. (1982a). Prevention of rhinovirus colds by human interferon alpha-2 from *Escherichia coli*. *Lancet*, **ii,** 186–7.

SCOTT, G. M., PHILLPOTTS, R. J., WALLACE, J., SECHER, D. S., CANTELL, K. & TYRRELL, D. A. J. (1982b). Purified interferon as protective against rhinovirus infection. *British Medical Journal*, **284,** 1822–5.

SCOTT, G. M., REED, S., CARTWRIGHT, T. & TYRRELL, D. A. J. (1980). Failure of human fibroblast interferon to protect against rhinovirus infection. *Archives of Virology*, **65,** 135–9.

SCULLARD, G. H., ALBERTI, A., WANSBROUGH-JONES, M. H., HOWARD, C. R., EDDLESTON, A. L., ZUCKERMAN, A. J., CANTELL, K. & WILLIAMS, R. (1979).

Effects of human leucocyte interferon on hepatitis B virus replication. *Journal of Clinical and Laboratory Immunology*, **1,** 277–82.

Scullard, G. H., Andres, L. L., Greenberg, H. B., Smith, J. L., Sawhney, V. K., Neal, A., Mahal, A., Popper, H., Merigan, T. C., Robinson, W. S. & Gregory, P. B. (1981a). Antiviral treatment of chronic hepatitis B virus infection: improvement in liver disease with interferon and adenine arabinoside. *Hepatology*, **1,** 228–32.

Scullard, G. H., Greenberg, H. B., Smith, J. L., Gregory, P. B., Merigan, T. C. & Robinson, W. S. (1982). Antiviral treatment of chronic hepatitis B virus infection II. Infectious virus cannot be detected in patient serum after permanent responses to treatment. *Hepatology*, **2,** 39–49.

Scullard, R. H., Pollard, R. B., Smith, J. L., Sacks, S. L., Gregory, P. B., Robinson, W. S. & Merigan, T. C. (1981b). Antiviral treatment of chronic hepatitis B virus infection: 1. Changes in viral markers with interferon combined with adenine arabinoside. *Journal of Infectious Diseases*, **143,** 772–83.

Smith, C. I., Kitchen, L. W., Scullard, G. H., Robinson, W. S., Gregory, P. B. & Merigan, T. C. (1982). Vidarabine monophosphate and human leukocyte interferon in chronic hepatitis B infection. *Journal of the American Medical Association*, **247,** 2261–5.

Solov'ev, V. D. (1969). The results of controlled observations on the prophylaxis of influenza with interferon. *Bulletin of the World Health Organization*, **41,** 683–8.

Stephen, E. L., Panner, W. L., Sammons, M. L., Baron, S., Levy, H. B. & Spertzel, S. (1975). Prophylaxis of yellow fever in rhesus monkeys by a nuclease-resistant derivative of Poly I-Poly C. *Federation Proceedings*, **34,** 960.

Stevens, D. A. & Merigan, T. C. (1972). Interferon, antibody and other host factors in herpes zoster. *Journal of Clinical Investigation*, **51,** 1170–8.

Strander, H., Cantell, K., Carlström, G. & Jakobsson, P. A. (1973). Clinical and laboratory investigations on man: systemic administration of potent interferon to man. *Journal of the National Cancer Institute*, **51,** 733–9.

Sugar, J., Kaufman, H. E., & Varnell, E. D. (1973). Effect of exogenous interferon on herpetic keratitis in rabbits and monkeys. *Investigative Ophthalmology*, **12,** 378–81.

Sundmacher, R., Cantell, K. & Neumann-Haefelin, D. (1978a). Combination therapy of dendritic keratitis with trifluorothymidine and interferon. *Lancet*, **ii,** 687.

Sundmacher, R., Cantell, K. & Neumann-Haefelin, D. (1981). Evaluation of interferon in ocular viral diseases. In *The Biology of the Interferon System*, ed. E. De Maeyer, G. Galasso & H. Schellekens, pp. 343–50. London, New York, Amsterdam: Elsevier/North-Holland.

Sundmacher, R., Cantell, K., Skoda, R., Hallermann, C. & Neumann-Haefelin, D. (1978b). Human leukocyte and fibroblast interferon in a combination therapy of dendritic keratitis. *Graefes Archiv für Klinische und Experimentelle Ophthalmologie*, **208,** 229–33.

Sundmacher, R., Neumann-Haefelin, D. & Cantell, K. (1976). Successful treatment of dendritic keratitis with human leucocyte interferon – a controlled clinical study. *Graefes Archiv für Klinische und Experimentelle Ophthalmologie*, **201,** 39–45.

Sundmacher, R., Neumann-Haefelin, D., Manthey, K. F. & Muller, O. (1976). Interferon in treatment of dendritic keratitis in humans: a preliminary report. *Journal of Infectious Diseases*, **133,** A160–A164.

Sutton, R. N. P. & Tyrrell, D. A. J. (1961). Some observations on interferon prepared in tissue cultures. *British Journal of Experimental Pathology*, **42,** 99–104.

Takano, K., Jensen, K. E. & Warren, J. (1963). Passive interferon protection in mouse influenza. *Proceedings of the Society for Experimental Biology and Medicine*, **114**, 472–5.

Taylor, P. E. & Zuckerman, A. J. (1968). Non-production of interfering substances by serum from patients with infectious hepatitis. *Journal of Medical Microbiology*, **1**, 217–19.

Tommila, V. (1963). Treatment of dendritic keratitis with interferon. *Acta Ophthalmologica*, **41**, 478–82.

Tyrrell, D. A. J. & Reed, S. E. (1973). Some possible practical implications of interferon and interference. In *Non-specific Factors Influencing Host Resistance*, ed. W. Braun & J. Unger, pp. 438–42. Basel: Karger.

Weimar, W., Heijtink, R. A., Schalm, S. W. & Schellekens, H. (1979). Differential effects of fibroblast and leucocyte interferon in HBsAg positive chronic active hepatitis. *European Journal of Clinical Investigation*, **9**, 151–4.

Weimar, W., Heijtink, R. A., Schalm, S. W., Van Blakenstein, M., Schellekens, H., Masurel, N., Edy, V. G., Billiau, A. & De Somer, P. (1977). Fibroblast interferon in HBsAg positive chronic active hepatitis. *Lancet*, **ii**, 1282.

Weimar, W., Heijtink, R. A., Tenkate, F. J. P., Schalm, S. W., Masurel, N., Schellekens, H. & Cantell, K. (1980a). Double-blind study of leucocyte interferon administration in chronic HBsAg-positive hepatitis. *Lancet*, **i**, 336–8.

Weimar, W., Schellekens, H., Lameijer, L. D. F., Masurel, N., Edy, V. G., Billiau, A. & De Somer, P. (1978). Double-blind study of interferon administration in renal transplant recipients. *European Journal of Clinical Investigation*, **8**, 255–8.

Weimar, W., Stitz, L., Billiau, A., Cantell, K. & Schellekens, H. (1980b). Prevention of vaccinia lesions in rhesus monkeys by human leucocyte and fibroblast interferons. *Journal of General Virology*, **48**, 25–30.

Weller, I. V. D., Bassendine, M. F., Murray, A. K., Summers, J., Thomas, H. C. & Sherlock, S. (1980). HBsAg-positive chronic liver disease: inhibition of viral replication by highly soluble adenine arabinoside 5′ monophosphate (Ara-AMP). *Gastroenterology*, **79**, 1129.

Wheelock, E. F. (1964). Interferon in dermal crusts of human vaccinia virus vaccinations: a possible explanation of relative benignity of variolation smallpox. *Proceedings of the Society for Experimental Biology and Medicine*, **117**, 650–3.

Wheelock, E. F., Schenker, S. & Combes, B. (1968). Absence of circulating interferon in patients with infections and serum hepatitis. *Proceedings of the Society for Experimental Biology and Medicine*, **128**, 251–3.

Yunde, H., Guoliang, M., Shuhua, W., Yuying, L. & Hantang, L. (1981). Effect of Radix *astragali seu hedysari* on the interferon system. *Chinese Medical Journal*, **94**, 35–40.

THE ANTITUMOUR EFFECTS OF INTERFERON

J. S. MALPAS

Department of Medical Oncology, St Bartholomew's Hospital, London, UK

INTRODUCTION

The interferon system was discovered more than 20 years ago when it was recognized that it played an important part in the defence against viruses (Isaacs & Lindenmann, 1957). The possible role of viruses in the aetiology of cancer suggested that interferon should be investigated for its antitumour activity. Support was given to this concept by a number of early animal experiments, including those in mouse leukaemia (Gresser, Coppey, Falcoff & Fontaine, 1967). Investigations in higher animals and man were inhibited by the absence of sufficient material for experimental studies. In 1966 Cantell showed that it was possible to produce large amounts of interferon from human leucocytes and subsequently the lymphoblasts and fibroblasts have acted as sources of interferons destined for human trial. Strander pioneered early work on human tumours (1974, 1977). More recently, DNA recombinant procedures have enabled much larger quantities of interferon to be produced, and this material is now being entered into clinical studies.

The case for the introduction of interferon into the clinic as a useful method of treating human tumours must be based, as for any other chemotherapeutic agent, on the evidence that it is as effective as current therapy, or that it has fewer side effects than conventional therapy. Sufficient material should be available, and it should not be prohibitively expensive. Therapeutic agents usually undergo extensive animal testing and are then introduced in a sequence of studies (Table 1). Tissue toxicity in animals is carefully investigated, and if the material is apparently safe and sufficient is available, it is then introduced into so-called Phase I studies in patients. The aim of Phase I studies is to define the optimal dose, the best method of administration, and the range of severity of toxicity. Activity against the malignant process is of secondary importance, but it is natural that response is noted. Having defined the optimal dose of the agent, and the best method of administration, patients with a variety

Table 1. *Clinical assessment of anti-cancer drugs*

Procedure	Aim
Phase I Studies	To establish maximum tolerated dose of drug, determine toxic effects and pharmacology
Phase II Studies	To study a number of tumours forming a broad spectrum of malignant disease. To detect response in 15–40 patients.
Phase III Studies	To test the new agent against existing drugs of known value

of different malignancies are treated in adequate number in so-called Phase II studies to determine whether the agents have any role in certain tumours. As many patients may already have been treated extensively with conventional therapies, a relatively low rate of response is acceptable; for example, one patient responding out of 14 would be quite a good response rate, and if good results are obtained, the agent can then be used in Phase III or randomized studies in which patients are treated comparing the best established regimens with the regimen including the new agent.

Such a logical progression was impossible to follow in introducing interferon, because of the shortage of the drug and its expense. This combination of circumstances made its introduction particularly vulnerable to false claims.

PHASE I STUDIES OF INTERFERON

Three Phase I studies are reported, two using human lymphoblastoid interferon (Priestman, 1980; Rohatiner, Balkwill, Malpas & Lister, 1981; Rohatiner, Balkwill, Griffin, Malpas & Lister, 1982), and one human fibroblast interferon study (MacPherson & Tan, 1980). Priestman reported the use of gradually increasing doses of human lymphoblastoid interferon given intramuscularly to 18 patients with various solid tumours. All patients had been treated extensively with chemotherapy or radiotherapy before starting interferon. Assessment of marrow, renal, hepatic and cardiac function was carried out during the administration of interferon. The dose of interferon was escalated from 0.3 megaunits per square metre of body surface area in successive groups of two or three patients until a dose of 3 megaunits per square metre was achieved. The maximum total dose given to these adult patients was therefore of the order of 4.5 to 6.0 megaunits. The dose-limiting side effect was pyrexia, which was reversible on stopping the administration of interferon. Having studied the effects of single injections, repeated administration for a period of 30 days was then carried out, and three treatment regimens were studied. In the first, 2.5 megaunits per square metre of body surface area were given daily intramuscularly for 30 days. Two other regimens involved escalation of the dose from 2.5 megaunits to 7.5 megaunits and 2.5 megaunits to 15 megaunits.

In the group as a whole, pyrexia remained the dose-limiting side effect of interferon therapy, but hypertension, hypotension, fatigue, anorexia, nausea and vomiting, anaemia, leukopaenia, thrombocytopenia and hepatotoxicity were also seen.

The level of circulating interferon in the preliminary analyses carried out in these patients suggested that at 5 megaunits per square metre, blood levels of 150 units of human lymphoblastoid interferon were found per millilitre of serum. Among the 18 patients treated, one with malignant melanoma showed more than 50% reduction in the size of deposits of tumour for about six weeks, but this patient died some time afterwards. Another patient showed a short-lived regression of a carcinoma of the stomach.

Raised serum levels of human lymphoblastoid interferon could be seen after intramuscular injections of the substance, either in a single injection or as a prolonged course. The maximum tolerated dose was of the order of 3 megaunits per square metre of body surface area, and there were no serious organ toxicities at this level.

Rohatiner *et al.* (1981) also used human lymphoblastoid interferon in a Phase I study predominantly in patients with haematological malignancies. They decided to give the material intravenously, and to study the pharmacokinetics of the drug. The same material was used as that studied by Priestman (1980). Gradually increasing doses of interferon were given to patients who had relapsed following conventional chemotherapy or had failed to respond to such therapy. Two patients were given bolus intravenous injections. The succeeding patients received interferon by continuous intravenous infusion at doses escalating from 5 to 200 megaunits per square metre of body surface area for 5, 7 or 10 days. A total of 39 patients were treated in this way. In addition to assessment of blood cell counts, renal function and liver function, serum sodium, potassium, chloride and bicarbonate levels were estimated, and the serum calcium and phosphate were also carefully monitored. Blood samples for estimation of interferon were taken before the start of the infusion at 2, 4, 8 and 12 h on the first day of treatment, and once or twice a day thereafter. The serum was stored at –20 °C for estimation later using a biological assay.

Both patients given a bolus intravenous injection developed rigors and pyrexia and one had an episode of hypotension from which he recovered. Patients receiving the infusion, almost without exception, became pyrexial; three became hypotensive. At 200 megaunits per square metre, toxicity in the central nervous system was evident with drowsiness and disorientation. This precluded further escalation of dose, and was a very serious dose-limiting effect. The blood count in those patients who were assessable (i.e. patients with a normal bone marrow) showed pancytopenia which usually recovered within a week of stopping interferon. Biochemical evidence of reversible hepatotoxicity was seen at dose levels of 50, 100 and 2000 megaunits of interferon. At 200 megaunits, marked metabolic disturbances became apparent, with severe hypocalcaemia (corrected serum calcium at the lowest levels was recorded at 1.35 mmol/l) and hyperkalaemia (maximum level 7.0 mmol/l).

The responses noted were a fall in the circulating blast counts in six patients with acute leukaemia, but with no effect on the bone marrow. In one patient a reduction in the percentage of myeloblasts within the bone marrow to less than 5% of the total marrow cell count occurred, but as the patient had tumour deposits elsewhere, this was counted as only a partial response. He could not be described as being in complete remission. The levels of interferon present in the serum were dose-related, with dose ranges of 10 to 50 megaunits per square metre giving rise to peak levels of 10^3 u/ml in the serum, and with doses greater than 50 megaunits per square metre, peak levels of between 10^3 and 10^4 u/ml were found in most patients.

This study showed that interferon could be given in high doses by continuous intravenous infusion. The same toxic side effects that had previously been seen in the lower dose range occurred, but new and unacceptable central nervous system and metabolic disturbances occurred with the high doses of 200 megaunits per square metre, and a recommendation was therefore made that doses of lymphoblastoid interferon not greater than 100 megaunits per square metre per day should be given intravenously.

A third Phase I study investigated fibroblast interferon (McPherson & Tan, 1980) in 12 patients with advanced cancer. Most of the patients received the drug by the intramuscular route, though two had fibroblast interferon by continuous intravenous infusion. Plasma levels of interferon were measured following injection. The patients were given 1 megaunit 3 times weekly for 2 weeks in the beginning, followed by 5 megaunits 3 times weekly for 2 weeks, and then 10 megaunits 3 times weekly for 2 weeks. Two patients received intravenous fibroblast interferon, one a bolus of 1 megaunit and the other, infusions increasing from 1.5 to 7.5 megaunits in an infusion over 2 h.

The study showed no dose-limiting toxicity in any of the patients. There was no evidence of biochemical disorder or abnormal liver or renal function. However, although some interferon was found in the plasma after intramuscular injection, levels of only 4–12 u/ml were seen. When the fibroblast interferon was given intravenously, levels of about 40 u/ml occurred. Thus it appears that fibroblast interferon is hardly absorbed from a muscular site, and this study was therefore unable to determine the upper level of fibroblast interferon dosage. The material was biologically active, for one patient with acute myeloblastic leukaemia showed clearance of blast cells from his peripheral blood, but this was the only response noted.

The information presently available on toxicity of interferon relates chiefly to the lymphoblastoid form, in which the upper limit of dosage has been defined for the intramuscular and intravenous routes. The data for fibroblast interferon is far less satisfactory, and that for leucocyte interferon is not available. However, leucocyte interferon has been in clinical use for a much longer period, and the side effects of long term therapy have been described (Ingimarsson, Cantell & Strander, 1979). Twenty nine patients, 27 with osteosarcoma and two with juvenile laryngeal papilloma, were treated with human leucocyte interferon by daily intramuscular injection on three days a week for periods of up to a year. The side effects noted were those of the acute response to short-term therapy, and it is notable that no other toxicities developed as a result of long-term exposure to the drug.

The remarkable consistency of the side effects noted using

leucocyte, lymphoblastoid and fibroblast interferon would argue that the toxicity is an inherent property of the interferon, and not due to contaminants.

PHASE II STUDIES OF INTERFERON

Phase II studies are designed to test the effects of an anticancer agent on a wide variety of tumours. They are normally based on information derived from Phase I trials, but because of the constraints on the use of interferon already mentioned, this has rarely been the case. It may be necessary to repeat many of the limited trials undertaken, with the new knowledge that has been made available with respect to tolerated dose levels, the alternative ways of scheduling, and the different modes of administration. The choice of tumours investigated has been somewhat arbitrary. Suggestions of viral aetiology of a tumour, some indication from *in vitro* effects on cultures of tumour cells, regression of tumours in experimental animals, and a failure of a human tumour to respond to any conventional therapy have all been influences determining the employment of interferon.

HUMAN LEUCOCYTE INTERFERON IFN-α

Human leucocyte interferon has been used more frequently than any other in the treatment of human malignant disease. There are now a number of reports on its use in lymphoma, leukaemia, myeloma, breast cancer, ovarian cancer, lung cancer, bladder cancer, osteosarcoma and neuroblastoma. It is usual to refer to responses seen in solid tumours as complete or partial, and currently acceptable definitions of these terms, which will be used subsequently, are given in Table 2.

The use of interferon in the lymphomas was reported by Merigan, Sikora, Breeden & Rosenberg (1978), who reviewed their further experience (Louie *et al.*) in 1981. Eleven patients with non-Hodgkin lymphoma received human leucocyte interferon in a dose of 5 megaunits twice daily for 30 days. One complete, three partial and three minimal responses were seen with a duration of 6–12 months in the responses. All patients had become refractory to conventional therapy, so that this response rate was satisfactory. It was noted that

Table 2. *Criteria for response in solid tumours*

Response	Criterion
Complete Response (CR)	Complete disappearance of all demonstrable disease
Partial Response (PR)	Greater than 50% reduction in the sum of the products of the longest perpendicular diameters of tumour with no disease progression elsewhere
No Response (NR)	No change or less than 50% reduction
Progression	Less than 50% increase

the responses occurred in the nodular or low-grade lymphomas, whereas no responses were seen in the diffuse histiocytic lymphomas, which progressed rapidly. Dose-limiting toxicity was leukopenia, which caused the discontinuation of therapy in three patients. Side effects were those noted previously, including fever, malaise, arthralgia and loss of appetite.

Gutterman *et al.* (1980) treated 11 patients with malignant lymphoma, of whom six had the nodular form of the disease, which was extensive, and four had been treated previously. One patient achieved a complete response, which was continuing without further therapy for nearly a year, and two others had partial responses. Again, one patient with a diffuse histiocytic lymphoma showed less than a partial response, with disease appearing at new sites during therapy.

Leucocyte interferon (IFN-α) given in adequate dosage will produce responses in previously treated patients without unacceptable side effects, and further investigation into the treatment of non-Hodgkin lymphoma with this material is justifiable.

In Hodgkin's disease the situation is less satisfactory as far fewer patients have been treated. One of the best documented (Blomgren *et al.*, 1976) was a 24 year old man with Stage IVB lymphocyte-predominant Hodgkin's disease, to whom leucocyte interferon was given intramuscularly. He had biopsy-proven parenchymal lung infiltration. He had not been treated previously with radiotherapy or chemotherapy. Resolution of his symptoms for a short time and a decrease in the size of lymphadenopathy and lung infiltration were seen following administration of interferon, but the patient only achieved less than a partial response. While on interferon, the disease started to progress clinically, and treatment with interferon was stopped. He was then treated with a combination chemotherapy regimen and achieved a complete response.

Response, which was not defined, was noted by Hill *et al.* (1979, 1980) in all of five patients with lymphoblastic leukaemia who were treated with leucocyte interferon. One out of three patients with acute non-lymphocytic leukaemia responded, and one patient with a blast crisis transformation of chronic granulocytic leukaemia also had a response. Doses of 0.5 megaunits per kilogram of body weight were given daily for up to two months. Toxicity was recorded as slight, but it is not clear whether complete remissions were seen in these patients, or simply a fall in the blast count. This latter seems more likely, as Rohatiner *et al.* (1982), using the lymphoblastoid form of α interferon, found a marked decrease in circulating blasts in six patients with acute leukaemia, but in only one patient did the bone marrow return to normal.

Breast cancer has been the subject of a number of trials of interferon (Gutterman *et al.*, 1980; Borden *et al.*, 1980). Gutterman *et al.* (1980) reported on 17 women with metastatic breast cancer who had received all the conventional modes of treatment including hormone therapy. They were treated with 3 or 9 megaunits of human leucocyte interferon daily for a minimum of four weeks. A total of seven responses were seen: six partial, and one less than partial.

Previous response to chemotherapy or hormonal therapy was a determinant for predicting response to interferon. Metastases in soft tissues responded better than those in the viscera. Disappearance of tumour cells in patients with malignant pleural effusions arising from breast cancer was reported by Jereb *et al.* (1977) using human leucocyte interferon directly instilled into the pleural cavity, but no indication of the dose used was given.

The studies by Gutterman *et al.* (1980) have prompted the American Cancer Society to undertake a more extensive trial of leucocyte interferon, and with larger numbers of patients, a better estimate of its effectiveness should be possible. Disappointingly, no long-term remissions or even stabilization of disease have so far been reported.

Myeloma has also been the subject of a number of studies with interferon. Mellstedt *et al.* (1979) reported on four cases of myeloma who had become resistant to Melphalan, and who received 3 megaunits daily intramuscularly for a period of 3–19 months. All four improved clinically, and disappearance of the paraproteinaemia was noted in two of the four. Gutterman *et al.* (1980), using similar treatment in 10 patients with myeloma, had three responses, both in clinical findings and laboratory test abnormalities.

Leucocyte interferon has been entirely ineffective in so-called solid tumours: Krown, Stoopler, Cunningham-Rundles & Oettgen (1980) showed it had no effect in 12 patients with carcinoma of the bronchus, and Freedman, Wharton & Rutledge (1981) recorded no response in 13 patients with ovarian carcinoma. A more recent study by Einhorn, Cantell, Einhorn & Strander (1982) showed one patient with ovarian carcinoma achieving a partial response out of five treated with interferon.

More disturbing is the failure to demonstrate any significant effect against overt metastatic osteosarcoma in view of the current debate about interferon's efficacy as an adjuvant treatment for osteosarcoma which has been pioneered in Sweden. Hidemoto *et al.* (1980) reported only slight diminution in pulmonary metastases in osteosarcoma treated with up to 3 megaunits of leucocyte interferon twice a week intramuscularly. Caparros, Rosen & Cunningham-Rundles (1982) treated 11 patients with 3 escalating to 10 megaunits of leucocyte interferon a day. None of the 10 evaluable patients showed any response.

Local application of leucocyte interferon has been reported to be successful in bladder papillomatosis, breast cancer and melanoma (Ikic *et al.*, 1981a), in head and neck tumours (Ikic *et al.*, 1981b) and in neuroblastoma (Sawada *et al.*, 1979, 1980), but these results will need confirmation.

There is good evidence, therefore, of activity by leucocyte interferon against some forms of human cancer. Long-term remissions are uncommon and no cure of disease has been reported. Leucocyte interferon has acceptable toxicity and side effects and is therefore a candidate for further studies. It would need further experience to say what its full role will be, but it may take its place alongside other chemotherapeutic agents as being of limited use in part of the spectrum of human cancer treatment.

HUMAN LYMPHOBLASTOID INTERFERON IFN-α

There is much less information on the response of human malignant disease to lymphoblastoid interferon. Results in early Phase I trials of the material (Priestman, 1980) showed responses in melanoma and gastric carcinoma. Rohatiner *et al.* (1981) had one patient with chronic lymphocytic leukaemia who showed a less than partial response of cervical lymphadenopathy, and six patients with acute leukaemia who showed a meaningful response, though not achieving a complete remission.

Nine patients with malignant melanoma were treated with lymphoblastoid interferon by Priestman, Retsas, Newton & Westbury (1981), where the material was given in a dose of 2.5 megaunits per square metre of body surface area intramuscularly daily for 30 days. Eight patients had previously received chemotherapy, one patient had refused chemotherapy but accepted interferon therapy. All patients had metastatic disease in the liver, skin, lung and lymph nodes. Most of the patients completed the course, which was moderately well tolerated. One patient achieved a partial remission, and another had a short period of stabilization of their disease. In the patient who responded, improvement was continuing at three months after starting therapy. It was noted that skin nodules responded, but visceral metastases did not.

Phase II studies are now currently being conducted in non-Hodgkin lymphomas, myeloblastic leukaemias, and other solid tumours using this material.

HUMAN FIBROBLAST INTERFERON IFN-β

There is relatively little information available on the use of this material in human malignant disease. Preliminary findings in the study by MacPherson & Tan (1980) suggested that although no responses were seen, the interferon did possibly arrest progress in cancer of the colon and renal cancer.

A single case report of a child with disseminated nasopharyngeal carcinoma who had previously been treated with chemotherapy and radiotherapy and who had a complete response to fibroblast interferon given in a daily dose of 4 megaunits intravenously for 35 days, and subsequently three times a week, was reported by Treuner *et al.* (1980). This child apparently was still continuing in complete remission six months later. This remarkable result is the only complete response to fibroblast interferon in a patient with a solid tumour of which the author is aware. Sixteen patients with myeloma have been treated in a Phase II study by Misset, Gastiaburu, De Vassal & Mathe (1981). They received the interferon in a dose of 6 megaunits weekly intravenously. None showed even a partial response, though the 'M' protein component was reduced by 25% in three patients. Such failure to respond was also noted by Billiau (1981) in three patients, Ogawa & Ezaki (1981) in two cases, and in the one patient treated by Furue, Kohita, Kobayashi & Hakozaki in 1980.

Fibroblast interferon can be given safely intrathecally and Misset, Gastiaburu, De Vassal & Mathe (1981) treated five patients with relapse of leukaemia or lymphoma in the central nervous system. One patient showed complete clearance of the lymphoblasts from the cerebrospinal fluid.

RECOMBINANT INTERFERON

Few reports of clinical trials using recombinant interferon have yet appeared. Horning *et al.* (1982) studied the effect of recombinant leucocyte A interferon on eight patients with advanced cancer. They found in a Phase I study that up to 198 megaunits could be given intramuscularly. The major side effects were similar to those seen with crude interferon. Tumour responses were seen in two patients with lymphoma and one with breast carcinoma; a patient with chronic myeloid leukaemia also showed a response. There are, however, a number of studies in progress looking at the effect of recombinant interferon on human malignancies. These have tended to concentrate on tumours that have not so far been exposed to the material. Preliminary experience in my own unit would suggest that it is effective in temporarily reducing the circulating white blood cell count in chronic granulocytic leukaemia, and reducing the size of the spleen in these patients. It is not yet evident whether it will be of any value in these conditions.

Although relatively small numbers of various tumour types have been treated, it is noteworthy that whereas some responses have been seen when leucocyte interferon has been used to treat myeloma, none has been observed with fibroblast interferon. This would suggest that the different types of interferon could have differing spectra of activity. It would caution against the dismissal of these agents without adequate trial and the results of therapy with recombinant interferon are awaited with interest.

Experience so far with Phase II trials in interferon suggest that toxicity is very similar between the interferons. The difference, if any, is of degree rather than qualitative. It is not possible to define their clinical value at the moment, as there are so many imponderables, but given that the material is no more expensive than other chemotherapeutic agents, they may achieve a modest place in the chemotherapeutic armamentarium.

Whether they will become the treatment of choice in some rare, non-malignant viral condition such as laryngeal papilloma (outside the scope of this chapter) is difficult to say. Complete regressions have been seen, but on cessation of therapy the papillomas return.

PHASE III STUDIES WITH INTERFERON

Phase III studies, in which an investigational agent is assessed in a comparative trial with well-established conventional agents, are not yet published. It is becoming apparent that some tumours, such as the non-Hodgkin lymphomas of low-grade malignancy, would be prime candidates for such studies. The chronic leukaemias, both lymphocytic and granulocytic, may be investigated in this way very soon. A clear advantage would have to be shown for interferon in the latter, as effective palliative therapy with oral medication is now available.

It would be appropriate to end this review where the whole subject started, with the prophylactic use of interferon in preventing the occurrence of metastases in the lungs of patients with osteosarcoma, pioneered by Strander *et al.* (1974). These workers found that the patients appeared to have an increased disease-free survival when given leucocyte interferon in a dose of 2.5 megaunits three times a week intramuscularly. Successive publications reported a definite improvement in disease-free survival over historical controls, and with non-randomized concurrent controls in patients with osteosarcoma treated at peripheral hospitals in Sweden. No randomization of these patients to therapy has been carried out in an acceptable manner, and consequently the results of these Swedish studies remain inconclusive. It is quite possible that this improvement in survival occurred by chance, because it is increasingly accepted that for any agent to be effective as an adjuvant therapy, an excellent response rate is always seen when it is used against overt disease, and no response at all has been seen in the largest trial on overt osteosarcoma (Caparros *et al.*, 1982).

While interferons remain of fundamental interest in the laboratory, and must of course continue to be studied, the results of their incursion into the field of clinical malignancy have so far been disappointing.

REFERENCES

Billiau, A. (1981). The clinical value of Interferons as antitumour agents. *European Journal of Clinical Oncology*, **17,** 949–67.

Blomgren, H., Cantell, K., Johansson, B., Lagergren, C., Ringberg, U. & Strander, H. (1976). Interferon therapy in Hodgkin's disease. *Acta Medica Scandinavica*, **199,** 527–32.

Borden, E., Dao, T., Holland, J., Gutterman, J. & Merrigan, T. (1980). Interferon in recurrent breast carcinoma. Preliminary report of the American Cancer Society clinical trials program. *Proceedings of the American Association for Cancer Research*, **21,** 187.

Caparros, B., Rosen, G. & Cunningham-Rundles, S. (1982). Phase II trial of interferon IFN in metastatic osteogenic sarcoma. *Proceedings of the American Association for Cancer Research*, **23,** 121.

Einhorn, N., Cantell, K., Einhorn, S. & Strander, H. (1982). Human leukocyte interferon therapy for advanced ovarian carcinoma. *American Journal of Clinical Oncology*, **5,** 167–72.

Freedman, R., Wharton, J. T. & Rutledge, F. (1981). Human leukocyte interferon (IFNα) in patients with epithelial ovarian carcinoma (a preliminary report). *Proceedings of the American Society of Clinical Oncology*, **22,** 372.

Furue, H., Kohita, T., Kobayashi, H. & Hakozaki, M. (1981). Clinical experience with human fibroblast interferon. *Proceedings: Conference on Clinical Potentials of Interferons and Malignant Tumours.* Published by the Japan Medical Research Foundation.

Gresser, I., Coppey, Y., Falcoff, E. & Fontaine D. (1967). Interferon and murine leukaemia. I. Inhibiting effects of interferon on development of Friend leukaemia in mice. *Proceedings of Society for Experimental Biology and Medicine*, **124,** 84–91.

Gutterman, J. V., Blumenschein, G. R., Alexanian, R., Yap, H. Y., Buzdar, A. U., Cabanillas, F., Hortobagyi, G. N., Hersh, E. M., Rasmussen, S. L., Harmon, M., Kramer, M. & Pestka, S. (1980). *Annals of Internal Medicine*, **93,** 399–406.

Hidemoto, I., Murakami, K., Yanagawa, T., Ban, S., Sawamura, H., Sakakidak, K., Matsuo, A., Imanishi, J. & Kishida, T. (1980). Effect of human leukocyte interferon on the metastatic lung tumour of osteosarcoma. *Cancer*, **46,** 1562–5.

Hill, N. O., Loeb, E., Khan A., Pardue, A. S., Hill, J. M., Aleman,C. & Dorn, G. (1980). Phase I human leukocyte interferon trials in leukaemia and cancer. *Proceedings of the American Society of Clinical Oncology*, **21,** 361.

HILL, N. O., LOEB, E., PARDUE, A. S., DORN, G. L., KHAN, A. & HILL, J. M. (1979). Response of acute leukaemia to leukocyte interferon. *Journal of Clinical Haematology and Oncology*, **9,** 137–49.

HORNING, S. J., LEVINE, J. F., MILLER, R. A., ROSENBERG, S. A. & MERIGAN, T. C. (1982). Clinical and immunologic effects of recombinant leucocyte A interferon in eight patients with advanced cancer. *Journal of American Medical Association*, **247,** 1718.

HOROSZEWICZ, J. S., LEONG, S. S., ITO, M., BUFFETT, R. F., KARAKOUSIS, C., HOLYOKE, E., JOB, L., DOLEN, J. G. & CARTER, W. A. (1978). Human fibroblast interferon in human neoplasia: clinical and laboratory study. *Cancer Treatment Reports*, **62,** 1899–906.

IKIC, D., NOLA,P., MARICIC, Z., SMUDJ, K., ORESIC, V., KNEZEVIC, M., RODE, B., JUSIC, D. & SOOS, E. (1981a). Application of human leucocyte interferon in patients with urinary bladder papillomatosis, breast cancer and melanoma. *Lancet*, **i,** 1022–4.

IKIC, D., PADOVAN, I., BRODAREC, I., KNEZEVIC, M. & SOOS, E. (1981b). Application of human leucocyte interferon in patients with tumours of the head and neck. *Lancet*, **i,** 1025–7.

INGIMARSSON, S., CANTELL, K. & STRANDER, H. (1979). Side effects of long term treatment with human leucocyte interferon. *The Journal of Infectious Disease*, **140,** 560–3.

ISAACS, A. & LINDENMANN, J. (1957). Virus interference: I. The interferon. *Proceedings of the Royal Society*, B**147,** 258–67.

JEREB, B., CERVEK, J., US-KRASOVEC, M., IKIC, D. & SOOS, E. (1977). Intrapleural application of human leukocyte interferon (HLI) in breast cancer patients with unilateral pleural carcinosis. In *Proceedings of the Symposium on Preparation, Standardisation and Clinical use of Interferon*, pp. 187–96. Yugoslav Academy of Sciences.

KROWN, S. E., STOOPLER, M. B., CUNNINGHAM-RUNDLES, S. & OETTGEN, H. F. (1980). Phase II trial of human leukocyte interferon (IF) on non-small cell lung cancer. *Proceedings of the American Association for Cancer Research*, **21,** 179.

LOUIE, A. C., GALLAGHER, J. G., SIKORA, K. LEVY, R., ROSENBERG, S. A. & MERIGAN, T. C., (1981). Follow up observation on the effect of human leukocyte interferon in non-Hodgkin lymphoma. *Blood*, **58,** 712–18.

MCPHERSON, T. A. & TAN, Y. H. (1980). Phase I. Pharmacotoxicology study of human fibroblast interferon in human cancers. *Journal of the National Cancer Institute*, **65,** 75–9.

MELLSTEDT, H., AHRE, A., BJORKHOLM, M., HOLM, G., JOHANSSON, B. & STRANDER, H. (1979). Interferon therapy in myelomatosis. *Lancet*, **i,** 245–7.

MERIGAN, T. C., SIKORA, K., BREEDEN, J. H. & ROSENBERG, S. A. (1978). Preliminary observations on the effect of human leukocyte interferon in non-Hodgkin lymphoma. *New England Journal of Medicine*, **299,** 1449–53.

MISSET, J. L., GASTIABURU, J., VASSAL, F. DE & MATHE, G. (1981). Phase II trial of interferon (IF) in malignant gammopathies, meningeal leukaemia and chronic lymphocytic leukaemia. *Proceedings of the American Society of Clinical Oncology*, **22,** 491.

OGAWA, M. & EZAKI, K. (1981). A clinical study of fibroblast interferon. *Proceedings of the Conference on Clinical Products of Interferons in Viral Disease and Malignant Tumours*. Japan Medicine Research Foundation.

PRIESTMAN, T. J. (1980). Initial evaluation of human lymphoblastoid interferon in patients with advanced malignant disease. *Lancet*, **ii,** 113–18.

PRIESTMAN, T. J., RETSAS, S., NEWTON, K. & WESTBURY, G. (1981). A Phase II evaluation of Wellcome Hu IFN in advanced malignant melanoma. In *The*

Biology of the Interferon System, ed. E. de Maeyer, G. Galasso & H. Schelleherz, pp. 421–4. Amsterdam, New York, Oxford: Elsevier/North-Holland Biomedical Press.

Rohatiner, A. Z. S., Balkwill, F. R., Griffin, D. B., Malpas, J. S. & Lister, T. A. (1982). A Phase I study of human lymphoblastoid interferon administered by continuous intravenous infusion. *Cancer Chemotherapy & Pharmacology*, **9,** 97–102.

Rohatiner, A. Z. S., Balkwill, F., Malpas, J. S. & Lister, T. A. (1981). A Phase I study of intravenous human lymphoblastoid interferon (HLBI). *Proceedings of the American Society for Clinical Oncology*, **22,** 371.

Sawada, T., Fujita, T., Kisonoki, T., Imanishi, J. & Kishida, T. (1979). Preliminary report on the clinical use of human leukocyte interferon in neuroblastoma. *Cancer Treatment Reports*, **63,** 2111–13.

Sawada, T., Takamatsu, T., Tanaka, T., Mino, M., Fujita, K., Kusunoki, T., Arizono, N., Fukuda, M. & Kishida, T. (1980). Effects of intralesional interferon on neuroblastoma. *Cancer*, **48,** 2143–6.

Strander, H. (1977). The interferon system and its possible use in the treatment of neoplastic disease. *Cancer Immunology and Immunotherapy*, **3,** abstract no. 86, S35.

Strander, H., Cantell, K., Jakobsson, P. A., Nilsonne, U. & Soderberg, G. (1974). Exogenous interferon therapy of osteogenic sarcoma. *Acta Orthopaedic Scandinavica*, **45,** 958–9.

Treuner, J., Niethammer, D., Dannecker, G., Hagmann, R., Neef, V. & Hofschneider, P. H. (1980). Successful treatment of nasopharyngeal carcinoma with interferon. *Lancet*, **i,** 817–18.

INDEX